Hamid R. Tizhoosh and Mario Ventresca (Eds.)

Oppositional Concepts in Computational Intelligence

Studies in Computational Intelligence, Volume 155

Editor-in-Chief

Prof. Janusz Kacprzyk
Systems Research Institute
Polish Academy of Sciences
ul. Newelska 6
01-447 Warsaw
Poland
E-mail: kacprzyk@ibspan.waw.pl

Hamid R. Tizhoosh
Mario Ventresca
(Eds.)

Oppositional Concepts in Computational Intelligence

 Springer

Prof. Hamid R. Tizhoosh
University of Waterloo
Systems Design Engineering
200 University Avenue West
Waterloo, Ontario N2L 3G1
Canada
Email: tizhoosh@uwaterloo.ca

Dr. Mario Ventresca
University of Waterloo
Systems Design Engineering
200 University Avenue West
Waterloo, Ontario N2L 3G1
Canada
Email: mventres@pami.uwaterloo.ca

ISBN 978-3-540-70826-1   e-ISBN 978-3-540-70829-2

DOI 10.1007/978-3-540-70829-2

Studies in Computational Intelligence   ISSN 1860949X

Library of Congress Control Number: 2008931300

ⓒ 2008 Springer-Verlag Berlin Heidelberg

Typeset & Cover Design: Scientific Publishing Services Pvt. Ltd., Chennai, India.

Printed in acid-free paper

9 8 7 6 5 4 3 2 1

springer.com

For free inquiry and coexistence of opposites

Preface

"The opposite of a correct statement is a false statement. But the opposite of a profound truth may well be another profound truth."

– Niels Bohr

This volume is motivated in part by the observation that opposites permeate everything around us, in some form or another. Its study has attracted the attention of countless minds for at least 2500 years. However, due to the lack of an accepted mathematical formalism for opposition it has not been explicitly studied to any great length in fields outside of philosophy and logic. This, despite the fact that we observe opposition everywhere in nature, our minds seem to divide the world into entities and opposite entities; indeed we use opposition everyday. We have become so accustomed to opposition that its existence is accepted, not usually questioned and its importance is constantly overlooked.

On one hand, this volume is a fist attempt to bring together researchers who are inquiring into the complementary nature of systems and processes and, on the other hand, provide some elementary components for a framework to establish a formalism for opposition-based computing. From a computational intelligence perspective, many successful opposition-based concepts have been in existence for a long time. It is not our intention to recast these existing methods, rather to elucidate that, while diverse, they all share the commonality of opposition - in one form or another, either implicitly or explicitly. To this end, we have attempted to provide rough guidelines to understand what makes concepts "oppositional".

The editors are convinced that in spite of the excellence of contributors, this volume cannot claim completeness simply due to the difficulties that each and every new scientific endeavor experiences. However, we profoundly believe that since the future of scientific research cannot ignore opposition, we can embrace the inescapable imperfections and are grateful to all colleagues who have contributed to this work.

Waterloo, ON, Canada
May 1, 2008

Hamid R. Tizhoosh
Mario Ventresca

Contents

List of Contributors

Florencio G. Asenjo
University of Pittsburgh,
P.O.Box 10244,
Pittsburgh, PA 15232-0244,
USA
fgasenjo@pitt.edu

Mohamed S. Kamel
Department of Electrical and
Computer Engineering,
University of Waterloo,
Waterloo, Ontario, N2L 3G1
Canada
mkamel@uwaterloo.ca

Masoud Mahootchi
Department of Systems Design
Engineering,
University of Waterloo,
Waterloo, Ontario, N2L 3G1
Canada
mmahootc@engmail.uwaterloo.ca

Alice Ralickas Malisia
Department of Systems Design
Engineering,
University of Waterloo,
Waterloo, Ontario, N2L 3G1
Canada
amalisia@alumni.uwaterloo.ca

Don L. McLeish
Department of Statistics and Actuarial
Science,
University of Waterloo,
Waterloo, Ontario, N2L 3G1
Canada
dlmcleish@math.uwaterloo.ca

Luís Moniz Pereira
Centro de Inteligência Artificial,
Universidade Nova de Lisboa,
2829-516 Caparica,
Portugal
lmp@di.fct.unl.pt

Alexandre Miguel Pinto
Centro de Inteligência Artificial,
Universidade Nova de Lisboa,
2829-516 Caparica,
Portugal
amp@di.fct.unl.pt

Kumaraswamy Ponnambalam
Department of Systems Design
Engineering,
University of Waterloo,
Waterloo, Ontario, N2L 3G1
Canada
ponnu@vlsi.uwaterloo.ca

Shahryar Rahnamayan
Faculty of Engineering and Applied
Science,
University of Ontario
Institute of Technology (UOIT),
Oshawa, Ontario, L1H 7K4 Canada
Shahryar.Rahnamayan@uoit.ca

Farhang Sahba
Department of Electrical and
Computer Engineering,
University of Toronto
Toronto, Ontario, M5S 3G4
Canada
farhang.sahba@utoronto.ca

Lei Shi
Department of Computer Science and
Engineering,
The Chinese University of Hong Kong,
Shatin, NT, Hong Kong,
China
shil@cse.cuhk.edu.hk

Maryam Shokri
Department of Systems Design
Engineering,
University of Waterloo,
Waterloo, Ontario, N2L 3G1
Canada
mshokri@engmail.uwaterloo.ca

Tse Guan Tan
Centre for Artificial Intelligence,
School of Engineering and Information
Technology,
Universiti Malaysia Sabah,
Locked Bag No. 2073, 88999 Kota
Kinabalu, Sabah,
Malaysia
tseguantan@gmail.com

Jason Teo
Centre for Artificial Intelligence,
School of Engineering and Information
Technology,
Universiti Malaysia Sabah,
Locked Bag No. 2073, 88999 Kota
Kinabalu, Sabah,
Malaysia
jtwteo@ums.edu.my

Hamid R. Tizhoosh
Department of Systems Design
Engineering,
University of Waterloo,
Waterloo, Ontario, N2L 3G1
Canada
tizhoosh@uwaterloo.ca

H. Jaap van den Herik
Tilburg centre for Creative Computing
(TiCC)
Faculty of Humanities,
Tilburg University P.O. Box 90153,
5000 LE Tilburg, The Netherlands
H.J.vdnHerik@uvt.nl

Mario Ventresca
Department of Systems Design
Engineering,
University of Waterloo,
Waterloo, Ontario, N2L 3G1
Canada
mventres@pami.uwaterloo.ca

Mark H.M. Winands
Maastricht ICT Competence Centre
(MICC)
Faculty of Humanities and Sciences,
Maastricht University
P.O. Box 616, 6200 MD Maastricht,
The Netherlands
m.winands@micc.unimaas.nl

1

Introduction

H.R. Tizhoosh and Mario Ventresca

Department of Systems Design Engineering, University of Waterloo, Canada
{tizhoosh,mventres}@pami.uwaterloo.ca

1.1 Introduction

Due to the omnipresence of opposition in the real world, regardless in what intensity and form we may encounter its diverse presence, the nature of entities and their opposite entities might be understood in different ways. A look at a whole set of words describing oppositeness is an indicator of this diversity: antipodal, antithetical, contradictory, contrary, diametrical, polar, antipodean, adverse, disparate, negative, hostile, antagonistic, unalike, antipathetic, counter, converse, inverse, reverse, dissimilar, divergent. All these words describe some notion of opposition and can be employed in different contexts to portray different relationships.

Recent theories based on our perceptions of the apparent contrary/opposite nature of the real world have begun to provide a formal framework within the neurological and behavioral sciences [18, 19, 20]. More specifically, these theories are concerned with how cognition and behavior arise from dynamical self-organization processes within the brain. Oppositional structures in linguistic information processing and communication play a pivotal role and any attempt to imitate human information processing without a framework for oppositional thinking would be futile. Furthermore, since much of computational intelligence is inspired from theories of how the brain works it seems reasonable to also examine the usefulness of opposition within the computational intelligence context.

The next section will provide a discussion regarding the understanding of opposition a long with some real world examples. Then, the following section provides a brief overview of attempts to formalize and explicitly utilize the concept of opposition. We conclude this introduction with a description of the relationship each chapter has with oppostional concepts.

1.2 What Is Opposition?

Opposition is concerned with the relationship between entities, objects or their abstractions of the same nature which are completely different in some manner. For instance, cold and hot both describe a certain temperature perception (are of the same kind),

H.R. Tizhoosh, M. Ventresca (Eds.): Oppos. Concepts in Comp. Intel., SCI 155, pp. 1–8, 2008.
springerlink.com

however, they are completely different since they are located at opposite spots of temperature scale. Moreover, these types of perceptions are relative to some observer or system implying that in many situations it may be very difficult to define a universally acceptable definition of opposite.

In general, the transition from one entity to its opposite can understandably establish rapid and fundamental changes. Social revolutions, for instance, mainly aim at attaining opposite circumstances, i.e. dictatorship vs. democracy. The goal is to alter the state or environment to one that is percieved to be more desirable. In the following we provide a short overview of some oppositional concepts from varying fields.

1.2.1   Oppositional Thinking

Many examples of opposition exist in philosophy. For example, ancient Egyptians based much of their lives around the notion of harmonizing opposites. These ideas likely influenced Pythagoras of Greece who through Aristotle gave us the Table of Opposites [17] which is presented in Table 1.1.

Table 1.1. Pythagoras's Table of Opposites, delivered through Aristotle

finite	$\Longleftrightarrow$	infinite
odd	$\Longleftrightarrow$	even
one	$\Longleftrightarrow$	many
right	$\Longleftrightarrow$	left
rest	$\Longleftrightarrow$	motion
straight	$\Longleftrightarrow$	crooked
light	$\Longleftrightarrow$	darkness
good	$\Longleftrightarrow$	evil
square	$\Longleftrightarrow$	oblong
male	$\Longleftrightarrow$	female

The more widely recognize concept of the Yin and Yang originates in ancient Chinese philosophy and metaphysics [16], which describes two primal opposing but complementary forces found in all things in the universe. Similiar ideas can also be found in Indian philosophy where the Gods Shiv and Shakti represent the two inseparable forces which are responsible for all forms of life in the universe [22] and Zoroastrianism which has the opposing spirits of Spenta Mainyu and Angra Mainyu representing life and destruction, respectively [22]. In fact, the main aspect of Zoroastrianism was believing in Ahura Mazda (God) who is permanently opposed by Ahriman (Satan), which may very well be the first time that a monotheistic religion introduced the duality in its philosophy.

The triad *thesis*, *antithesis*, and *synthesis* describe the Hegelian dialectic [15] to explain historical events. The thesis is a proposition, and the antithesis is its negation (opposite). The synthesis solves the conflict between the thesis and antithesis by reconciling their common truths, and forming a new proposition.

Immanuel Kant defines in *The Critique of Pure Reason* [14] the term *antithetic*:

"Thetic is the term applied to every collection of dogmatical propositions. By antithetic I do not understand dogmatical assertions of the opposite, but the self-contradiction of seemingly dogmatical cognitions (thesis cum antithesis), in none of which we can discover any decided superiority. Antithetic is not, therefore, occupied with one-sided statements, but is engaged in considering the contradictory nature of the general cognitions of reason and its causes."

In natural language, opposition can be detected at different levels: Directional opposition (north-south, up-down, left-right), adjectival opposition (ugly-handsome, long-short, high-low), and prefix opposition (thesis vs. anti-thesis, revolution vs. counter-revolution, direction vs. opposite direction) [13]. Further, one can distinguish complements (mutually exclusive properties:dead-alive, true-false), antonyms (two corresponding points or ranges of a scale: long-short, hot-cold), directional converses (two directions along an axis: east-west, north-south), and relational converses (the relative positions of two entities on opposite sides: above-below, teacher-pupil) [13]. Human communication without utilizing linguistic opposition is unimaginable. It seems as though opposites are needed to understand and describe the state of a system, otherwise the system would likely not warrant explanation.

In psychology, reinforcement is a crucial learning mechanism for both humans and animals [12]. Rewards vs. punishments, and positive vs. negative reinforcements have been intensively investigated [12]. Every reinforcement signal has its own distinct influence, which may not be easily achieved with the opposite reinforcement signal. Learning, at its heart, requires opposite types of feedback for the development of useful behavior [12].

In physics, antiparticles are subatomic particles having the same mass as one of the elementary particles of ordinary matter but with opposite electric charge and magnetic moment [11]. The positron (positive electron), hence, is the antiparticle of the electron. As another physical example, electric polarization is defined as a slight relative shift of positive and negative electric charge in opposite directions within an insulator or dielectric, induced by an external electric field. Polarization occurs when an electric field distorts the negative cloud of electrons around positive atomic nuclei in a direction opposite the field [1]. Opposition seems to be an inherent characteristic of the physical universe.

In mathematics we have concepts close to opposition, but tending to focus on its implicit interpretation and practical usefulness. For instance, the bisection method for solving equations makes use of $\pm$ sign change in order to shrink the search interval [7]. In probability theory, the probability of the contrary situation is calculated by $1 - p$ if the initial event occurs with probability p. Similarly, opposite numbers between $[-\infty, \infty]$ are represented by the pair $(x, -x)$. In Monte Carlo simulation, antithetic random numbers are used to reduce variance [9, 8] (this concept is possibly the only subject in mathematics that is an explicit implementation of what we will introduce in this volume as opposition-based computing). And of course, we have the concept of proof by contradiction which begins with an untrue statement and shows it must be false, therefore the opposite of the original statement is true.

These examples show that opposition plays a central role in our lives. It seems to be required for reasoning, learning, describing things and is prevalent in many fields ranging from theology to linguistics to physics. In spite of all these examples, however, it should be mentioned that understanding and defining opposition may not be trivial in most cases.

1.2.2 Formalizing and Understanding Opposition

Numerous proposals for understanding and formalizing opposition can be seen throughout history and are continuously fine-tuned to incorporate new knowledge or views. There is no particular accepted "theory of opposition", rather different opinions, some more specific to certain domains than others. Most of this work is based in the area of logic.

An early classification of opposites in nature is given by the Pythagorean Table of Opposites (delivered through Aristotle) seen in Table 1.1. Where the question of *one vs. many* has been of particular interest for the last two millennia.

Aristotle later provided his Square of Opposites (Figure 1.1) which aimed to describe the ways in which propositions of a system are logically related. The terms A (universal affirmative), E (universal negative), I (particular affirmative) and O (particular negative) are connected by relationships which show the type of opposition that relates them; contrary statements cannot both be true at the same time, subcontrary statements can both be true but not both false, and subalternate statements which concern the implied nature between a particular and a universal statement. The underlying assumption behind this concept is that every affirmative statement has a single contradictory statement, of which one must always be true.

By allowing for non-existent entities, Boole [6] alleviated the need for contrary, subcontrary and subalternate terms from Aristotles Square of Opposites. What remains is simply those propositions which are contradictory.

De Morgan reasoned that all kinds of opposition are formally equivalent, "*Every pair of opposite relations is indistinguishable from every other pair, in the instruments of operation which are required*" [5]. Thus, a mathematical $\pm$ could be used in an attempt to find an algebraic algorithm for reasoning involving any kind of opposition. Along

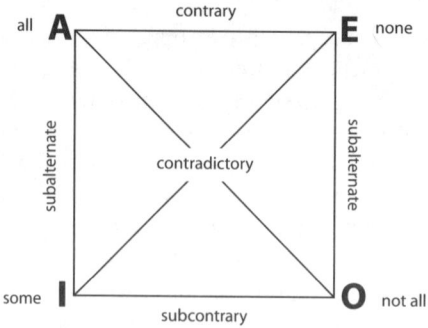

Fig. 1.1. Aristotle's Square of Opposition [21]

this line of thinking Leibniz, De Morgan and Sommers' natural language has a logic that makes use of two kinds of expressions, the signs of opposition [4].

More recently, the Square of Opposites has been generalized through n-Opposition Theory which is based on the relations between modal logic and geometry [3]. The generalized form expands to shapes of higher dimensions, which seem to be hinting towards a deep, complex underlying theory of oppositional logic.

Abstract mathematics has also attempted to formalize the concept of opposition. An example is dual or opposite category of the branch of mathematics known as category theory [2]. Given some grouping of entities according to a relationship, the dual category is formed by considering the opposite relationship. For example, reversing the direction of inequalities in a partial ordering.

There are too many alternative logics and mathematical dualities and interpretations thereof to include a comprehensive overview here, but the reader should realize that the problem of understanding opposition is still an active area. Furthermore, these logics and dualities form or have the potential to form the basis of a vast number of application areas, computational and otherwise.

1.3 About This Book

This volume attempts to provide a collection of different perspectives on opposition in computational intelligence. The lack of a unifying umbrella is certainly the driving motivation for this book, even though such an umbrella still needs time to be built. The authors of this volume may have a different understanding of opposition. However, they all employ oppositional thinking in diverse ways whose relationship may not be apparent at first sight.

We have divided this book into four parts. The first, Motivations and Theory (Chapters 2–4), focuses on motivating principles or theoretical findings related to oppositional concepts. The next three parts examine the topics of (a) Search and Reasoning (Chapters 5–6) (b) Optimization (Chapters 7–9) and (c) Learning (Chapters 10–12), although in many cases these may overlap. The final part of the volume keys in on real-world applications using oppositional algorithms (Chapters 13–14).

1.3.1 Motivations and Theory

In "Opposition-Based Computing" H.R. Tizhoosh, M. Ventresca and S. Rahnamayan attempt to provide a framework for the classification and study of techniques which employ oppositional principles. Initial definitions, basic theorems and experimental evidence for specific uses of opposition are provided as proof-of-concept evidence.

The use of antithetic variates is presented by D.L. McLeish in "Antithetic and Negatively Associated Random Variables and Function Maximization". The idea behind antithetic variates is that extreme values found during simulation algorithms can be made less harmful by counteracting them with other, negatively correlated, extreme values. This approach aims to provide a wider sampling of the space in order to minimize the covariance between the two samples to reduce the variance of the sampling procedure.

F.G. Asenjo describes a different view of opposition from a logical perspective in "Opposition and Circularity". The roots of this work extend back to 1902 and the discovery of sets which are not members of themselves. This has eventually led to a redefining of some basic logical assumptions and aspects of our thought process and reasoning can be described using these antinomic logics.

1.3.2   Search and Reasoning

In "Collaborative vs. Conflicting Learning, Evolution and Argumentation", Pereira and Pinto provide a logic-based approach for reconciling opposite arguments. It is often the case that arguments are deemed incompatible. Nevertheless, it may be possible to find common ground for a resolution between these opposite arguments. The overall goal of this work is to combine antagonistic and collaborative argumentation in order to produce a common ground alternative.

Game trees, which are inherently oppositional, are examined in "Proof-Number Search and Its Variants" by H. Japp van den Herik and M.H.M. Winnands. Specifically, they deal with Proof-Number search which is a special case of adversarial search algorithm. The goal is to reduce computational overhead of searching game trees while still discovering a quality decision.

1.3.3   Optimization

Opposition-based ant colony algorithms are extended in "Improving the Exploration Ability of Ant-Based Algorithms" by A. Malisia. Integrating antipheromone into the traditional pheromone update and decision making processes is shown to have potential for improving different ant-based optimization approaches.

Through the use of antichromosomes S. Rahnamayan and H.R. Tizhoosh are able to improve the differential evolution algorithm in "Differential Evolution via Exploiting Opposite Populations". The effect of antichromosomes for both population initialization and the actual optimization process is explored. Additionally, the generation jumping scheme is introduced which yields a higher convergence rate to higher quality solutions by jumping between opposite populations.

In "Evolving Opposition-Based Pareto Solutions: Multiobjective Optimization Using Competitive Coevolution", T.G. Tan and J. Teo extend ideas of opposition-based differential evolution to the multiobjective domain. They are able to better approximate the true Pareto front through the use of opposition.

1.3.4   Learning

L. Shi in "Bayesian Ying-Yang Harmony Learning for Local Factor Analysis: A Comparative Investigation" provides a comparison for local factor analysis model selection. The concept of Ying-Yang Harmony Learning is based on the balance between an external observation and its inner representation.

M. Shokri, H.R. Tizhoosh and M. Kamel in "The Concept of Opposition and its Use in Q-learning and Q(λ) Techniques" suggest the use of non-Markovian updates to improve reinforcement learning techniques. The additional updates accelerate the convergence based on knowledge about the action and opposite actions which are exploited to assign reward/opposite reward to corresponding actions.

In "Two Techniques for Improving Gradient-Based Learning Algorithms", M. Ventresca and H.R. Tizhoosh explore properties of a specific adaptive transfer function, the opposite transfer function. Two variations on the concept are presented and shown to improve the accuracy, convergence and generalization ability of backpropagation-based learning algorithms.

1.3.5 Real World Applications

The problem of image segmentation is tackled by F. Sahba and H.R. Tizhoosh in "Opposite Actions in Reinforced Image Segmentation". By defining opposite actions during image thresholding, reinforcement learning is extended to extract prostate boundaries in ultrasound scans.

Using a real-world application, M. Mahootchi, H.R. Tizhoosh and K. Ponnambalam in "Opposition Mining for Reservior Management" demonstrate how to conduct opposition mining in a pratical context. This concept is used to perform additional reinforcement learning updates resulting in increased modelling performance.

As mentioned above, these chapters in no means summarize all possible uses and applications of opposition. These are merely some successful examples of much broader range of oppositional concepts in computational intelligence.

References

1. Encyclopædia Britannica, The New Encyclopædia Britannica (2007)
2. Pierce, B.C.: Basic Category Theory for Computer Scientists. MIT Press, Cambridge (1991)
3. Moretti, A., Beziau, J.Y., Costa-Leite, A., Facchini, A.: Aspects of Universal Logic, Geometry for Modalities? Yes: Through n-Opposition Theory, Cahiers de logique - Universite de Neuchatel (2004)
4. Sommers, F.: The calculus of terms. Mind 79, 1–39 (1970)
5. De Morgan, A., Heath, P.: On the Syllogism and Other Logical Writings. Routledge and Kegan Paul (1966)
6. Boole, G.: The Calculus of Logic. Cambridge and Dublin Mathematical Journal 3, 183–198 (1848)
7. Hamming, R.: Numerical Methods for Scientists and Engineers. Dover Publications (1987)
8. McLeish, D.: Monte Carlo Methods and Finance. Wiley, Chichester (2005)
9. Hall, P.: Antithetic resampling for the bootstrap. Biometrika 76(4), 713–724 (1989)
10. Wood, C., Jena, D.: Polarization Effects in Semiconductors: From Ab Initio Theory to Device Applications. Springer, Heidelberg (2007)
11. Cottingham, W.N., Greenwood, D.A.: An Introduction to the Standard Model of Particle Physics. Cambridge University Press, Cambridge (1998)
12. Pearce, J.M.: Animal Learning and Cognition. Psychology Press (1997)
13. Cruse, D.A.: Lexical Semantics. Cambridge University Press, Cambridge (1986)
14. Kant, I.: Critique of Pure Reason. Cambridge University Press, Cambridge (1999)
15. Hegel, G.W.F.: Phenomenology of Spirit, SOS Free Stock (1998)
16. Chan, W.: A Source Book in Chinese Philosophy. Princeton University Press, Princeton (1963)
17. Aristotle, Sachs, J.: Aristotle's Metaphysics. Green Lion Press (1999)
18. Kelso, J.A.S., Engstrm, D.A.: The Complementary Nature. MIT Press, Cambridge (2006)

19. Kelso, J.A.S., McDaniel, R., Driebe, D.: Uncertainty and Surprise, The Complementary Nature of Coordination Dynamics: Toward a Science of the in-Between, vol. 3. Springer, Heidelberg (2005)
20. Grossberg, S.: The Complementary Brain – Unifying Brain Dynamics and Modularity. Trends in Cognitive Sciences 4(6), 233–246 (2000)
21. Thompson, M.: On Aristotle's Square of Opposition. The Philosophical Review 62(2), 251–265 (1953)
22. Jones, L.: Encyclopedia of Religion. MacMillan Reference Books (2005)

Part I

Motivations and Theory

2

Opposition-Based Computing

H.R. Tizhoosh[1], Mario Ventresca[1], and Shahryar Rahnamayan[2]

[1] Department of Systems Design Engineering, University of Waterloo, Canada
  {tizhoosh,mventres}@pami.uwaterloo.ca
[2] Faculty of Engineering and Applied Science, University of Ontario Institute of Technology (UOIT), Canada
  Shahryar.Rahnamayan@uoit.ca

Summary. Diverse forms of opposition are already existent virtually everywhere around us but the nature and significance of oppositeness is well understood only in specific contexts within the fields of philosophy, linguistics, psychology, logic and physics. The interplay between entities and opposite entities is apparently fundamental for maintaining universal balance. However, it seems that there is a gap regarding oppositional thinking in engineering, mathematics and computer science. Although many opposition-based techniques exist in these fields, the oppositional properties they employ are not usually directly studied. A better understanding of opposition could potentially establish new search, reasoning, optimization and learning schemes with a wide range of applications. For instance, improving convergence rate for hyperdimensional problems could be improved through the use of oppositional strategies.

A large number of problems in engineering and science cannot be approached with conventional schemes and are generally handled with intelligent techniques such as evolutionary, neural, reinforcing and swarm-based techniques. These methodologies, however, suffer from high computational costs. In this work, the outlines of opposition-based computing, a proposed framework for computational intelligence, will be introduced. The underlying idea is simultaneous consideration of guess and opposite guess, estimate and opposite estimate, viewpoint and opposite viewpoint and so on in order to make better decisions in a shorter time, something that all aforementioned techniques could benefit from. The goal is to better understand the role of opposition within a computational intelligence context with the intention of improving existing or developing more powerful and robust approaches to handle complex problems.

Due to its diversity, it is difficult, if not impossible, to uniquely and universally define the nature and the scope of what we call *opposition*. However, there is no doubt that we need a mathematical formalism if we are going to, at least to some degree, exploit the oppositional relationships in real-world systems. Hence, we attempt to establish a generic framework for computing with opposites in this chapter, a framework which may not mathematically capture the very essence of oppositeness in all systems accurately but it will be, as we hope, a point of departure for moving toward opposition-based computing. The ideas put forward in this chapter may not be shared by all those who work on some aspect of opposition (and usually without labeling it opposition), nevertheless, we believe that this framework can be useful for understanding and employing opposition.

2.1 Introduction

The idea of explicitly and consiously using opposition in machine intelligence was introduced in [21]. The paper contained very basic definitions and three application

H.R. Tizhoosh, M. Ventresca (Eds.): Oppos. Concepts in Comp. Intel., SCI 155, pp. 11–28, 2008.
springerlink.com   © Springer-Verlag Berlin Heidelberg 2008

examples providing preliminary results for genetic algorithms, reinforcement learning and neural networks. The main scheme proposed was to not only to look at chromosomes, actions and weights but simultaneously consider anti-chromosomes, counter-actions and opposite weights. The paper argued that by considering these opposites and by simultaneous analysis of quantities and their opposite quantities we can accelerate the task in focus by increasing the chances of discovering the basins of attraction for optimal (or high quality) solutions.

In this chapter, the framework of opposition-based computing (OBC) will be provided, new definitions will be established and a rudimentary classification of oppositional concepts in computational intelligence will be provided. In section 2.2 the nature of opposition will be briefly discussed. Section 2.3 provides introductory definitions which we propose to be employed to understand opposition. We also show an example based on low-discrepancy sequences, of the type of improvements opposition can potentially deliver. In Section 2.4 we formally describe opposition-based computing using our definitions from Section 2.3. In section 2.5 a rough guideline or classification of OBC algorithms is provided. Section 2.6 summarizes the work with some conclusions.

2.2   What Is Opposition?

According to the American Heritage Dictionary *opposite* is defined as, "being the other of two complementary or mutually exclusive things" and *oppositional* as "placement opposite to or in contrast with another" [1]. Generalizing these definitions we can infer that opposites are two complementary elements of some concept or class of objects/numbers/abstractions, and the relationship which defines their complementarity is the notion of opposition[1]. More formally, two elements $c_1, c_2 \in C$ are considered opposite relative to some mapping $\varphi \colon C \to C$, where φ is a one-to-one function and C is some non-empty set of concepts.

The class C can represent any form of concept, whether it be a physical entity (i.e. a state or estimate) or more abstract notion (i.e a hypothesis or viewpoint). The mapping between the opposite elements is also very robust, indeed the very notion of opposite is defined relative to this mapping, which can essentially be any function determining which entities are opposites. Therefore, changing φ consequently changes how the concepts are percieved.

Consider the following two simple situations, letting C be a continuous interval between $[0, 1]$ and $X \in C \colon$.

1. Let $\varphi(X) = (X + 0.5)\%1$. The modulo operator guarantees a constant distance of 0.5 between $X = x$ and its opposite $\breve{x} = \varphi(x)$[2].
2. Let $\varphi(X) = 1 - X$. Under this definition as $X = x \to 0.5$, $\breve{x} \to 0.5$ (the center of the search interval). Therefore φ is less concerned with physical distance within the natural ordering of C and more related to similarity.

These situations highlight an important facet of oppositional thinking. Namely, the natural ordering of concepts from class C is not necesarily the only manner in which

---

[1] See the introduction, Chapter 1, for a more general motivation on opposition.
[2] We denote the opposite of x with $\breve{x}$.

opposites can be defined. More often than not the problem definition may indicate or hint towards something more appropriate. In fact, the second scenario contains a paradox in that if $x = 0.5$ then $\breve{x} = 0.5$, leaving a debate as to whether a concept can be its own opposite or whether it is possible for the concept to not have an opposite at all[3]. This situation must also be appropriately addressed through the definition of φ.

The above examples were restricted to a single dimensional problem. A logical next step is to examine how oppositional thinking changes as the problem dimensionality increases. Let $C =< c_1, c_2, \ldots, c_n >$ be an arbitrary concept in C and c_i be the ith dimensional component of the n-dimensional system. We can then define $\Phi(C) =< \varphi^1(c_1), \varphi^2(c_2), \ldots, \varphi^n(c_n) >$ where φ^i defines opposition solely for dimension i and Φ is the resultant mapping over all n single dimension mappings. That is, the opposite of a multi-dimensional system is defined with respect to the opposite of each component. Also, as with the single dimension situation, Φ must be a one-to-one function.

Let S be an arbitrarily selected subset of integers from $\{1, 2, \ldots, n\}$, corresponding to one of $\{\varphi^1, \varphi^2, \ldots, \varphi^n\}$. In total there exist 2^n possible subsets, of which only one has all n integers (denoted S^*). Then, for every S we can calculate

$$\tau(S) = \frac{|S|}{|S^*|} = \frac{|S|}{n}, \tag{2.1}$$

to represent the *degree of opposition* between the concepts represented by S and the true opposite S^*, where $|\cdot|$ is the cardinality of the respective set. It is important to notice that in this case each component is given equal weighting, meaning that there does not exist a component which contributes more "oppisitional tendency" than any other (i.e. each dimension of the problem is equally important). However, allowing unequal weighting allows for greater flexibility and freedom when employing opposition in computational intelligence (as will be shown throughout this book).

By introducing the idea of varying degrees of oppositeness we also can ask whether opposition is a static or dynamic concept. This refers to the redefinition of opposition as undesirable elements of C are excluded through a search, optimization or learning process (i.e. $|C|$ can vary at each iteration). We denote a system with a dynamic concept of opposition as Φ_t, indicating there exists a time component to the mapping (where each component is also time varying and denoted by φ_t^i).

Exploiting *symmetry* is in many cases a good example for incorporating opposition within learning and search operations. In its essence, considering oppositional relationships is nothing but integration of *a-priori knowledge*. There exist a large number of works on using a-priori or expert knowledge in computational intelligence, why then, one may ask, should we concern ourselves with opposition since we are integrating a-priori knowledge anyway? The knowledge of oppositeness in context of a given problem is a very special type of a-priori knowledge, which, if we comprehend the oppositional relationships between quantities and abstractions, its processing bears virtually no uncertainty. If, for instance, we know that the quantity/abstraction X has a high evaluation in context of the problem in focus, then its opposite $\breve{X}$ can be confidently dismissed as delivering a low evaluation. By recognizing oppositional entities we

---

[3] This is a very old debate in logic and philosophy which everybody may know via the example "is the glass half full or half empty?".

acquire a new type of a-priori knowledge that, if processed as general, non-oppositional knowledge, its contributions to problem solving remain widely unexploited.

A final major issue to address is whether opposite concepts are required to have an antagonistic nature (mutually exclusive), or can they interact in a constructive manner (balance maintenance). Is the system constantly deciding between which of two concepts is most desirable or are these two concepts used together to produce a more desirable system? Furthermore, is this interaction a constant or dynamic phenomena? That is, can competitive and cooperative relationships exist between opposites at different times? Most likely antipodal answers to these questions are true for different problem settings.

Most of the ideas behind these notions can be found scattered in various research areas. Chapters 3 and 4 further discuss some of the statistical and logical motivations and interpretations of opposition, respectively.

2.3   Computing with Opposites

In this section we attempt to establish a generic framework for computing with opposites which will be used in subsequent sections when discussing opposition-based computing. We have left the definitions abstract to provide flexibility and robustness in their practical application. First we provide important definitions followed by a simple example of using the outlined concepts.

2.3.1   Formal Framework

In this subsection we will provide abstract definitions pertaining to type-I and type-II opposition. For each definition we discuss its implications and provide examples where appropriate.

Definition 1 (Type-I Opposition Mapping).   *Any one-to-one mapping function, $\Phi\colon \mathcal{C} \to \mathcal{C}$, which defines an oppositional relationship between two unique[4] elements C_1, C_2 of concept class $\mathcal{C}$ is a type-I opposition mapping. Furthermore, the relationship is symmetric in that if $\Phi(C_1) = C_2$, then $\Phi(C_2) = C_1$.*

This mapping is understood to define the opposite nature between C_1, C_2 with respect to the problem and/or goal of employing the relationship. In some cases this may be directly related to solving the problem at at hand (i.e. with respect to minimizing an evaluation function). Another possible use is to implcitly aid in solving the problem, for example to provide a more diverse range of solutions.

Definition 2 (Type-I Opposition).   *Let $C \in \mathcal{C}$ be a concept in n-dimensional space and let $\Phi\colon \mathcal{C} \to \mathcal{C}$ be an opposition mapping function. Then, the type-I opposite concept is determined by $\check{C} = \Phi(C)$.*

---

[4] The assumption is that an element cannot be its own opposite, although it is technically possible to relax this condition.

For convenience, we often omit the explicit reliance on Φ. For example, if $\breve{x}_i = \Phi(x_i) = a_i + b_i - x_i$, we simply write this as $\breve{x}_i = a_i + b_i - x_i$. Analogously, we can also denote type-I opposites as $\breve{x}_i = -x_i$ or $\breve{x}_i = 1 - x_i$ depending on the range of the universe of discourse.

Type-I opposites capture *linear opposition* and are simple and easy to calculate. For instance, if we regard *Age* as an integer set $A = \{0, 1, \ldots, 100\}$ and define $\Phi(age) = 100 - age$, then the opposite of a 10 year old child is an 90 year old adult, where age is some random element of A (the sex could provide additional oppositional attributes).

Definition 3 (Type-I Super-Opposition). *Let C be an n-dimensional concept set C. Then, all points $\breve{C}^s$ are type-I super-opposite of C when $d(\breve{C}^s, C) > d(\breve{C}, C)$ for some distance function $d(\cdot, \cdot)$.*

Type-I super-opposition plays a role in systems where the opposite is defined such that $|\breve{C}^s| \geq 1$ corresponding to the existence of at least one point further in distance from concept C, but not with respect to logical meaning. Consider $\Phi(X) = -X$ where $-\infty < X < \infty$. For $X = x = 0.1$, $\breve{x} = -0.1$, but there exists an infinite number of values further from x than its opposite $\breve{x}$. These extreme (or super) points are in super-opposition to C. Note that $|\breve{C}^s| \geq 0$ for any Φ (i.e. super-opposition is not required to exist). For instance, if we regard the *age* as an integer set $A = \{0, 1, \ldots, 100\}$, then the super-opposites of a 10 years old are all those members with an age above 90 years. On the other hand, consider an adult of exactly 100 years; his/her opposite is an unborn child (exactly equal to age 0). So in this case when $age = 100$ there is no super-opposition (i.e. $|\breve{C}^s| = 0$).

As another example of super-opposition, let $C = \mathbb{R}$ and use $\breve{x} = a + b - x$ as the relationship between opposites. Then for $x \in [a, b]$

$$\breve{x}^s \in \begin{cases} [a, \breve{x}) & \text{for} \quad x > (a+b)/2 \\ [a, b] - \{x\} & \text{for} \quad x = (a+b)/2 \\ (\breve{x}, b] & \text{for} \quad x < (a+b)/2 \end{cases} \qquad (2.2)$$

represents the corresponding super-oppostion relationship. In other words, for $x = \frac{a+b}{2}$ the entire interval except x becomes the super-opposite of x. This means that for $x \to \frac{a+b}{2}$ the type-I opposites converge to the same value and the range of super-opposite values increases. Using our above child/adult age example, then a member with an age of 50 years has all elements $\neq 50$ as its super-opposites, including the 10 year old and the 90 year old, simultaneously.

Definition 4 (Type-I Quasi-Opposition). *Let C be an n-dimensional concept in set C. Then, all points $\breve{C}^q$ are type-I quasi-opposite of C when $d(\breve{C}^q, C) < d(\breve{C}, C)$, for some distance function $d(\cdot, \cdot)$.*

Similar to super-opposition, quasi-opposition will not enter into all systems. For example, consider a binary problem where each $x_i \in \{0, 1\}$. In this case, neither quasi- nor super-opposition are of any relevance. However, for a more practical situation consider the single dimensional hypothecial case where $\breve{x} = a + b - x$, then

$$\breve{x}^q \in \begin{cases} (\breve{x}, \breve{x} + \min[d(a, \breve{x}), d(\breve{x}, c)]) & \text{for} \quad x > (a+b)/2 \\ \emptyset & \text{for} \quad x = (a+b)/2 \\ (\breve{x} - \min[d(b, \breve{x}), d(\breve{x}, c)], \breve{x}) & \text{for} \quad x < (a+b)/2 \end{cases} \qquad (2.3)$$

defines the set of quasi-oppositional values with $c = \frac{a+b}{2}$.

Whereas the super-opposition is naturally bounded by the extremes of the universe of discourse, the quasi-opposites can be bounded differently. A convenient way is to mirror the same distance that the opposite has to the closest interval limit in order to set a lower bound for quasi-opposites. This is illustrated in Figure 2.1. Of course, one could define a degree of quasi-oppositeness and determine the slope of gradual increase/decrease based on problem specifications. This, however, will fall into the scope of similarity/dissimilarity measures, which has been extensively investigated for many other research fields [5, 10, 19, 20].

We have defined how type-I opposition deals with the relationship between concepts based on features of the concepts without regard to the actual quality of the concept. Although, there does exist a relationship between these two cases, they are at heart, different. Quality is measured with respect to the question being asked of the opposites, in contrast to the physical structure of the two concepts. We can summarize that type-I opposition models/captures oppositional attributes in a linear or almost linear fashion, is easy to compute and, most likely, can be regarded as an approximation of the real opposites (type-II opposition) for non-linear problems.

Definition 5 (Type-II Opposition Mapping). *Any one-to-many function* $\Upsilon \colon f(\mathcal{C}) \to f(\mathcal{C})$ *which defines an oppositional relationship between the evaluation* f *of a concept*

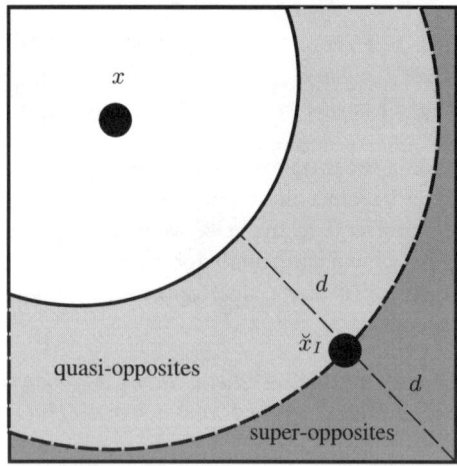

Fig. 2.1. Simplified representation of quasi- and super-opposites in a 2-dimensional space. The distance d of the type-I opposite $\breve{x}_I$ from the closest corner can be used to define the quasi- and super-oppositional regions. The boundary between these two region (bolded dashed circle segment) can be regarded quasi-opposite for a strict interpretation.

$C \in \mathcal{C}$ *to a set S of all other evaluations of concepts also in $\mathcal{C}$ such that $C \in \Upsilon(s)$* $\forall s \in S$.

The focus on evaluation f of concepts, instead of focusing on the concepts themselves, leads to loss of information regarding the true relationship of opposite concepts (i.e. type-II is a phenotypical relationship as opposed to a structural one). Given some concept C and its associated evaluation $f(C)$, the type-II opposition mapping will return a set of opposite evaluations which can then lead to concepts (although the specific true type-I opposite concept cannot be determined with certainty). Here, Υ returns a set because f is not restricted to any functional form. For example, if f is a periodic function such as the sine wave defined over $[-2\pi, 2\pi]$, then for any $f(x)$ there will be at least two opposites since a horizontal plane will cut the wave twice at each of $-3/2\pi, 1/2\pi$ and $-1/2\pi, 3/2\pi$, respectively (5 points at $-2\pi, -\pi, 0, \pi, 2\pi$ and will cut the wave at 4 points elsewhere).

Definition 6 (Type-II Opposition). *Let concept C be in set $\mathcal{C}$ and let $\Upsilon : f(\mathcal{C}) \rightarrow f(\mathcal{C})$ be a type-II opposition mapping where f is a performance function (such as error, cost, fitness, reward, etc.) Then, the set of type-II opposites of C are completely defined by Υ. Type-II opposition can also be coined* non-linear opposition.

An example of type-II opposition is in the following. Let $y = f(x_1, x_2, \cdots, x_n) \in \mathbb{R}$ be an arbitrary function where $y \in [y_{\min}, y_{\max}]$ are the extreme values of f. Then, for every point $C = (c_1, c_2, \cdots, c_n)$ we can define the type-II opposite point $\check{C} = (\check{c}_1, \check{c}_2, \cdots, \check{c}_n)$ according to

$$\check{C} = \Upsilon(C) = \{c \mid \check{y} = y_{\min} + y_{\max} - y\}. \tag{2.4}$$

We can assume that f is unknown, but that $y_{\min}$ and $y_{\max}$ are given or can be reasonably estimated.

Alternatively, at sample t we may consider a type-II opposite with respect to $t - 1$ previous samples. Using Equation (2.4), this temporal type-II opposite can be calculated according to

$$\check{c}_i(t) = \left\{ c \mid \check{y}(t) = \min_{j=1,\ldots,t} y(j) + \max_{j=1,\ldots,t} y(j) - y(t) \right\}. \tag{2.5}$$

Figure 2.2 depicts the difference between type-I and type-II opposition for a simple case where we are trying to find the extremes of a non-linear function. As apparent, type-II opposition can provide more reliable information.

In general, a type-II opposition scenario implies a deep understanding of the tartget concept, which is typically difficult to attain for real-world applications. However, type-II opposition mappings can be approximated online as the learning/search is in progress. We term this situation *opposition mining*.

Definition 7 (Opposition Mining). *Let $C^* \in \mathcal{C}$ be a target concept and Υ a type-II opposition mapping. Then, opposition mining refers to the online discovery of $\bar{\Upsilon}$, which represents an approximation to Υ.*

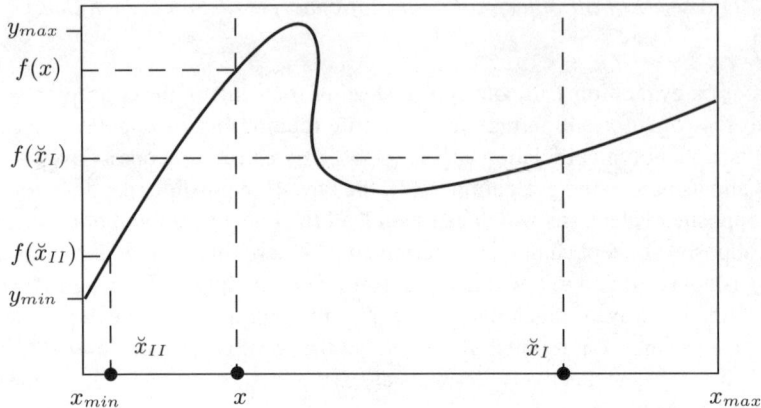

Fig. 2.2. Type-I versus type-II opposition. A nonlinear function $f : x \rightarrow [y_{min}, y_{max}]$ has the value $f(x)$ at the position x. The type-I opposite $\breve{x}_I = x_{max} + x_{min} - x$ generates the function value $f(\breve{x}_I)$. The type-II opposite $\breve{x}_{II}$, however, corresponds to the real opposite value $f(\breve{x}_{II}) = \breve{f}(x) = y_{max} + y_{min} - f(x)$. Evidently, $f(\breve{x}_I) \rightarrow f(\breve{x}_{II})$ for a linear or almost linear f.

The definitions for super- and quasi-oppostion can also be extended to the type-II domain, although we have not provided definitions here. Similarily, we can discuss the notion of a degree of opposition for both type-I and type-II scenarios (we will assume, without loss of generality a type-I situation).

Definition 8 (Degree of Opposition). *Let Φ be a type-I opposition mapping and $C_1, C_2 \in C$ arbitrary concepts such that $C_1 \neq \Phi(C_2)$. Then, the relationship between C_1 and C_2 is not opposite, but can be described as partial opposition, determined by a function $\tau : C_1, C_2 \rightarrow [0, 1]$. As $\tau(C_1, C_2)$ approaches 1 the two concepts are closer to being opposite of each other.*

So, it is possible to compare elements $C_1, C_2 \in C$ within the framework of opposition. Calculating the degree of opposition $\tau(C_1, C_2)$ will depend on how concepts are represented for the problem at hand. Nevertheless, this definition allows for the explanation of the relationship between elements that are not truly opposite. For example, consider the statements

$$s_1 = \text{"Everybody likes apples"}$$
$$s_2 = \text{"Nobody likes apples"}$$
$$s_3 = \text{"Some people like apples"}$$
$$s_4 = \text{"Some people do not like apples"}$$

We can describe the relationship between these concepts with respect to s_1 using the degree of opposition. Doing so, we can determine that $\tau(s_1, s_4) < \tau(s_1, s_3) < \tau(s_1, s_2)$, where the exact value of each relationship is ignored but is related to the logical meaning of the statements. It is important to distinguish between the degree of opposition and super- and quasi-opposition. The latter relationships are related to physical distance whereas the former is concerned with the logical relationship. In some cases these may overlap, and others there may be no such similarity.

As an example of how to combine the degree of opposition with a temporal situation as described above, assume only the evaluation function $g(X)$ is available. This is accomplished by adding the temporal variable t to the definition of the τ function. Then, using the same scenario as Equation (2.5) the type-II temporal opposite for variables X_1 and X_2 could then be computed according to

$$\tau(X_1 = x_1, X_2 = x_2, t) = \frac{|g(x_1) - g(x_2)|}{\max\limits_{j=1,\ldots,t} g(x_j) - \min\limits_{j=1,\ldots,t} g(x_j)} \in [0, 1]. \qquad (2.6)$$

A more practical example of employing the degree of opposition can be found in [22], where opposite actions were utilized for a reinforcement learning (RL) algorithm. In order to make additional updates to the Q matrix for the Q-learning algorithm, an RL agent has to know how to find opposite actions and opposite states. Clearly, this is an application-dependant problem although in some applications such as control and navigation a good definition of opposite actions is generally straightforward (*increase temperature* is the opposite of *decrease temperature*, and *move left* is the opposite of *move right*). Nonetheless, general procedures may be defined to facilitate the situation where a good opposite action may not be so apparent.

In [22] this was accomplished by defining the degree of opposition τ as a measure in how far two actions a_1 and a_2 are opposite of each other in two different states s_i and s_j, respectively, such as

$$\tau(a_1|_{s_i}, a_2|_{s_j}) = \eta \times \left[1 - \exp\left(-\frac{|Q(s_i, a_1) - Q(s_j, a_2)|}{\max\limits_{k}\left(Q(s_i, a_k), Q(s_j, a_k)\right)} \right) \right], \qquad (2.7)$$

where the Q matrix contains the accumulated discounted rewards (which guide learning), and η is the *state similarity* and can be calculated based on state clustering or by a simple measure such as

$$\eta(s_i, s_j) = 1 - \frac{\sum\limits_{k} |Q(s_i, a_k) - Q(s_j, a_k)|}{\sum\limits_{k} \max\left(Q(s_i, a_k), Q(s_j, a_k)\right)}. \qquad (2.8)$$

In [22] limited examples are provided to intuitively verify the usefulness of this measure in context of the reinforcement learning problems. However, online opposition mining based on this measure, embedded inside the parent algorithm, has not been investigated.

2.3.2 Example for Using Opposition

There are numerous manifestations of oppositional relationships in both physical world and the human society, too many to show an example for each. Instead, we will restrict ourselves to a random sampling scenario. We assume at least one optimal solution exists within a given interval $[a, b]$, and that we are given feedback regarding the distance of a guess to the closest optimal solution. To make comparing the approaches easier we group guesses into pairs.

It has been shown that for a problem with a single optima, where distance feedback is given as evaluation criteria that guesses made in a specific dependent manner (in contrast to purely random) have a higher probability of being closer to the optimal solution [17,18].

Theorem 1 (Searchig in n-Dimensional Space). *Let* $y = f(X^n)$ *be any unknown function with* $X^n = (x_1, \ldots, x_n)$ *such that* $x_i \in [a_i, b_i]$, *where* $a_i, b_i \in \mathbb{R}$. *Also, let* y *have some extreme values* $S = \{s_1, \ldots, s_n\}$, *without loss of generality assumed the minima of* y. *Generate* X, Y *based on a uniform distribution with respect to the interval* $[a_i, b_i]$ *and let* $\check{X} = a_i + b_i - X_i$. *Then*

$$Pr\left(\|\check{X}, S\| < \|Y, S\|\right) > Pr\left(\|Y, S\| < \|\check{X}, S\|\right)$$

where $\| \cdot \|$ *denotes the Euclidean distance.*

Similarily, if $\check{X} = (0.5 + X)\%1$ we can also achieve a similar expected behavior. Indeed any low discrepancy sequence (see Chapter 3) should yield an improvement, in terms of probability of making guesses closer to the optima, over purely random sampling. This concept has been employed in quasi-Monte Carlo methods [8,9,15,23], evolutionary algorithms [7,11,12,17], particle swarms [13,14,16] and neural networks [3,6], to name a few. The underlying concept is similar in all cases. Although in relation to opposition-based computing, typically the focus is on pairs of samples *(guess, opposite guess)*.

Table 2.1 shows probabilities of randomly generating a solution in 2, 5 and 10 dimensions over two opposition mappings, both compared to purely random sampling. For each dimension we assume the interval is $[0, 1]$, without loss of generality. The first method, called "Mirror", uses $\check{x}_i = 1 - x_i$, and the second method, "Modulo", is defined as $\check{x}_i = (0.5 + x_i)\%1$. To make comparing the approaches simpler, we examine pairs of guesses. The table shows triples $< P(\min(X, Y) = \min(\check{X}, Y)), P(\min(X, \check{X}) < \min(X, Y)), P(\min(X, \check{X}) > \min(X, Y) >$ which compare the probability of two guesses being equal, opposition-based technique is more desirable and random guesses are better, respectively. The data was gathered by randomly selecting 1000 solutions, where 10,000 guessing pairs were considered. The value provided represents the average over these experiments.

In all experiments, the non-random guessing yields a higher probability of being closer to one of the optimal solutions. Although, as the number of optimal solutions increases, this benefit decreases, especially for the Mirror opposition mapping. But, the Mirror function seems to be the best of the two mappings when there only exists one optimal solution.

In general for distance-based problems, given m optimal solutions represented as the set $S = \{S_1, ..., S_m\}$, the goal is to devise an opposition mapping such that if we let $g_1 = d(X, s^{1,*})$, $g_2 = d(Y, s^{2,*})$ and $g_3 = d(\Phi(X), s^{3,*})$ then we desire the expected value relationship,

Table 2.1. Generating guesses where $< P(\min(X,Y) = \min(\check{X},Y)), P(\min(X,\check{X}) < \min(X,Y)), P(\min(X,\check{X}) > \min(X,Y) >$ for variables X, Y. In most cases, a Modulo function outperforms the Mirror function, but both show an improvement over purely random sampling. See Chapter 3 for more on low-discrepancy sequences.

Method	\multicolumn{3}{c}{Number of optimal Solutions}		
	1	2	5
\multicolumn{4}{c}{2 Dimensions}			
Mirror	$< 0.358, 0.358, 0.283 >$	$< 0.344, 0.343, 0.313 >$	$< 0.338, 0.338, 0.321 >$
Modulo	$< 0.364, 0.364, 0.273 >$	$< 0.355, 0.355, 0.291 >$	$< 0.344, 0.344, 0.312 >$
\multicolumn{4}{c}{5 Dimensions}			
Mirror	$< 0.361, 0.361, 0.278 >$	$< 0.346, 0.346, 0.309 >$	$< 0.334, 0.334, 0.332 >$
Modulo	$< 0.359, 0.359, 0.282 >$	$< 0.353, 0.354, 0.293 >$	$< 0.347, 0.347, 0.307 >$
\multicolumn{4}{c}{10 Dimensions}			
Mirror	$< 0.361, 0.362, 0.277 >$	$< 0.345, 0.345, 0.310 >$	$< 0.332, 0.332, 0.336 >$
Modulo	$< 0.344, 0.360, 0.296 >$	$< 0.343, 0.358, 0.299 >$	$< 0.340, 0.356, 0.304 >$

$$E[\min(g_1, g_3)] < E[\min(g_1, g_2)] \tag{2.9}$$

where $d(\cdot, \cdot)$ is a distance function and $s^{1,*}, s^{2,*}, s^{3,*}$ are the solutions closest to each of X, Y and $\Phi(X)$, respectively. Thus, the most desirable opposition map Φ^* will be one that maximizes the difference

$$\Phi^* = \operatorname*{argmax}_{\Phi} E[\min(g_1, g_2)] - E[\min(g_1, g_3)], \tag{2.10}$$

although improved performance should be observed for any Φ satisfying equation 2.9. Notice though that it is possible to design poor opposition mapping functions which actually decrease the performance of the algorithm.

2.4 Opposition-Based Computing

Having introduced the foundational definitions for opposition-based computing, we will now provide its definition as well as a first attempt at a classification strategy for opposition-based computing methods in the following section.

Definition 9 (Opposition-Based Computing). *We speak of opposition-based computing, when a computational method or technique implicitly or explicitly employs oppositional relations and attributes either at its foundation (purely oppositional algorithms) or to improve existing behavior of some parent algorithm (opposition-based extensions of existing algorithms).*

Situations in which opposition is inherent to the algorithm existence (for instance switching between opposite alternatives in a decision-making scenario) is generally termed *implicit* because the underlying concept tends to be overlooked in favor of improving its behavior by other means. For example, adversarial searching using game tree searching typically follows a turn-based rule which alternates between two play-ers. The focus of research in this area tends towards decreasing computational time via improving the decision as to which nodes are to be explored instead of changing the un-derlying adversarial structure of the algorithm. A similar concept also applies to other competition-based methods and paradigms.

In contrast to implicit opposition, explicit forms occur when the goal is to utilize the non-apparent or imperceptible opposition as a means of improving an existing, usually non-oppositional algorithm. An example of this is the use of antithetic sampling to re-duce the variance and improve convergence rate of Monte Carlo integration methods. The integration method works without the use of any opposites, but by explicitly us-ing antithetic variates we can observe improved behavior in terms of reduced variance leading to faster convergence.

In both implicit and explicit algorithms it is possible to further determine the man-ner in which opposites interact with respect to the scale $[0 = cooperative, \ldots, 1 = competitive]$.

Definition 10 (Degree of Competition). *Let $C_1, C_2 \in C$ be arbitrary concepts where $C_1 = \Phi(C_2)$. With inherent respect to the algorithm employing opposition, Σ, the degree at which C_1, C_2 compete for use in Σ is given by the function $\zeta \colon C_1, C_2|\Sigma \to [0, 1]$. Values of ζ closer to 1 indicate a more competive situation. For readability we often drop the Σ, though it is implied.*

Basically, this definition allows us to measure how cooperative or competitive the con-sideration of two opposite guesses/concepts are. For a game tree search $\zeta \to 1$ (totally competitive), and in case of antithetic variates we take an average of the results such that $\zeta \to 0$ (totally cooperative).

We can also speak of the degree of competition between concepts which have a degree of opposition < 1. The above definition changes slightly in that the requirement of $C_1 = \Phi(C_2)$ is dropped and we adopt a subscript ζ_τ to represent the case that the concepts are not true opposites.

Whether wishing to describe the degree of opposition or the degree of competition between opposites, there may be a time componet to the interaction. In such cases the goal is to describe the degree of opposition between aribitrary concepts $C_1, C_2 \in C$ as time (or algorithm iterations) t varies,

$$\Upsilon_t(C_1, C_2) = \frac{d\Upsilon(C_1, C_2)}{dt}. \tag{2.11}$$

In a similar fashion we can describe the behavior of the degree of competition between concepts with respect to algorithm Σ at time t as

$$\zeta_t(C_1, C_2) = \frac{d\zeta(C_1, C_2)}{dt}. \tag{2.12}$$

These definitions are very broad and there are many possible manifestations of opposition that can arise. We provide possibilities below, but this is by no means a comprehensive list. Of course, some of these areas overlap, and can be hybridized.

2.5   OBC Algorithms

Opposition can be embedded into existing algorithms implicitly or explicitly. For instance, the Proof Number Search (Chapter 6) is an example of implicit opposition usage, whereas the ODE algorithm (Chapter 8) is an explicit implementation. In this section we establish a general framework for explicit employment of OBC within existing methodologies. For this purpose we begin general definitions of *OBC algorithms*.

Definition 11 (Implicit OBC Algorithms). *Any algorithm that incorporates oppositional concepts without explicitly using type-I or type-II opposites is an implicit OBC algorithm, or short an I-OBC algorithm.*

The range of I-OBC techniques is large since they have been in use since quite some time. For example, the well-known bisection method [4] for solving equations can be considered an I-OBC algorithm since it shrinks the search interval by looking at positive versus negative sign change. As well, using Bayes theorem [2] is an implicit usage of oppositional relationship between the conditional probabilities $p(A|B)$ and $p(B|A)$[5]. A different, stronger version of implicit incorporation of oppositional concepts is Bayesian Yin-Yang Harmony Learning which is based on the alternative viewpoints of Bayes Rule (Chapter 10).

Definition 12 (Explicit OBC Algorithms). *Any algorithm dealing with an unknown function $Y = f(X)$ and a known evaluation function $g(X)$ (with higher values being more desirable) extended with OBC to calculate type-I or type-II opposite $\check{X}$ of numbers, guesses or estimates X for considering $\max\left(g(X), g(\check{X})\right)$ in its decision-making is an explicit OBC algorithm, or short an E-OBC algorithm.*

Generally, three classes of E-OBC algorithms can be distinguished:

1. **Initializing E-OBC Algorithms** – The concept of opposition is only used during the initialization. The effect is a better start (initial estimates closer to solution vicinity). Further, since it is done before the actual learning/searching begins, this creates no additional overhead.
2. **Somatic E-OBC Algorithms** – This approach is based on modification of the body of an existing algorithm. For instance, changing the weights in a neural network during the learning based on integration of opposites weights changes the way the algorithm works in every step. The effect of OBC will be much more visible since opposition is considered in every iteration/episode/generation.
3. **Initializing and Somatic E-OBC Algorithms** – This class uses OBC both during initialization and actual learning/search by combining the two previous approaches.

Algorithm 1 provides the generic structure of E-OBC algorithms. However, it seems that the selection of an appropriate OBC algorithm and its specific steps directly depend on the problem at hand and cannot be universally established.

---

[5] The underlaying idea of Bayes theorem is looking at events from opposite perspectives when analyzing a cause-effect problem. For example, if we have the prior probability of cancer, $p(\text{cancer})$, and on smoking, $p(\text{smoking})$, then the conditional posterior probability $p(\text{cancer}|\text{smoking})$ can be determined when we make an antipodal observation for the conditional probability $p(\text{smoking}|\text{cancer})$.

Algorithm 1. General pseudo code of initializing and somatic E-OBC algorithms. Different schemes may/shall be defined for specific parent algorithms.

1: Establish evaluation function $g(\cdot)$ (e.g. error, fitness, reward etc.)
2: Initialize parameters P=$\{p_1, \cdots, p_n\}$ (e.g. weights, chromosomes etc.)
3: Find opposites $\breve{P} = \{\breve{p}_1, \cdots, \breve{p}_n\}$

4: **if** First option **then**
5:     Establish $M = P \cup \breve{P}$
6:     Sort M in descending order with respect to $g(\cdot)$
7:     Choose the n first parameters $m_1, \cdots, m_n$ from M
8: **end if**

9: **if** Second option **then**
10:     Run the algorithm with P
11:     Run the algorithm with $\breve{P}$
12:     Choose the better one with respect to $g(\cdot)$
13: **end if**

14: **for** each learning/search loop of the parent algorithm **do**
15:     Calculate the estimate X based on schemes provided by the parent algorithm
16:     **if** OBC-condition satisfied **then**
17:         Calculate $\breve{X}$
18:         Perform the parent algorithm's steps for $\breve{X}$
19:         Take X or $\breve{X}$ based on $\max(g(X), g(\breve{X}))$
20:     **end if**
21: **end for**

Line 1 in Algorithm 1 controls the frequency of relaying on opposition while the parent algorithm runs. The following cases can be distinguished:

1. **OBC condition always true** – OBC can be applied during the entire learning/search, meaning that in all iterations/episodes/generations, proper opposite quantities are calculated and comparisons with original quantities are conducted. It could appear that in this way we can fully exploit the benefit of oppositional concept. However, in some cases the additional overhead for calculation of opposite entities can exceed its advantage; the extended algorithm becomes less effective than its original version in terms of convergence speed (the accuracy will be at least as high as the regular call without OBC due to $\max(g(X), g(\breve{X}))$).
2. **OBC condition as a threshold** – If we decide to activate OBC only for a fraction of time, then the most straightforward condition for opposite estimates is to set a threshold θ. If number of iterations/episodes/generations is less than θ, for instance, then opposite estimates will be calculated and considered. This will eliminate the overhead complexity but adds a degree of freedom which needs empirical knowledge to be filled.
3. **OBC condition as a probabilistic constraint** – Opposite consideration could be triggered by a probability function, which ideally is coupled with the evaluation function $g(X)$. Empirical and uniform probabilities may be used as well.

Generally, probabilistic control of opposition frequency should be either a decreasing or increasing function. For former, we rely on opposition rather at the begining stages of learning/search and do not use it towards the end to suppress the computational overhead. For latter case, we increase the opposition frequency as learning/search progresses since, perhaps, we do not know the true opposites initially. This would correspond to dynamic/online opposition mining.

In following subsections we try to categorize different OBC algorithms.

2.5.1   Search and Reasoning

Generally speaking, a search algorithm explores possible solutions to a problem in a systematic manner with the goal of discovering some solution. Although, in many instances the goal may not be obtainable in a reasonable amount of time. A popular approach to deal with this scenario is to incorporate heuristic knowledge about the problem or solution to guide the searching algorithm, instead of an uninformed search which does not use any such knowledge. Reasoning is a process of searching for some explanation for a belief or action. Therefore search and reasoning are closely coupled processes. Examples include:

1. Implicit: adversarial search and reasoning (see Chapter 6).
2. Explicit: understanding or compromising between opposing arguments, considering opposite reasonings (see Chapter 5).

2.5.2   Optimization

The purpose of optimization algorithms is to maximize or minimize some real-valued function. Without loss of generality, consider a minimization problem where we are given a function $f \colon A \to \mathbb{R}$ from some set A we seek to find an element $a^* \in A$ such that $f(a^*) \leq f(a) \ \forall a \in A$. Of course, this generalizes all searches, but the reader should keep important aspects such as constraint satisfaction, multiple objectives, etc in mind. Examples include:

1. Implicit: Competitive coevolution (see Chapter 9).
2. Explicit: Using a low-discrepancy paired guessing strategy to improve diversity and/or convergence rate (see Chapter 3).

2.5.3   Learning

Learning refers to the process of modifying behavior as a result of experiences and/or instructions. In other words, given some initial hypothesis, learning allows for its modification as more data are observed. This contrasts with search and reasoning in that learning is not concerned with explaining why the data is observed, only to perform some action when it occurs. Often learning can be seen as an optimization problem, however, here we distinguish them because they do not always imply one another. Examples include:

1. Implicit: model selection using opposing viewpoints of a problem (see Chapter 10).
2. Explicit: considering two opposing actions or states during active learning (see Chapter 11).

2.6   Summary and Conclusions

In this work, the general framework of opposition-based computing was introduced. Basic definitions of type-I and type-II opposites along with categorization of oppositional algorithms were provided as well. Generally, OBC provides a straightforward framework for extension of many existing methodologies as this has been demonstrated for neural nets, evolutionary algorithms, reinforcement learning and ant colonies. The OBC extensions of existing machine learning schemes seem to primarily accelerate search, learning and optimization processes, a characteristic highly desirable for hyperdimensional complex problems.

Opposition-based computing, however, encompasses multiple challenges as well. First and foremost, a solid and comprehensive formalism is still missing. Besides, the definition of opposition may not be straightforward in some applications such that the need for opposition mining algorithms should be properly addressed in future investigations. On the other side, the cost of calculating opposite entities can exceed the time saving so that the OBC-extended algorithm becomes slower than the parent algorithm. Extensive research is still required to establish control mechanisms to regulate the frequency of opposition usage within the parent algorithms.

Due to our imperfect understanding of interplay between opposite entities, this work will most likely be a preliminary investigation. Hence, more comprehensive elaborations with a solid mathematical understanding of opposition remain subject to future research.

The motivation of this chapter was not to just provide a framework for oppositional concepts claiming to cover all possible forms of opposition (due to the diversity of oppositeness this seems to be extremely difficult, if not impossible), but also to establish a general umbrella under which other, existing techniques can be classified and studied. This should not be understood to just put a new label on existing methodologies and algorithms but to consciously comprehend the oppositional nature of problems and exploit a-priori knowledge in light of inherent antipodality and complementarity of entities, quantities and abstractions.

2.7   Contributions and Acknowledgements

The idea of opposition-based computing and its applications for genetic algorithms, neural nets, reinforcement learning and ant colonies has been introduced by Dr. Hamid R. Tizhoosh. Mario Ventresca has designed and developed OBC versions of backpropagation algorithm and simulated annealing, and provided some formalism to establish the OBC definitions. Dr. Shahryar Rahnamayan has introduced opposition-based differential evolution (ODE) with practical and theoretical contributions to some of the formalism in this work. The authors would like to thank Dr. Don McLeish for valuable discussions with respect to the relationship between antithetic variates and opposite quantities. We also are grateful to Maryam Shokri and Dr. Farhang Sahba for their contributions to the OBC research. As well, we would like to thank Alice Malisia and Masoud Mahootchi who have provided examples how OBC can be practically implemented.

References

1. The American Heritage Dictionary of the English Language. Houghton Mifflin (2000)
2. Carlin, B.P., Louis, T.A., Carlin, B.: Bayes and Empirical Bayes Methods for Data Analysis. Chapman and Hall/CRC, Boca Raton (2000)
3. Cervellera, C., Muselli, M.: Deterministic Design for Neural Network Learning: An Approach based on Discrepancy. IEEE Transactions on Neural Networks 15(3), 533–544 (2004)
4. Chapra, S.C., Canale, R.P.: Numerical Methods for Engineers. McGraw-Hill Education, New York (2005)
5. Cheng, V., Li, C.H., Kwok, J.T., Li, C.K.: Dissimilarity learning for nominal data. Pattern Recognition 37(7), 1471–1477 (2004)
6. Jordanov, I., Brown, R.: Neural Network Learning Using Low-Discrepancy Sequence. In: Foo, N.Y. (ed.) Canadian AI 1999. LNCS, vol. 1747, pp. 255–267. Springer, Heidelberg (1999)
7. Kimura, S., Matsumura, K.: Genetic Algorithms using Low-Discrepancy Sequences. In: Genetic and Evolutionary Computation Conference, pp. 1341–1346 (2005)
8. Kucherenko, S.: Application of Quasi Monte Carlo Methods in Global Optimization. In: Global Optimization, pp. 111–133 (2006)
9. Kucherenko, S., Sytsko, Y.: Application of Deterministic Low-Discrepancy Sequences in Global Optimization. Computational Optimization and Applications 30, 297–318 (2005)
10. Li, M., Chen, X., Li, X., Ma, B., Vitanyi, P.M.B.: The similarity metric. IEEE Transactions on Information Theory 50(12), 3250–3264 (2004)
11. Maaranen, H., Miettinen, K., Makela, M.M.: Quasi-random Initial Population for Genetic Algorithms. Computers and Mathematics with Applications 47(12), 1885–1895 (2004)
12. Maaranen, H., Miettinen, K., Penttinen, A.: On Initial Populations of a Genetic Algorithm for Continuous Optimization Problems. Journal of Global Optimization 37, 405–436 (2007)
13. Nguyen, Q., Nguyen, X., McKay, R., Tuan, P.: Initialising PSO with Randomised Low-Discrepancy Sequences: The Comparative Results. In: Congress on Evolutionary Computation, pp. 1985–1992 (2007)
14. Nguyen, X., Nguyen, Q., McKay, R.: Pso with Randomized Low-Discrepancy Sequences. In: Genetic and Evolutionary Computation Conference, p. 173 (2007)
15. Niederreiter, H., McCurley, K.: Optimization of Functions by Quasi-random Search Methods. Computing 22(2), 119–123 (1979)
16. Pant, M., Thangaraj, R., Grosan, C., Abraham, A.: Improved Particle Swarm Optimization with Low-Discrepancy Sequences. In: Congress on Evolutionary Computation (to appear, 2008)
17. Rahnamayan, S., Tizhoosh, H.R., Salamaa, M.: A Novel Population Initialization Method for Accelerating Evolutionary Algorithms. Computers and Mathematics with Applications 53(10), 1605–1614 (2007)

18. Rahnamayan, S., Tizhoosh, H.R., Salamaa, M.: Opposition versus randomness in soft computing techniques. Applied Soft Computing 8(2), 906–918 (2008)
19. Sarker, B.R., Saiful Islam, K.M.: Relative performances of similarity and dissimilarity measures. Computers and Industrial Engineering 37(4), 769–807 (1999)
20. Stewart, N., Brown, G.D.A.: Similarity and dissimilarity as evidence in perceptual categorization. Journal of Mathematical Psychology 49(5), 403–409 (2005)
21. Tizhoosh, H.R.: Opposition-Based Learning: A New Scheme for Machine Intelligence. In: International Conference on Computational Intelligence for Modelling Control and Automation, vol. 1, pp. 695–701 (2005)
22. Tizhoosh, H.R.: Opposition-based reinforcement learning. Journal of Advanced Computational Intelligence and Intelligent Informatics (JACIII) 10(4), 578–585 (2006)
23. Yaochen, Z.: On the Convergence of Sequential Number-Theoretic Method for Optimization. Acta Mathematicae Applicatae Sinica 17(4), 532–538 (2001)

3

Antithetic and Negatively Associated Random Variables and Function Maximization

Don L. McLeish

Department of Statistics and Actuarial Science, University of Waterloo, Canada

Summary. A pair of antithetic random numbers are variates generated with negative dependence or correlation. There are several reasons why one would wish to induce some form of negative dependence among a set of random numbers. If we are estimating an integral by Monte Carlo, and if the inputs are negatively associated, then an input in an uninformative region of the space tends to be balanced by another observation in a more informative region. A similar argument applies if our objective is Monte Carlo maximization of a function: when the input values are negatively associated, then inputs far from the location of the maximum tend to be balanced by others close to the maximum. In both cases, negative association outperforms purely random inputs for sufficiently smooth or monotone functions. In this article we discuss various concepts of negative association, antithetic and negatively associated random numbers, and low-discrepancy sequences, and discuss the extent to which these improve performance in Monte Carlo integration and optimization.

3.1 Antithetic and Negative Association

Antithetic random numbers are normally used to generate two negatively correlated copies of an unbiased estimator so that on combination, we have reduced variance. They are one of a large number of variance-reduction techniques common in Monte-Carlo methods (for others, see [12]). For example, suppose we wish to estimate an integral

$$\int_{-\infty}^{\infty} h(x)f(x)dx$$

where $f(x)$ is a probability density function and h is an arbitrary function for which this expected value exists. One very simple method for doing this, usually known as "crude Monte Carlo", is to generate independent observations $X_i, i = 1, 2, ..., n$ from the probability density function $f(x)$ and then estimate the above integral with a sample mean

$$\frac{1}{n} \sum_{i=1}^{n} h(X_i).$$

In order that the above estimator be unbiased, i.e. that

$$E\left[\frac{1}{n} \sum_{i=1}^{n} h(X_i)\right] = \int_{-\infty}^{\infty} h(x)f(x)dx$$

H.R. Tizhoosh, M. Ventresca (Eds.): Oppos. Concepts in Comp. Intel., SCI 155, pp. 29–44, 2008.
springerlink.com

we do not need the assumption of independence, but only that each of the random variables X_i have probability density function $f(x)$ so it is natural to try and find joint distributions of the random variables X_i which will improve the estimator, i.e. make its variance $var\left(\frac{1}{n}\sum_{i=1}^{n}h(X_i)\right)$ smaller. One of the simplest approaches to this problem is through the use of antithetic random numbers. This is easier to understand in the case $n = 2$, for then, if $h(X_1)$ and $h(X_2)$ have identical distributions and therefore the same variance,

$$var\left(\frac{1}{2}\sum_{i=1}^{2}h(X_i)\right) = 2[var(h(X_1)) + cov(h(X_1), h(X_2))] \qquad (3.1)$$

Of course if X_1 and X_2 are independent random variables, in order to minimize this quantity while fixing the common marginal distribution of X_1, X_2, we need to minimize the covariance between the two, $cov(h(X_1), h(X_2))$, subject to the constraint that X_1, X_2 have the required probability density function $f(x)$. The solution is quite simple. Suppose that X_1 and X_2 both have finite variance,

$$F(x_1, x_2) = P(X_1 \le x_1, X_2 \le x_2)$$

is the joint cumulative distribution function and $F_1(x_1) = P(X_1 \le x_1), F_2(x_2) = P(X_2 \le x_2)$ are the marginal cumulative distribution functions of X_1, X_2 respectively. Then in [10] it is proved that

$$cov(X_1, X_2) = \int_{-\infty}^{\infty}\int_{-\infty}^{\infty}(F(x_1, x_2) - F_1(x_1)F_2(x_2))dx_1 dx_2 \qquad (3.2)$$

and [11] gives a simple proof of this identity. In [3] it is proved that if $\alpha(X_1)$ and $\beta(X_2)$ are functions of bounded variation on the support of the probability distribution of the random vector (X_1, X_2) with finite first moments and $E(|\alpha(X_1)||\beta(X_2)|) < \infty$, then

$$cov(\alpha(X_1), \beta(X_2)) = \int_{-\infty}^{\infty}\int_{-\infty}^{\infty}(F(x_1, x_2) - F_1(x_1)F_2(x_2))\alpha(dx_1)\beta(dx_2)$$

Now suppose that we wish to minimize the variance (3.1). Since we are constrained to leave the marginal distribution alone, we can only change the dependence between X_1 and X_2 and thus vary the term $F(x_1, x_2)$ in (3.2). Our problem becomes the minimization of $F(x_1, x_2)$, if possible for all values of x_1, x_2 with a constraint on the marginal distributions of X_1 and X_2.

Suppose we generate both X_1 and X_2 by inverse transform, that is by generating (continuous) uniform random numbers U_1 and U_2 of the interval $[0, 1]$ and then solving $F(X_1) = U_1, F(X_2) = U_2$ or, equivalently by using the inverse of the function F, $X_1 = F^{-1}(U_1), X_2 = F^{-1}(U_2)$[1] Obviously we build some dependence between X_1 and X_2 by generating dependence between U_1 and U_2 so let us try an extreme form of dependence $U_2 = 1 - U_1$, for example. In this case, the joint cumulative distribution function is

---

[1] This definition requires that $F(x)$ be continuous and strictly increasing. The definition of inverse transform that applies to a more general c.d.f. is $X = \inf\{x; F(x) \ge U\}$.

$$\begin{aligned}
F(x_1, x_2) &= P(X_1 \le x_1, X_2 \le x_2) \\
&= P(F^{-1}(U_1) \le x_1, F^{-1}(1 - U_1) \le x_2) \\
&= P(U_1 \le F(x_1), 1 - U_1 \le F(x_2)) \\
&= P(U_1 \le F(x_1) \text{ and } U_1 \ge 1 - F(x_2)) \\
&= \max(0, F(x_1) - 1 + F(x_2)).
\end{aligned}$$

There is an elegant inequality due to Fréchet [5], [4] which states that for all joint cumulative distribution functions $F(x_1, x_2)$ which have marginal c.d.f. $F(x)$,

$$\max(0, F(x_1) + F(x_2) - 1) \le F(x_1, x_2) \le \min(F(x_1), F(x_2))$$

and so indeed the smallest possible value for $F(x_1, x_2)$ is achieved when we generate both X_1 and X_2 from the same uniform random variate and put $X_1 = F^{-1}(U_1)$, $X_2 = F^{-1}(1 - U_1)$. This strategy for generating negative dependence among random variables is referred to as the use of *antithetic* (or opposite) *random numbers*.

Definition 1. *Random numbers X_1 and X_2 with continuous cumulative distribution functions F_1, F_2 respectively are said to form an antithetic pair if for some uniform[0,1] random variable U, we have $X_1 = F_1^{-1}(U)$, and $X_2 = F_2^{-1}(1 - U)$.*

The concept of antithetic random numbers as a tool for reducing the variance of simulations appears in the early work of Hammersley (see [8] and [7]). The extent of the variance reduction achieved with antithetic random numbers is largely controlled by the degree of symmetry in the distribution of the estimator. In the best possible case, when $h(F^{-1}(U))$ is a *linear function* of U, the use of paired antithetic variates can entirely eliminate the Monte Carlo variance so that the estimator with

$$\frac{1}{2}[h(X_1) + h(X_2)]$$

$X_1 = F^{-1}(U_1)$, $X_2 = F^{-1}(1 - U_1)$ is perfect in the sense that it is exactly equal to $\int_{-\infty}^{\infty} h(x)f(x)dx$ and its variance, $var(\frac{1}{2}[h(X_1) + h(X_2)])$ is zero. When two random variables come in antithetic pairs, e.g. $X_1 = F^{-1}(U_1)$, and $X_2 = F^{-1}(1 - U_1)$, we will later use the notation $X_2 = X_1^-$ to indicate this dependence.

For a more general function, of course, the variance of the estimator is unlikely to be zero but as long as the function h is monotonic, we are guaranteed that the use of antithetic random numbers is at least as good in terms of variance as using independent random numbers, i.e. in this case

$$var\{\frac{1}{2}[h(X_1) + h(X_2)]\} \le \frac{1}{2}var\{h(X_1)\}$$

The problem of generating negatively associated random variables in more than two dimensions has a richer variety of solutions. One possibility is to generate n (assuming n is even) points using $n/2$ independent uniform random numbers $U_1, ..., U_{n/2}$ and then their antithetic partners, $1 - U_1, ..., 1 - U_{n/2}$. In general random variables $(X_1, ..., X_n)$ are said to be negatively dependent if for any two functions h_1 and h_2

where h_1 is a function of a subset of the $X_i's$ say $\{X_{i_1}, X_{i_2}, ..., X_{i_k}\}$ which is non-decreasing in each of the k components, and h_2 is a similar function of a subset of the remaining $X_i's$ i.e. a function of $\{X_j; j \neq i_1, i_2, ..., i_k\}$ then

$$cov(h_1(X_{i_1}, X_{i_2}, ..., X_{i_k}), h_2\{X_j; j \neq i_1.i_2, ..., i_k\}) \leq 0$$

Such structures of positive and negative dependence were introduced by [1]. Random variables $(X_1, ..., X_n)$ achieve the extreme antithesis property if they are exchangeable (that is the joint distribution of any permutation of $(X_1, ..., X_n)$ is the same as that of $(X_1, ..., X_n)$) and if they achieve the minimum possible correlations for the given marginal distribution. Notice for exchangeable random variables X_i all with marginal c.d.f. $F(x)$, variance σ^2 and correlation coefficient ρ, that

$$var(\sum_{i=1}^{n} X_i) = \sigma^2 n[1 + (n-1)\rho] \geq 0. \tag{3.3}$$

This implies that $\rho \geq -\frac{1}{n-1}$ so that the smallest possible value for the covariances among n exchangeable random variables is $-\frac{1}{n-1}$. This is not always achievable but it is possible to get arbitrarily close to this value for uniformly distributed random variables. The following iterative Latin hypercube sampling algorithm (see [2]) allows us to get arbitrarily close in any dimension n:

Iterative Latin Hypercube Sampling Algorithm

1. *Begin with $t = 0$. Define $U^{(0)} = (U_1^{(0)}, U_2^{(0)}, ...U_n^{(0)})$ all independent $U[0, 1]$.*
2. *Draw a random permutation π_t of $\{1, 2, ..., n\}$ independent of all previous draws.*
3. *Define*

$$\mathbf{U}^{(t+1)} = \frac{1}{n}(\pi_t + \mathbf{U}^{(t)})$$

4. *Set $t = t + 1$. If $t > t_{max}$ exit, otherwise return to step 2.*

In [2] it is shown that

$$cor(U_i^{(t)}, U_j^{(t)}) = -\frac{1}{n-1}(1 + \frac{1}{n^{2t}}) \text{ for } i \neq j$$

which very rapidly approaches the minimum possible value of $-\frac{1}{n-1}$ as t grows. Running this algorithm for a few (e.g. $t_{max} = 10$) steps then results in a vector $\mathbf{U}^{(t)}$ whose components are pairwise very close to the minimum possible correlation $-\frac{1}{n-1}$.

Because antithetic random numbers and negative dependence control the covariance between two monotone functions, it is a device suited best to reducing the variance of Monte Carlo integrals such as

$$var\{\frac{1}{n}\sum_i h(X_i)\}.$$

We might also wish to induce dependence in a sequence of values X_i however if we wanted to determine the location of the minimum or the maximum of the function h, and this is the theme of the next section.

3.2 Monte Carlo Maximization Using Antithetic and Dependent Random Numbers

In this section, we will discuss the general problem, that of the design of points $u_1, ..., u_m$ in one or more dimensions with a view to improvement over the case of independent random variables. The basic questions are: how should we select the points u_i and how much improvement can we expect over a crude independent sample? We begin with a general result. We wish to approximately maximize a function $g : [0,1]^d \to \Re$ and to that end determine a set of points $(u_1, u_2, ...u_m)$ which are marginally uniformly distributed (i.e. each u_1 has a uniform distribution on $[0,1]^d$) but possibly dependent so that the probability of improvement over crude

$$p_{im} = P[\max(g(U_1), g(U_2), ...g(U_n)) < \max(g(u_1), g(u_2), ..., g(u_m))] \qquad (3.4)$$

is maximized for smooth functions g. Here each of the random vectors $U_1, ..., U_n$ are independent uniform on $[0,1]^d$ and each of the u_i are also uniform$[0,1]^d$ but the latter may be dependent.

Properties of the maximum are closely tied to the Hessian H of the function $g(x)$ at its maximum, a matrix H where

$$H_{ij} = \frac{\partial^2 g(x)}{\partial x_i \partial x_j}.$$

Assume first that this matrix H has all eigenvalues equal to $-\lambda$ so that, after an orthogonal transformation we can assume that $H = -\lambda I_d$, where I_d is the $d-$dimensional identity matrix. Suppose D is a random variable defined as the closest distance from a random point in $[0,1]^d$, say U to a point in the set $\{u_1, u_2, ..., u_m\}$, i.e.

$$D^2 = \min\{(U - u_j)'(U - u_j), j = 1, 2, ..., m\}.$$

Then there is a simple relationship between the probability of improvement and the distribution of D^2 valid for reasonably large values of n (large enough that edge effects do not play a significant role),

$$p_{\text{imp}} \simeq E\{\exp(-nV(D))\} = E\{\exp(-c_d nD^d)\} \qquad (3.5)$$

where $V(D)$ is the volume of a ball of radius D in dimension d and c_d is defined by $V(D) = c_d D^d$. Particular values are:

d	1	2	3	4	5	6	7	8
c_d	2	π	$\frac{4}{3}\pi$	$\frac{\pi^2}{2}$	$\frac{8\pi^2}{15}$	$\frac{\pi^3}{6}$	$\frac{16\pi^3}{105}$	$\frac{\pi^4}{24}$

Proof of formula (3.5). Let D be the distance between the point $u^* = \arg\max g(u)$ and the closest point in the set $\{u_1, u_2, ..., u_m\}$. Let $V_{u^*}(D)$ be the volume of the intersection of the unit cube $[0,1]^d$ with a ball of radius D around the point u^*. Then the probability that no point among $\{U_1, ..., U_n\}$ is inside this ball is the probability that every one of the points U_i falls outside the ball:

$$
\begin{aligned}
p_{\text{imp}} &= E[(1 - V_{u^*}(D))^n] \\
&\simeq E[(1 - c_d D^d)^n] \\
&\simeq E[\exp(-c_d n D^d)] \text{ for } n \text{ large and } D^d \text{ small.}
\end{aligned}
\tag{3.6}
$$

Note that the approximation in (3.6) holds provided that D is small, (*i.e.* m is large), because then the edge effect which occurs when u^* is within d units of the boundary of the cube is negligible.

This approximation (3.5) is remarkably accurate for reasonably large n and shows that in order to maximize the probability of improvement, we need to choose a point set u_i which is as close as possible to a randomly chosen point in $[0,1]^d$. The potential improvement over independent U_i is considerable. To generate such as set of u_i we will use the notation $x \bmod 1$ to represent the fractional part of the real number x. Suppose $d = 1$ and we choose points equally spaced such as

$$
u_i = (u_0 + \frac{i}{n}) \bmod 1, i = 1, 2, ..., n
\tag{3.7}
$$

for u_0 uniform[0,1]. For example in Figure 3.1, $n = 10$, the value of u_0 was 0.815 and the closest distance from the random point U to a point in the set $\{u_0, u_1, ...u_9\}$ is denoted D.

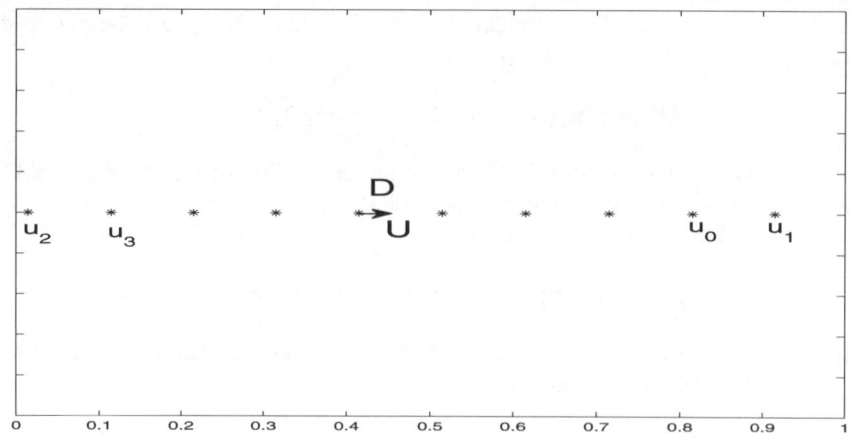

Fig. 3.1. The points (6) when n=10

It is easy to see that D has a uniform $[0, \frac{1}{2n}]$ distribution and so

$$p_{imp} \simeq E\{\exp(-2nD)\} = 2n \int_0^{1/2n} e^{-2nx} dx = 1 - e^{-1} \simeq 0.632.$$

In this case the probability of improvement over the independent case can be computed exactly and a correction to this approximation obtained since

$$p_{imp} = 2n \int_0^{1/2n} (1 - 2x)^n dx = \frac{n}{n+1} (1 - (1 - \frac{1}{n})^{n+1})$$

$$\simeq (1 - e^{-1}) - \frac{1}{n}(1 - \frac{5}{2}e^{-1}) + O(n^{-3}) = 0.63212 - \frac{0.0803}{n} + O(n^{-3})$$

The term $\frac{0.0803}{n}$ above renders the approximation nearly exact. For example when $n = 2$,

$$p_{imp} = 4 \int_0^{\frac{1}{4}} (1 - 2x)^2 dx = \frac{7}{12} \simeq 0.58333$$

whereas the first two terms give $0.63212 - \frac{0.0803}{2} = 0.59197$. The closeness of these two answers leads us to believe that use of the first two terms will provide an accurate approximation for any $n > 2$.

There is nothing particularly unique about the placement of points suggested by (3.7) that we used to generate dependent values u_i. This is simply a device used for stratification of the sample, so that every subinterval of length $1/n$ contains a point. There are a variety of other potential algorithms for filling space evenly so that the sizes of holes are small, often referred to as *low-discrepancy sequences* under the general topic *Quasi-Monte Carlo Methods*. See for example [14]. We will discuss briefly a special case of these methods, the Halton sequence later.

We might, however, also try using antithetic random numbers as a device to get close to an optimum. Suppose u_0 is uniform on $[\frac{1}{2}, 1]$ and $u_0^- = 1 - u_0$. Then the distance D from U, a uniform$[0, 1]$ random variable to the closer of u_0 and u_0^- has the same distribution as that of the distance between two independent random points on the interval $[0, \frac{1}{2}]$, with cumulative distribution function $F(x) = 1 - (1 - 2x)^2$ for $0 < x < \frac{1}{2}$ and probability density function $F'(x) = 4(1 - 2x)$, for $0 < x < \frac{1}{2}$. In this case

$$p_{imp} = E[(1 - V(D))^n] = E[(1 - 2D)^n] \tag{3.8}$$

$$= 4 \int_0^{1/2} (1 - 2x)^n (1 - 2x) dx = \frac{2}{n+2} \tag{3.9}$$

which is $\frac{1}{2}$ in the case $n = 2$. When $n = 2$, two antithetic random numbers are no better or worse than two completely independent random numbers. In fact for a particular function, it is possible that a pair of antithetic random numbers are worse than a pair of independent random numbers. Consider, for example, a function with maximum at $u = \frac{1}{2}$ which is symmetric about its maximum (for example the function $g(u) = (1 - u)$). Then $P(\max(g(U_1), g(U_2)) < \max(g(u_0), g(u_0^-)))$ is the probability that u_0 is closer to $\frac{1}{2}$ than either U_1 or U_2 and this probability is $\frac{1}{3}$.

However if the function g is monotone, it is a somewhat different story, because in this case the antithetic random numbers are more likely to be closer to the maximum, at one of the two ends of the interval. In particular, for one-dimensional monotone functions g and continuous random variables X_i, the use of antithetic inputs provides for an improvement over independent inputs with probability $\frac{7}{12}$.

Theorem 1. *For strictly monotone g,*

$$P(\max(g(U_1), g(U_2)) < \max(g(u_0), g(u_0^-))) = 7/12 \qquad (3.10)$$

Proof. By the monotonicity of g (we assume it is increasing),

$$P(\max(g(U_1), g(U_2)) < \max(g(u_0), g(u_0^-))) = P(\max(U_1, U_2) < \max(u_0, u_0^-))$$

$$= \int_0^1 [P(U_1 < \max(x, 1-x))]^2 dx$$

$$= \int_0^1 (\max(x, 1-x))^2 dx = \frac{7}{12}$$

Theorem 1 means that, at least for monotone functions g, the candidates U_1, U_2, u_0, u_0^- for maximizing the function $g(u)$ maximize it with probabilities $\frac{5}{24}, \frac{5}{24}, \frac{7}{24}, \frac{7}{24}$ respectively.

For the validity of Theorem 1, we do not need to require that u_0 and u_0^- are antithetic in the usual sense that $u_0^- = 1 - u_0$ or that the function g is monotonic provided that we can define a suitable partner u_0^- for u_0. The ideal pairing is one for which, when $g(u_0)$ is small, then $g(u_0^-)$ will tend to be large. Let us suppose for example we have established some kind of pairing or "opposite" relationship (u, u^*) so that for some value $g_{1/2}$,

$$g(u) > g_{1/2} \text{ if and only if } g(u^*) < g_{1/2}. \qquad (3.11)$$

For example for the function $g(u) = u(1 - u)$, the standard use of antithetic random numbers fails to provide variance reduction but in this case we can define $u^* = \frac{1}{2} - u$ for $u < \frac{1}{2}$, and $u^* = \frac{3}{2} - u$ for $u \geq \frac{1}{2}$ and then (3.11) holds with $g_{1/2} = \frac{3}{16}$. Moreover we suppose that the transformation $u \to u^*$ is *measure-preserving*[2] so that, for example, if u is uniformly distributed on some set A, then u^* is uniformly distributed on the corresponding set A^*. Let us replace antithetic random numbers such as u_0^- by u_0^* in (3.10). We denote the cumulative distribution function of $g(u_0)$, by $F(x)$ and its derivative $f(x) = F'(x)$, its probability density function (which we will assume exists, to keep things simple).

Theorem 2. *Suppose g is strictly monotone g, the transformation $u \to u^*$ is measure-preserving and for some value $g_{1/2}$ (3.11) holds. Then*

$$P(\max(g(U_1), g(U_2)) < \max(g(u_0), g(u_0^*))) = 7/12 \qquad (3.12)$$

---

[2] A probability or measure space together with a measure-preserving transformation on it is referred to as a *measure-preserving dynamical system*.

Proof. By the measure-preserving property, if u is uniformly distributed over the set A then u^* is uniformly distributed over A^*. Furthermore, since $u \to u^*$ is measure-preserving and that the distribution of $g(u)$ is continuous, $P(g(u) > g_{1/2}) = P(g(u) < g_{1/2}) = \frac{1}{2}$. Then

$$P[\max(g(U_1), g(U_2)) < \max(g(u_0), g(u_0^*)] = \int [F(\max(g(u), g(u^*)))]^2 du$$

$$= \int_{\{u; g(u) > g_{1/2}\}} F^2(g(u)) du + \int_{\{u; g(u) < g_{1/2}\}} F^2(g(u^*)) du$$

$$+ \int_{\{u; g(u) = g_{1/2}\}} F^2(g(u^*)) du$$

$$= \int_{g_{1/2}}^{\infty} F^2(x) f(x) dx + \int_{g_{1/2}}^{\infty} F^2(x) f(x) dx + 0 \qquad (3.13)$$

$$= \frac{2}{3}(1 - F^3(g_{1/2})) = \frac{7}{12}$$

This shows that for approximating a maximum or minimum, the preferred notion of "opposites" is one which maps large values of the function g into small values, not one which maps large values of u into small values of u, unless it is known that the function g is monotone.

Return to the context of Theorem 1. If we compare the more favorable of a random number and its antithetic with a **single** random variable X_1, then the probability that the better of the antithetic pair is larger is $\frac{3}{4}$, which means that the best is determined by u_0, u_0^-, U_1 with frequencies 37.5%. 37.5% and 25% respectively.

Theorem 3. *For strictly monotone g,*

$$P(g(U_1) < \max(g(u_0), g(u_0^-))) = \frac{3}{4}$$

Proof. Assume without loss of generality that g is monotonically increasing. By the monotonicity of g,

$$P[g(U_1) < \max(g(u_0), g(u_0^-))] = P[U_1 < \max(u_0, u_0^-)]$$

$$= \int_0^1 P[U_1 < \max(x, 1 - x)] dx$$

$$= \int_0^1 \max(x, 1 - x) dx = \frac{3}{4}$$

We can carry out a similar calculation in $d = 2$ dimensions. The easiest way of distributing m^2 points at random in the unit square is with a *shifted lattice*, i.e. begin with a point $z = (z_1, z_2)$ uniform in the unit square $[0, 1]^2$ and then define points on a lattice

$$u_{ij} = ((z_1 + \frac{i}{m}) \bmod 1, (z_2 + \frac{j}{m}) \bmod 1), i, j = 0, 1, ..., m - 1$$

See Figure 3.2 for such a set of points in the case $m = 10$.

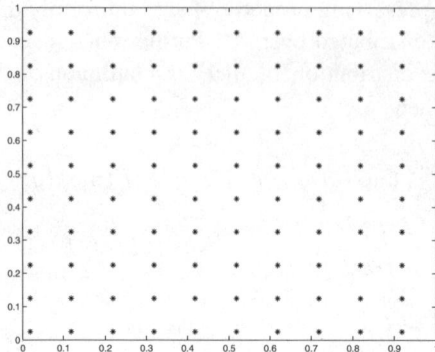

Fig. 3.2. A shifted lattice with $n = 100$

If D = the distance between a random point in the unit square and this point set $\{u_{ij}\}$, then we can determine $P(D \leq x)$ from a simple geometric argument. When $x < \frac{1}{2m}, P(D \leq x) = \pi x^2 m^2$. When $x > \frac{1}{2m}$, see the area of the shaded region in Figure 3.3.

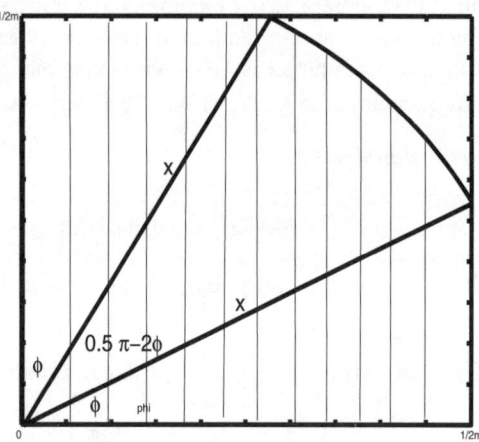

Fig. 3.3. Finding the distance between a random point in a square and the lower left corner

This area is the sum of the pie-shaped region, $x^2(\frac{\pi}{4} - \phi)$ where $\phi = \cos^{-1}(\frac{1}{2mx})$ plus the sum of the area of two triangles, each with area $\frac{1}{2}(\frac{1}{2m})\sqrt{x^2 - \frac{1}{4m^2}}$. Thus

$$P(D \leq x) = \begin{cases} \pi x^2 m^2 & \text{for } x \leq \frac{1}{2m} \\ 4m^2 x^2(\frac{\pi}{4} - \cos^{-1}(\frac{1}{2mx})) & \text{for } \frac{1}{\sqrt{2}m} > x > \frac{1}{2m} \\ +(4m^2 x^2 - 1)^{1/2} & \\ 1 & \text{for } x > \frac{1}{\sqrt{2}m} \end{cases}$$

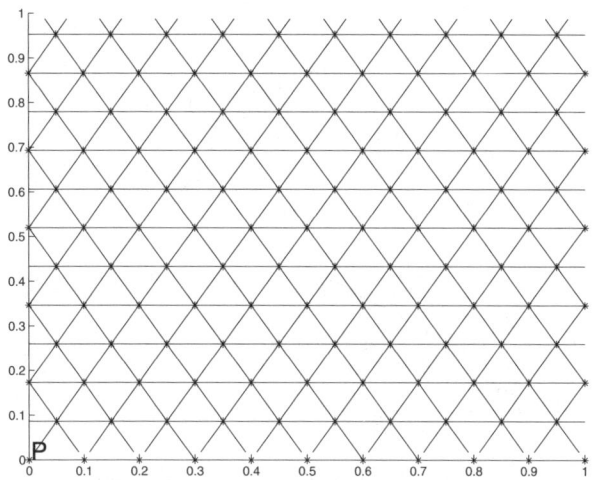

Fig. 3.4. Triangulation or Tessellation of the unit square

It remains to determine

$$p_{\text{imp}} \simeq E[(1 - \pi D^2)^n] \simeq E[\exp(-\pi n D^2)]$$

which, although not the prettiest integral in the world, is feasible under the above distribution. We will compare various other possibilities in 2-dimensions with this one.

A more efficient design for filling spaces in 2 dimensions is with a *triangular tessellation*. Suppose the points $(u_1, u_2, \ldots u_n)$ provide an equilateral triangulation or *tessellation*[3] of the unit square. By this we mean that each vertex u_i (except possibly for those on the boundary) is the corner of 6 triangles in the triangularization (see Figure 3.4). Suppose the side of each triangle is of length y. Then there are approximately $1/y$ points on a given row, and the vertical distance between rows is the height of a triangle or $y\sqrt{3}/2$ so there are approximately $\frac{2}{y\sqrt{3}}$ rows of points giving about $n = \frac{2}{y^2\sqrt{3}}$ points in total. We use this relationship to determine y from the number of points n.

Again in this case it is not difficult to determine the distribution of the squared distance D^2 from a random point in the unit square to the nearest vertex of a triangle. Assume the triangle has sides of length y. We want to determine the distribution of the distance between a random point in the triangle and the closest vertex. The sides of the triangle are as in Figure 3.5: we wish to determine the distribution of the squared distance of the point P to the closest corner of the large equilateral triangle.

By symmetry we can restrict to a point in the smaller right-angle triangle ABC. If we pick a point P at random uniformly in ABC (the point C has coordinates $(\frac{y}{2}, \frac{y}{2\sqrt{3}})$)

[3] Weisstein, Eric W. "Regular Tessellation." From MathWorld–A Wolfram Web Resource. http://mathworld.wolfram.com/RegularTessellation.html

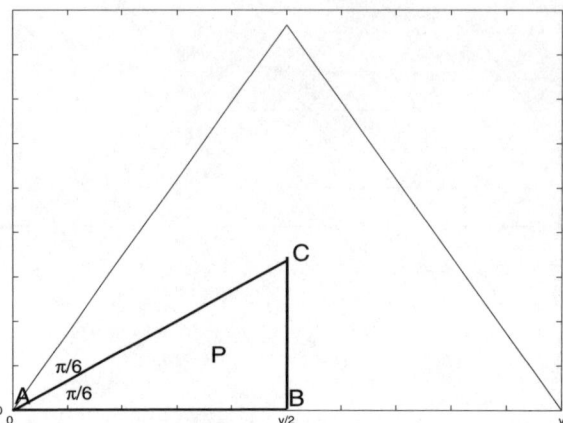

Fig. 3.5. Determining the c.d.f. for a triangular tessellation

then for $0 < z < \frac{y^2}{3}$, the probability that the squared distance to this random point P is less than z is

$$\frac{8\sqrt{3}}{y^2} \int_0^{y/2} \int_0^{x/\sqrt{3}} I[x^2 + y^2 \leq z] dy dx = \frac{8\sqrt{3}}{y^2} \int_0^{\pi/6} \int_0^{\min(\sqrt{z}, y\sec(\theta)/2)} r dr d\theta$$

$$= \frac{4\sqrt{3}}{y^2} \int_0^{\pi/6} \min(z, \frac{y^2 \sec^2(\theta)}{4}) d\theta$$

$$= \begin{cases} \frac{2}{y^2} \frac{\pi}{\sqrt{3}} z & \text{if } z \leq \frac{y^2}{4} \\ \sqrt{\frac{12z}{y^2} - 3} \\ + \frac{4\sqrt{3}}{y^2} z(\frac{\pi}{6} - \arccos(\frac{y}{2\sqrt{z}})) & \text{if } \frac{y^2}{3} > z > \frac{y^2}{4} \\ 1 & \text{if } z \geq \frac{y^2}{3} \end{cases}$$

where $I[x^2 + y^2 \leq z] = 1$ or 0 as $x^2 + y^2 \leq z$ or $x^2 + y^2 > z$ respectively.

From this we can determine the probability density function of D^2 and the value of

$$p_{imp} \simeq E\{\exp(-\pi n D^2)\} \tag{3.14}$$

There is another method for generating a uniform point set that is commonly used, the method of low discrepancy sequences. A simple example of this is the *Halton* sequence. We begin with a special case in one-dimension, the Van der Corput sequence (see for example [12, Section 6.4]), obtained by reversing the digits in the representation of the sequence of integers in a given base. Consider the binary (base $b = 2$) representation of the sequence of natural numbers:

$$1, 10, 11, 100, 101, 110, 111, 1000, 1001, 1010, 1011, 1100, 1101, \ldots$$

We map these into the unit interval $[0, 1]$ so that the integer $\sum_{k=0}^t a_k b^k$ is mapped into the point $\sum_{k=0}^t a_k b^{-k-1}$ using the displayed algorithm for producing $n_{\max}$ Van der Corput random numbers.

Van Der Corput Algorithm

1. *Begin with $n = 1$.*
2. *Write n using its binary expansion. e.g. $13 = 1(8) + 1(4) + 0(2) + 1(1)$ becomes 1101*
3. *Reverse the order of the digits. e.g. 1101 becomes 1011.*
4. *Determine the number in [0,1] that has this as its binary decimal expansion. e.g. $1011 = 1(\frac{1}{2}) + 0(\frac{1}{4}) + 1(\frac{1}{8}) + 1(\frac{1}{16}) = \frac{11}{16}$.*
5. *Set $n = n + 1$ and if $n > n_{\max}$ exit, otherwise return to step 2.*

Thus 1 generates $\frac{1}{2}$, 10 generates $0(\frac{1}{2}) + 1(\frac{1}{4})$, 11 generates $1(\frac{1}{2}) + 1(\frac{1}{4})$ etc. The intervals are recursively split in half in the sequence $\frac{1}{2}, \frac{1}{4}, \frac{3}{4}, \frac{1}{8}, \frac{5}{8}, \frac{3}{8}, \frac{7}{8}, \ldots$ and the points are fairly evenly spaced for any value for the number of nodes.

The Halton sequence is the multivariate extension of the Van der Corput sequence. In higher dimensions, say in d dimensions, we choose d distinct primes, $b_1, b_2, \ldots b_d$ (usually the smallest primes) and generate, from the same integer m, the d components of the vector using the method described for the Van der Corput sequence. For example, when $d = 2$ we use bases $b_1 = 2$, $b_2 = 3$. The first few points, $(\frac{1}{2}, \frac{1}{3}), (\frac{1}{4}, \frac{2}{3}), (\frac{3}{4}, \frac{1}{9}), (\frac{1}{8}, \frac{4}{9}), (\frac{5}{8}, \frac{7}{9}), (\frac{3}{8}, \frac{2}{9}), (\frac{3}{8}, \frac{2}{9}), (\frac{7}{8}, \frac{5}{9}), (\frac{1}{16}, \frac{8}{9}), (\frac{9}{16}, \frac{1}{27}), \ldots$ are generated as in Table 1 using the bold values in pairs:

We plot the first 1000 points of this sequence in Figure 3.6. Since it appears to fill space more uniformly than independent random numbers in the square, it should also reduce the value of D^2 and consequently increase the probability of improvement over crude Monte Carlo.

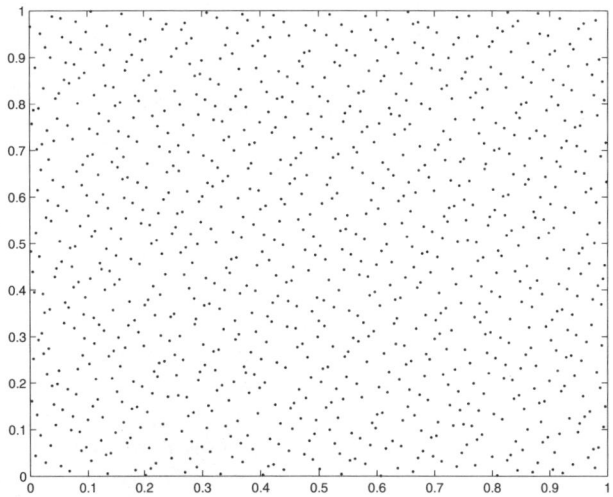

Fig. 3.6. The first 1000 points of the Halton sequence of dimension 2

Table 1. Generating a 2-dimensional Halton sequence

m	representation base 2	first component	representation base 3	second component
1	1	1/2	1	1/3
2	10	1/4	2	2/3
3	11	3/4	10	1/9
4	100	1/8	11	4/9
5	101	5/8	12	7/9
6	110	3/8	20	2/9
7	111	7/8	21	5/9
9	1000	1/16	22	8/9
10	1001	9/16	100	1/27

Table 2. Comparison of p_{imp} for various methods of placing u_i

n	crude	antithetic pairs	Halton	shifted lattice	triangular tessellation
4	$\frac{1}{2}$	$\frac{1}{2}$	0.54	0.66	0.61
16	$\frac{1}{2}$	$\frac{1}{2}$	0.59	0.60	0.64
64	$\frac{1}{2}$	$\frac{1}{2}$	0.59	0.61	0.65
256	$\frac{1}{2}$	$\frac{1}{2}$	0.58	0.62	0.63

Table 2 provides an approximation to p_{imp} in the case $d = 2$ for non-monotonic functions for various methods for distributing the points u_i and for various values of n. As we have seen, we can improve on the performance of these methods if the function to be optimized is known to be monotonic.

3.2.1 Other Optimization Methods Using Monte Carlo

There is a very large number of different routines, based on Monte-Carlo, for optimizing functions. The methods described above are largely black-box methods that do not attempt to adapt to the shape of the function. However there are alternatives such as Markov Chain Monte Carlo methods (see [6] and [15]) that sequentially adapt to the function shape, drawing more observations in regions for example near the maximum and fewer where the function is relatively small.

We wish to maximize a non-negative function $g(x)$ over a domain which might well be high-dimensional. The idea is essentially quite simple-we move around the space with probabilities that depend on the value of the function at our current location and at the point we are considering moving to. If we are currently (at time t) at location x_t in our space then we start by proposing a move to a new location x_{t+1}^*, a point randomly chosen in the domain. This new location is chosen according to some distribution or probability density function $q(x_{t+1}^*|x_t)$ that may depend on our current location x_t. For example we might pick a proposed location x_{t+1}^* at random inside a ball or small radius around x_t. Having chosen a proposed move to x_{t+1}^*, whether the move is actually carried out must depend on the the the size of the function both at x_t and at x_{t+1}^*. In particular, we "accept" the move with probability

$$\min\left(1, \frac{g(x_{t+1}^*)q(x_t|x_{t+1}^*)}{g(x_t)q(x_{t+1}^*|x_t)}\right) \tag{3.15}$$

and if the move is accepted then $x_{t+1} = x_{t+1}^*$, otherwise $x_{t+1} = x_t$ (we stay at the last location). Of course the process of "accepting" (or not) the proposed move requires generating an additional uniform random number U_t and accepting the move if $U_t \leq \frac{g(x_{t+1}^*)q(x_t|x_{t+1}^*)}{g(x_t)q(x_{t+1}^*|x_t)}$. It is apparent from the form of (3.15) that proposed moves to higher ground are favoured (more likely to be accepted) because then $\frac{g(x_{t+1}^*)}{g(x_t)}$ is larger. The other terms in the ratio, for example $q(x_{t+1}^*|x_t)$ are to compensate for a possible imbalance or asymmetry in the proposal distribution. For example if we are very likely to move to a particular point x_{t+1}^* from x_t then the term $q(x_{t+1}^*|x_t)$ in the denominator adjusts for this. If we run this algorithm over a long period of time then it is possible to show that the limiting distribution of x_t is proportional to the distribution $g(x)$ so that values of x_t will tend to be clustered near the maximum of the function. If the contrast between large and small values of g is insufficient to provide a clear separation, or if the function g is not non-negative, then the algorithm can be applied instead with $g(x)$ replaced by a function $e^{Tg(x)}$ for some user-selected "temperature" parameter T. For large values of the "temperature" T, the resulting density is very much peaked around its maxima. This algorithm is often referred to as the Metropolis-Hastings algorithm (see [15], [13] and [9]).

3.3 Conclusion

There are clear advantages to using a dependent sequence of values $u_0, \ldots u_m$ over a crude independent sequence. For moderate sample sizes e.g. 16-64, there is some advantage to the more complex triangular tessellation of the unit square but for very small or very large sample sizes, it appears that in 2 dimensions, the simple shifted lattice provides for improvement over crude Monte Carlo with probability around 62%. For higher dimensions a Halton sequence or another low-discrepancy sequence is a reasonable compromise between ease of use and efficiency. If we wish to adapt our sequence to the function g to be optimized there is a very rich literature on possible methods, including Markov-Chain Monte Carlo algorithms such as the Metropolis-Hastings algorithm (see [15]), and in this case, at the expense of some additional coding, we can achieve much greater efficiencies.

References

1. Alam, K., Saxena, K.M.L.: Positive dependence in multivariate distributions. Comm. Statist. A – Theory Methods 10(12), 1183–1196 (1981)
2. Craiu, R.V., Meng, X.-L.: Multiprocess Parallel Antithetic Coupling for Backward and Forward Markov Chain Monte Carlo. Ann. Statist. 33, 661–697 (2005)
3. Cuadras, C.M.: On the covariance between functions. J. Multivariate Anal. 81, 19–27 (2002)

4. Fréchet, M.: Les Probabilitiés Associées à un Système d'Événments Compatibles et Dépendants (Hermann & Cie, Paris), vol. Premiere Partie (1940)
5. Fréchet, M.: Surles tableaux de corrélation dont les marges sont données. Ann. Univ. Lyon. Sect. A 14(3), 53–77 (1951)
6. Gilks, W.R., Richardson, S., David Spiegelhalter, S.D. (eds.): Markov Chain Monte Carlo in Practice. Chapman & Hall/CRC, London (1996)
7. Hammersley, J.M., Handscomb, D.C.: Monte Carlo Methods. Methuen, London (1965)
8. Hammersley, J.M., Morton, K.V.: A New Monte Carlo technique: Antithetic variates. Proc. Cambridge Philos. Soc. 52, 449–475 (1956)
9. Hastings, W.K.: Monte Carlo Sampling Methods Using Markov Chains and Their Applications. Biometrika 5, 97–109 (1970)
10. Hoeffding, W.: Masstabinvariante korrelations-theorie. Schr. Math. Inst. University Berlin 5, 181–233 (1940)
11. Lehmann, E.L.: Some concepts of dependence. Ann. Math. Statist. 37, 1137–1153 (1966)
12. McLeish, D.L.: Monte Carlo Simulation and Finance. Wiley, New York (2005)
13. Metropolis, N., Rosenbluth, A.W., Rosenbluth, M.N., Teller, A.H., Teller, E.: Equations of State Calculations by Fast Computing Machines. Journal of Chemical Physics 21, 1087–1092 (1953)
14. Niederreiter, H.: Random Number Generation and Quasi-Monte Carlo Methods. In: CBMS-NSF. Applied Mathematics, vol. 63. SIAM, Philadelphia (1992)
15. Robert, C.P., Casella, G.: Monte Carlo Statistical Methods, 2nd edn. Springer, New York (2004)

4

Opposition and Circularity

Florencio G. Asenjo

Department of Mathematics, University of Pittsburgh, USA
fgasenjo@pitt.edu

Summary. From the middle of the twentieth century contradictory logics have been growing in number, scope, and applications. These are logics which allow for some propositions to be true and false at the same time within a given formal system. From psychology to physics one finds many examples of areas of the mind and of physical reality which exhibit an unmistakable degree of antinomicity. Quantum mechanics is a prime example. Now, it is commonly thought that antinomies depend essentially of negation. This is not so. In fact, there are many antinomies whose contradictory character is created by the fusing of opposite assertions into one. Indeed, antinomicity should be considered an instance of the more general notion of opposition. A logic of opposition is essential to deal with this general case.

But what about terms? The nouns, adjectives, and verbs that form part of statements? Terms are neither true nor false, they have to be fed to a predicate in order to obtain a true or false sentence. Yet, terms can be antinomic in a very important sense. We often define a term A in terms of an opposite term B, but only to find that in turn B has to be defined in terms of A. This "vicious circle" is most of time inevitable, and it makes of B part of the meaning of A and vice versa. I sustain that circularity should be looked at positively and put to good use. An example: "one" and "many" are opposite concepts. Now, if we attempt to go beyond the naive approach of taking one and many neutrally as simple ideas absolutely complete each in itself, we realize that aware of it or not we think of "one" as "one of many," and of "many" as "many ones." One and many are inextricably interwoven concepts, and a logic of "antinomic terms" is necessary to deal with this regular semantic phenomenon that is found in all categories of thought as they combine with their opposite counterparts.

This chapter deals with (i) the relations that exist between antinomicity, opposition, and circularity; with (ii) ways of dealing explicitly with the antinomic character of certain terms, a character which becomes obvious every time we think of them reflectively; then with (iii) the answer to the question of what changes are necessary to make in the usual constructions of logical models to enable us to expose antinomic terms explicitly: formalization must undergo substantial changes if we are to make systematically obvious the fact that, whether we know or not, we do think regularly in terms of opposites.

4.1 Purpose

The study of contradictory logics has grown considerably since the middle of the twentieth century. Nowadays there are many such logics developed from different starting points, all with a positive attitude toward contradictions, and a number of them having significant applications, most notably to quantum mechanics, a discipline whose

H.R. Tizhoosh, M. Ventresca (Eds.): Oppos. Concepts in Comp. Intel., SCI 155, pp. 45–57, 2008.
springerlink.com © Springer-Verlag Berlin Heidelberg 2008

contradictory conclusions keep growing in extraordinary and puzzling ways, notable examples: nonlocality and the Bose-Einstein Condensate. Classical formal logics are really not adequate to deal with such conclusions. Here I shall use the words "contradiction," "paradox", and "antinomy" as synonymous. These words are often given somewhat distinct connotations with respect to one another but for the purpose of this work such differences are not essential and will be disregarded.

One of the oldest antinomies is that of the liar: "I am not telling the truth." If we apply the sentence to itself, it is obviously false if assumed to be true, and true if assumed to be false. Many examples exist of this kind of paradoxical statement given in ordinary or technical languages. It is often believed that antinomies depend essentially on the use of negation. This is not the case. Antinomies can be obtained without negation just by putting together into a single conjunction two opposite statements: antinomies generated by negation are only a special case. The logic of antinomies should be broadly understood as a special logic of opposition [1]. In this chapter I do not want to revisit the ideas of the work I just referred to, nor to look into the opposition of sentences in particular; rather, I shall look into the opposition of terms in themselves, the terms that occur as subject of a sentence. I shall be interested especially in the kind of opposition that involves circularity, that is, the cases - much more common than one would think - in which opposite terms hinge semantically on one another to the point of making each term unintelligible without thinking of its opposite counterparts at the same time.

4.2   Vicious Circles

Our inveterate desire for simplicity pushes us to constantly analyze wholes into parts. However, there are situations and processes in which to disassemble the complex only obfuscates the picture. There are properties that emerge when small parts are assembled into a larger whole and which were not present in any of the parts taken in isolation, properties that become lost when we insist in splitting the whole into its smallest possible components. This unfortunate attitude is present also in the way we comprehend ideas, especially our most basic concepts, the so-called categories of thought. We like to keep categories simple so that our thinking is kept simple as well. But simple thoughts are often not adequate for the understanding of a complex reality. Clarity can be deceiving. Even in the realm of pure thought we cannot escape complexity if we really want to comprehend how our mind works in order to go effectively beyond platitudes.

Charles Peirce remarked that on first approximation we do take ideas at a neutral state, simple, unadorned [18]. We all start at this uncomplicated first approximation. One is one and many is many, and we do not care to go beyond this clear-cut approach. But when the mind begins to function reflectively the picture changes. Any attempt at defining a concept as a function of another, and then do it vice versa ends in what is pejoratively called a 'vicious circle". But such "vicious circles" are not only inevitable in any striving toward real comprehension, they are also essential, indeed positive components of sound thought. We need to become explicitly aware of this semantic situation and incorporate circularity as a regular way of building our repertoire of basic ideas. Nothing can be defined thoroughly without resorting to circularity at some point. Categories, in particular, can only be defined in circular ways. This should be accepted as a normal way our mind functions. In the present considerations we want to focus especially on circular concepts formed by the semantic conjunction of an idea and its opposite. Let us call " antinomic term" a circle made up of two or more opposite nouns or gerunds together, and let us call " antinomic predicate" a circle made up of two or more opposite predicates taken together, keeping in mind that relations are predicates calling for two or more terms in a given order to complete a well-formed sentence.

Take the example of "one" and its opposite "many", the latter in the sense of "more than one". Both categories emerge from our naive perceptive experience, and we usually take them as neutral, simple ideas. We identify objects one by one, and consider each of them as a "one-in-itself". Similarly, we identify a plurality of objects as a "many". We perform these identifying acts of perception without any pondering whatsoever, yet, if we stop to think on what we are doing, the "one" of an object is not a primitive concept in itself, but the result of a choice, of the act of addressing our attention to one object among many. "One" implies the rejection of other "ones", it is a "one-of-many". This means that if we were to define this more realistic idea of "one", we could not do it without bringing along the idea of "many". Vice versa, it would not be possible to conceive of "many" at this reflective level in any other way than as "many ones". We have then that what in logic is labeled a "vicious circle" is in effect an unavoidable fact of semantics, as a consequence of which, we cannot extract any component of an antinomic term without automatically hiding its deeper meaning, the fact that the component originates in an inextricable circularity.

Matter-and-energy is another antinomic term. Energy is characterized by its effect on matter, and matter is known to be energy waiting to be released, the source of nuclear power, for example. Whole-and-parts constitutes one more example of a very fruitful opposition, indeed, a very productive antinomy. And perhaps one of the most obvious cases of antinomicity is that of identity-and-difference, an antinomic "vicious circle" which is of the essence of becoming. Becoming clearly implies change, but change is never absolute: this would be tantamount to chaos; something always remains identifiably the same during the most radically transforming process. Real-and-potential, universal-and-particular, activity-and-passivity also add to the endless list of antinomic terms.

But the complexity of thought does not stop at the first conjunction of opposite concepts. Let us look again at the one-and-many circle: it involves a second circle. "One" is chosen from many ones, "many" is a gathering of ones. Choosing-and-gathering is another circle hidden inside the meaning of one-and-many, a circle composed of two active gerunds, pointing at two opposite movements of the mind acting in conjunction.

The semantic situation we have just described has been recognized already by a number of authors (see [7] and [10]), although it has not been taken as a logical theme by itself, nor has it been analyzed, and much less acknowledged as taking sometimes the form of a cascade of antinomic terms in successive inclusion deep inside. The linguist Jost Trier, who introduced the notion of field in semantics, put it this way: "Every uttered word reminds us of the one of opposite sense ... The value of a word is recognized only if it is bounded against the value of the neighboring and opposing words" [7]. The antinomy is the force that makes up the semantic field [10]. And Charles Bally before Trier: "Logical notions exist together in our mind with the opposite concepts, and these oppositions are ready to present themselves to the consciousness of the speaking subject. It can be said that the contrary of an abstract word is part of the sense of that word" [11].

4.3   Hidden Circles as Special Cases of Pregnant Meaning

Implicit circularity is a particular case of a more general semantic phenomenon which we have labeled " pregnant meaning". By the last expression we mean the condition of certain concepts, statements, or paragraphs of obscurely containing other significant connotations than those set in clear explicitness. The implicit content cries to be uncovered and may consist of related ideas, situations, or memories buried under the surface of our consciousness, important some times, insignificant occasionally. Marcel Proust has left several classical descriptions of the semantic phenomenon we are referring to and of the process of recovering those half-forgotten remembrances that suddenly claim for our attention promoted by a casual reference or event [19]. This is a common occurrence: often words and sentences are gravid with related meanings whose whole import escapes us momentarily but that we can usually uncover with a well-directed reflection. The concept of "one" is pregnant with the concept of "many", and vice versa [6].

4.4   Antinomic Terms Versus Antinomic Predicates

Affirmation and negation, pertains to sentences; the logic of opposition allows for the conjunction of two opposite sentences to be both asserted together, and the conjunction to be true and false at the same time. We want to focus now on the parallel treatment of antinomic terms and predicates. The antinomic term "one-and-many" has already been discussed. Predicates are subject to a similar kind of circularity. We all accept the meaning of the word "bittersweet" without stopping to reflect that it is in fact an example of antinomic predicate. A situation or a memory may be bitter and sweet at the same time, which can clearly be expressed by an antinomic sentence that states such opposition by the propositional conjunction of the two sentences stating bitterness and sweetness separately. But the predicates of bitterness and sweetness combine by themselves to form the antinomic predicate "bittersweet". The same applies to the relation of membership in set theory. The set of all sets which are not members of themselves is and is not a member of itself. This is the well-known Russells antinomy, which in turn induces the antinomic relation "to be and not to be a member of" [5].

   We want now to introduce symbols to systematize this new situation. We already noted in passing that terms as well as predicates may have more than one opposite. The binary predicate "less than" has the opposites "greater than", "neither less than nor greater than", and "less than and greater than". In order to represent opposition for both terms and predicates we shall use the negation symbol "$\neg$"-extending its usual logical meaning to opposition in general - negation being after all a special case of opposition. The expressions "$\neg t$" and "$\neg P$" indicate the term and predicate opposite respectively to the term t and the predicate P, if there is only one opposite in each case. If t and P have more than one opposite, we shall list them respectively thus: $\neg_1 t, \neg_2 t,...$, and $\neg_1 P, \neg_2 P, ...$ The symbol "$\neg$" has now three possible applications: to sentences, to terms, and to predicates. No confusion can be derived from this multiple use of the same symbol, syntactically or semantically.

In a similar way we shall extend the use of the connective "&" (and) used to represent the conjunction of sentences. Here we shall use the same symbol to indicate the conjunction of terms and predicates. Thus, "t & $\neg t$" and "P & $\neg P$" are formal expressions to represent, say, "one-and-many" and "bittersweet". Again, no confusion should derive from the use of the same symbol "&" to stand for the three kinds of conjunction.

Semantically, sentences become true or false when interpreted in a given structure, which then becomes a model of the sentence or its negation. A sentence may be true in one structure and false in another. Constant terms are interpreted in classical model theory each by a specific individual in the domain of interpretation of the given structure - the universe of discourse. Variable terms range over such domain. Of course, different constant terms have different interpretation in different domains.

For us, terms will not all be simple, or in other words, atomic. Complex terms of the forms t & $\neg t$, or t & $\neg_1 t$ & $\neg_2 t$, etc., are to be taken as irreducible complex entities - such as complex numbers in mathematics are irreducible pairs of real numbers - and each complex term interpreted by a different single or complex individual according to the way the domain of interpretation is defined: in any case, opposition necessarily enlarges the universe of discourse by forcing the membership of complex individuals in it.

As for predicates, classically each is assigned a subset of the domain of interpretation if the predicate calls for only one term. This subset is precisely the collection of all individuals that satisfy the predicate. An antinomic predicate P & $\neg P$ should be assigned the intersection of the two subsets that interpret P and $\neg P$ respectively, an intersection that is always empty in classical logic. If the predicate is an n-ary relation with n greater than or equal to 2, the interpretation of such relation is, of course, the collection of all n-tuples of individuals that satisfy the relation.

4.5   A Term Is Made Up of All the Predicates That Apply to It

A term that represents a concept or an actual entity - a thing - is never an atom. Each is the sum of all the predicates, including all the relations, that the term satisfies by itself or occurring in pairs, triples, etc. In the parallel real world, concrete things can be said equally to be made up of all the properties and relations they exhibit. Relations, in particular, incorporate at the core of each actual entity the presence of other objects to which the entity is related, a presence that may transcend time and location. To sum up: as we try to analyze a thing, what we do is in effect to mentally unfold property after property and relation after relation as though we were peeling off an onion. We shall then accept as a given that a term that represents a concrete or abstract object is composed entirely of predicates unary, binary, ternary, etc. This conduces us to state that in addition to predicate formulas of the form $P(t)$ - read "the predicate P applies to the term t" - we should now introduce " term formulas" of the form $t(P)$ - read "the term t displays the predicate P" - both formulas eventually becoming true or false when specifically interpreted in a fixed structure.

In order to make clear the difference between $P(t)$ and $t(P)$ - expressions which superficially may sound synonymous - let us set up a sequence of logical types or levels.

Type 0 is the level of terms and predicates by themselves, neither predicate formulas nor term formulas occur at this level. Type 1 is the level of formulas of the form $P(t), P(t_1, \cdots, t_n)$, etc. These formulas may contain P or t as variables; if not, or if P and t do not occur free, the formulas are called sentences. The usual propositional connectives, as well as quantifiers over terms belong to type 1, where the classical and the antinomic predicate logics can both be developed with their associated semantic interpretations, models, etc.

We want now to place the term formulas $t(P)$ in a different type. For this purpose, we will make use of the notion of negative types introduced by Hao Wang with an entirely different objective [21]. He used negative types only as a device to show the independence of the axiom of infinity in set theory, not to analyze terms. In fact, Wang states that "it may be argued that in logic and mathematics there is no reason for us to discuss individuals [12]. In his Theory of Negative Types "there are no more individuals, and all types are on an equal footing [12]. Here, we want to use negative types precisely to formalize the structure of terms and of their interpretations. Since thought begins and ends with individuals, and since logic and mathematics originate in abstractions performed on individuals, it is reasonable to make room for a more concrete way of considering terms by looking into their inside, that is, the predicates that make them. Type-1 is the first level in which to look at terms in such a way, and to this level belong term formulas such as $t(P)$.

The domain of interpretation of these term formulas contains in type-1 all the sets of individuals of the domain of interpretation of terms in type 1. Each predicate $P_1, P_2...$ of the language of all formulas is interpreted in type 1 by one such set, precisely the set of individuals that satisfy that predicate in type 1. Each term t in turn is then interpreted in type-1 by the set of all the interpreted predicates P for which $P(t)$ is true in type 1. Given that in type-1 predicates are not interpreted by point-like individuals but each by a fixed set of individuals, each term becomes in turn associated with a collection of sets in succession, like a collection of folds in an onion. There are examples in mathematics of such a kind of universe of discourse. The so-called Riemann surfaces to represent functions of complex variables is one such example [8]. In this type of representation an indefinite number of planes, or just regions, are superimposed successively each over the next to represent the inverse values of such functions in a single-valued manner. We shall interpret here each term by a Riemann surface of sorts, that is, by the collection of folds that successively interpret the predicates that satisfy t. Each term t is then interpreted in type-1, by its own Riemann surface, its own "space of onions". Each predicate P that satisfies t is in turn interpreted by exactly one fold of the onion; again, each fold is the set of individuals that satisfies P, in type 1, which naturally includes t itself as an individual from type 1. Negative types are then constructed on the basis of what is already built in the corresponding positive types.

If the term under consideration is antinomic, say $t \,\&\, \neg t$, then its corresponding Riemann surface is such that each fold corresponds to a predicate P that satisfies t and its opposite, that is, the set-theoretic intersection of the fold for t and the fold for the $\neg t$. In antinomic models such interjection is not empty because there are predicates that satisfy t and $\neg t$. See example at the end of the next section.

Syntactically, the chief difference between $P(t)$ and $t(P)$ is the following. In type 1, the predicate formula $P(t)$, with P constant and t variable, is quantified over all terms in expressions such as $(\forall t)P(t)$ and $(\exists t)P(t)$. In contrast, in type-1 the term formula $t(P)$, with t constant and P variable, is quantified over all predicates in expressions such as $(\forall P)t(P)$ and $(\exists P)t(P)$. Parallel to the predicate calculus in type 1, there is a calculus of terms in type-1.

4.6   More on the Semantics of Terms

In type 1 a term is never identified as the sum of its predicates. The focus is on predication, and the individuals that interpret the terms as well as the terms themselves are devoid of content, points in the universe of discourse. Yet, it is the term that exhibits the predicates, not vice versa. In type-1 the term is not fed to a predicate: the predicate is unfolded in the term. In type 1 only predicate formulas can be true, false, true and false, or neither true nor false. At such level, it does not make sense to speak of the truth of a term. Yet, we can assert with Christian Wolff that everything that exists is in a sense true ("Omne ens est verum") [22]. Nature provides individuals, not genera, the latter being a product of the mind. These individuals come assembled in folded stages, leaf over leaf; we analyze them accordingly, leaf by leaf. Concrete terms are therefore true because they always have real, identifiable properties, and because they display the impact of all the other real terms to which they relate. This makes true predication possible, and the sentence $t(P)$ true. No real predication can be asserted of a unicorn.

If a term is antinomic, say t & $\neg t$, some properties apply to t and to its opposite $\neg t$, others only to t and not to its opposite. In such cases, respectively, $(t$ & $\neg t)(P)$ is true, $t(P)$ is true, and $(\neg t)(P)$ false. There is also another way in which a term can be antinomic: when a predicate P holds for the term t, and so does as well the opposite predicate $\neg P$. Then, both $t(P)$ and $t(\neg P)$ are true; t is antinomic by exhibiting both a predicate P and its opposite $\neg P$, making $t(P$ & $\neg P)$ also true. No real object represented by a term t can be false, this would mean that for all predicates $P, t(P)$ is false. Since everything that exists exhibits properties and relations, the "truth of things" - veritas rei - is inescapable in the real world. If $(t\&\neg t)$ stands for matter-and-energy and P for capacity for action, $(t\&\neg t)(P)$ is true. If t stands for a villain that performs altruistic actions and P for goodness and $\neg P$ for badness, $t(P\&\neg P)$ is true.

4.7   "Vicious Circles" as a Kind of Concept Expansion

The notion of concept expansion originates in the way ideas in mathematics, the sciences, etc., keep enlarging the scope of their application, and therefore that of their meaning as well [4]. Thus, the concept of number started with thinking about the natural numbers, then it was successively extended to the integers, the rational numbers, the real numbers, the complex numbers, etc. Given that the truth of a sentence depends on the model in which the sentence is interpreted, it is not surprising that specific true sentences related to an expanded concept may not be true in the expanded models without effecting important reinterpretations of their meaning, if that is at all possible.

When two concepts originally separated by opposition merge by conjunction into one complex idea in a merger, each concept becomes an example of concept expansion but in a different sense. This applies to all circles, not only antinomic ones. What distinguishes circles from other types of concept expansion is that the said expansion does not take place by a broadening of the scope of application of each of its components - horizontally, so to say - but by deepening its original meaning - vertically. The expansion consists in an intrinsic enlargement of the meaning of the concepts involved which adds new semantic levels and resonances to each of them. Through renewed antinomic actions like choosing-and-gathering, identifying-and-distinguishing, etc., we can continue such process of expansion up to the point when we become satisfied with the state of our comprehension.

4.8   The Inescapable Vagueness of Thought

There is a bias among many mathematicians toward a belief in the existence of Platonic Ideas, a heaven of essences eternally identical to themselves waiting to be apprehended periodically as we need them. This bias obfuscates our understanding of the true nature of thought. Predicates do not float by themselves in a heaven of Ideas, they exist in an individual, or nowhere. We think necessarily with relatively vague concepts, concepts whose contour we seldom can circumscribe clearly. Kant saw this situation well and described it as follows: "The limits of the concept never remain within secure boundaries . . .., I can never be certain that the clear representation of a given concept has been effected completely . . . The integrity of the analysis of my concept is always in doubt, and a multiplicity of examples can only make it probable [14].

Mathematical concepts are of course the exception to the primacy of vagueness. But in general, we make our way toward definiteness in a universe of thoughts which remain to the end incurably vague. In Peirce's words: "Vagueness is an ubiquitous presence and not a mark of faulty thinking, no more to be done away with than friction in mechanics [16]. The apparent definiteness of language deceives us into believing that we have captured reality in the flesh when we describe it with words. The truth is that, as Russell said, "all language is vague [20] despite appearances to the contrary. Language is finite and discrete, reality is the opposite. We approach reality with atomic thoughts, but there are no atoms in either world or mind.

The realizations just pointed out lead us to reflect that opposition itself must not be taken as being always clear-cut except in the most abstract thought. Between a term and its opposite there is a continuous spectrum of possible contrasts in between. There is in actuality a degree of opposition, a degree that goes continuously from total opposition, to no opposition at all. Accordingly, there are also degrees of circularity: t may depend on its opposite $\neg t$ more than $\neg t$ on t, etc. In fuzzy mathematics the degree of membership of an element e to a set S is measured by a real number a between 0 (no membership) and 1 (full-fledged membership); the expression $e \in_a S$ symbolizes this fact [15]. In fuzzy set theory we have not one, as in classical set theory, but an infinite number of predicates of membership: for each real number a such that $0 \leq a \leq 1$, there is one binary predicate of membership $\in_a$.

Following the pattern of fuzzy set theory, we can now deal with degrees of opposition. Let us do it in two alternative ways. First, the expression $\neg_a t$ shall indicate that $\neg_a t$ is a term opposite to the term t with a degree of opposition a. In an additional way, we can introduce a binary predicate of opposition "Opp", and make the expression $t_1 Opp_a t_2$ indicate in symbols that the term t_1 is in opposition to the term t_2 with degree of opposition a. The predicate Opp is symmetric, that is, if $t_1 Opp_a t_2$, then $t_2 Opp_b t_1$ as well, but the opposition of t_2 to t_1 may have a different degree b. Both approaches open the door to a fuzzy logic of opposition which we shall not pursue here.

In type 1 specific systems usually deal with a finite number of constant predicates and an infinite number of terms. In classical set theory for example, we have only one two-place constant predicate, membership, in fuzzy set theory, we have as many predicates of membership as membership has degrees. In a parallel way, in type-1, if we allow for degrees of opposition, for each term t we have as many opposite terms as degrees of opposition, in principle infinite. The same applies to predicates. In both types 1 and -1 we can allow for an infinite number of predicates of the form $\neg_a P$ representing a predicate opposed to P with degree a.

4.9   Contradictions Are Not Contagious

To feed an antinomic term such as t & $\neg t$ to a nonantinomic predicate P in type 1 does not necessarily produce an antinomic sentence: $P(t$ & $\neg t)$ can be simply true or simply false. Similarly, to feed an antinomic predicate as P & $\neg P$ to a term t in type-1 does not make $t(P$ & $\neg P)$ antinomic, the latter can be simply true or simply false. The conjunction of n predicates generates a compound predicate P_1 & P_2 & $\cdots$ & P_n in type 0. In type 1 $(P_1$ & P_2 & $\cdots$ & $P_n)(t)$ is a well-formed predicate formula, and in type-1, $t(P_1$ & P_2 & $\cdots$ & $P_n)$ is a well-formed term formula. Thus, for example, if T is the set of all sets that are not members of themselves, and ϵ & $(\neg \epsilon)$ the corresponding antinomic predicate, then $(\epsilon$ & $(\neg \epsilon))(T)$ is true, not antinomic. Correspondingly, $T(\epsilon$ & $(\neg \epsilon))$ is true in type-1, not antinomic.

Both t & $\neg t$ and P & $\neg P$ are not to be considered "unitary" in any sense. We subconsciously tend to think that there is a supremacy of unity over multiplicity in thought, that unity is generally a desirable objective to reach. But complex numbers for instance do not form a unity; they are each a single entity, made up of an irreducible pair, a multiple object - a many taken as a many, not as a one. So are circles of any kind also irreducible multiplicities, including the conjunction of antinomic mental processes such as identifying-and-distinguishing, etc. Now, whereas it makes no difference to write t & $\neg t$ or $\neg t$ & t on the one hand, or P & $\neg P$ or $\neg P$ & P on the other, we do obtain very different entities if we commute the order of the mental operations in a conjunction of gerunds. We know already that it makes a substantial difference to gather distinguishable items first and then choose from them, or to identify items to then gather them into a many - we end up with opposite objects of thought. We usually pay no attention to the order in which we carry out our movements of the mind, but they do branch creatively in different, often opposite directions toward opposite ends.

4.10   More on the Differences between $P(t)$ and $t(P)$

Satisfiability can be defined in type-1 as follows: the constant predicate P satisfies the constant term t if and only if $P(t)$ is true in type 1 in the usual model-theoretic sense. But given the different role P plays in a formula in type-1, the situation is more complex there. I think it better to explain such situation with an example. To say that Mary is good (in symbols: $P(t), t$ for Mary, P for goodness) means that Mary is selected from all the good persons I know and kept in isolation. To say, as we do in type-1, that goodness is one of Mary's traits (in symbols: $t(P)$), means that goodness is one of the personal qualities that Mary possesses, but in addition, given that goodness is correlated to the assemblage of all the good persons in my knowledge, Mary is good together with all the other good persons I know. The predicate that satisfies the term automatically conveys in type-1 the set of all the other terms that the predicate satisfies as well. Only an abstract way of thinking makes us think that Mary can exhibit goodness in isolation of what all the other good persons, exhibit. Of course, we think in abstract terms and predicates routinely, we need to move on, and we circumvent reflection out of necessity. And yet, implicitly we do know that to say that Mary exhibits goodness as one of her traits places her in the company of all the other persons I know of whom the same can be said. This is the ultimate meaning of saying "No person is an island", that is, a mere isolated object with attributes: Indeed, "Life is always with others", and as part of many personal circles we are often defined by others as much as we define them.

4.11   Two Kinds of Relations

We have so far put the emphasis on properties, qualities that an object can have, that is to say, on one-place predicates, predicates that call for exactly one term, simple or antinomic, to yield a complete predicate formula in type 1. But at least equally important are relations, i.e., n-place predicates with n greater or equal to two. Binary, ternary relations, etc., yield a complete predicate formula in type 1 when given the required number of terms in the appropriate order. This reduction of relations to n-place predicates was introduced by Norbert Wiener, the inventor of cybernetics, and made the so-called logic of relations superfluous as an independent chapter of logic [17]. But this applies only to what can be called "external relations", relations that are attached to the correct number of terms after the terms have already been given. As Bradley and Whitehead have remarked, there is another kind of relation which they labeled "internal", relations that intrinsically affect the terms related, become of the essence of the terms, are part of the terms, and independent of the terms for reason of being genetically prior, like chemical valences acting as relations ready to link [2]. This way of viewing relations as constitutive of the terms cannot be described in type 1. In type-1, in contrast, relations become the subject of term formulas as they are fed to a term. We can say that external relations belong to type 1 and internal relations to type-1. A relation that is external in type 1 reemerges as internal in type-1.

In an effort to persuade the reader of the reality of internal relations and give a sense of their omnipresent concreteness let us look at an eminently antinomic experience that we constantly have in daily life, the phenomenon of intersubjectivity, the fact that we regularly put ourselves in other peoples place, that we think with their minds, feel with their hearts, and will their will. We have at least a subconscious realization of this fact of life; often we are even conscious that we think as though we were somebody else thinking, more: that we are momentarily somebody else thinking, that the other mind is an opposite consciousness introjected into my own. Literally, the other person looks at me from within, and I see myself with the eyes of others, sometimes better than I see myself - just as the others see themselves through my eyes. Rightly or wrongly, we often define ourselves in terms of other persons, have other persons as active agents within in many considerations and decisions. This way person's minds are part of one another is intersubjectivity, a term introduced by Husserl to describe such presence and primacy of personal relations in the constitution of each persons consciousness [9]. We look inside each individual and we find ... relations, internal relations that literally bring the outside inside. There is opposition and circularity in the way I live my life from within through the life of others. Within and without are extremely relative words, indeed. Relations that effectively bridge what seems to be in with what seems to be out are present in us as part of our make up as subjects, as terms. As we reflect on it, all this becomes obvious to us: there is not a figment of solipsism in real life. Personal relations are, as a norm, much more than a mere external accretion; as we introject the others we make of social life a network of circular oppositions intrinsic to us, and a factor in the outcome of our actions. Internal relations, functioning like what in mathematics are called independent variables act inside the terms.

4.12   Final Remarks

It should be clear that the semantic constructions in type-1 are not the mirror image of those of type 1, the way negative integers are a mirror image of the natural numbers, and therefore an alternative model of the latter. The same rules that apply to the propositional and predicate calculuses in type 1 apply in type-1, except that predicate formulas are replaced by term formulas and universal and existential quantification - as already mentioned - range not over domains of terms, but over domains of predicates. The expression

$$t(P_1) \Rightarrow t(P_2)$$

reads "if the term t exhibits the property or relation P_1, then necessarily exhibits the property or relation P_2." We should distinguish between $\neg(t(P))$ and $(\neg t)(P)$; in the first expression, the symbol "$\neg$" is the logical propositional operator of negation that yields a compound statement; in the second expression, $\neg t$ is the term opposite to t, $(\neg t)(P)$ is then atomic. If both $t(P)$ and $(\neg t)P$ are satisfied by P, then P satisfies $(t$ & $\neg t)(P)$, and hence P belongs to the folds that interpret $(t$ & $\neg t)$.

A classical example of contradictory opposition generated by negation is Russell's antinomy of the set T of all sets which are not members of themselves, a set which necessarily belongs and does not belong to itself, as we already pointed out. Propositionally, this is stated by $(T \in T)$ & $\neg(T \in T)$. But still in type 1, we also have available from type 0 the antinomic binary relation $(\epsilon$ & $(\neg\epsilon))$, where the symbol "$\neg$" stands for negation applied to the relation ϵ, "$\neg\epsilon$" reads "is not a member of". We can then express Russell's antinomy alternatively as $(\epsilon$ & $(\neg\epsilon))(T, T)$, an atomic rather than a compound predicate formula: ϵ and T are both constant and $(\epsilon \ \& \ (\neg\epsilon))(T, T)$ is a simply true sentence. In type-1, on the other hand, the antinomy becomes $T(\epsilon \ \& \ (\neg\epsilon))$. While not all antinomic predicates and relations engender antinomic terms, $(\epsilon \ \& \ (\neg\epsilon))$ does.

In the haste to get rid of the antinomy, no effort has been made to investigate the nature of T itself. "To be and not to be a member of" is an unsettling idea which points to a radical ambiguity of thought, an ambiguity which subverts our leanings toward clarity, simplicity, and definiteness. We want our thoughts about ideas, and about reality itself, to be exactly one way and not any other. Then we find T, an ambiguous entity fit for an ambiguous universe. And then we find the many unsettling conclusions that quantum mechanics keeps adding daily to our conception of the world. This makes us conclude that ambiguity is ultimately one of the characteristics of mind and reality. There are regions of the mind and of the world in which things are not just one way, and in which the coexistence of opposites reigns supreme and is the rule.

The conceptual and real ambiguity we have pointed out establishes an uncertainty principle of thought and fact. There are unsurpassable barriers to our deep desire for achieving certainty, barriers that keep receding but cannot disappeared altogether. Bradley said: "nothing in the end is real but the individual" [3], but "individuality means the union of sameness and diversity with the qualification of each by the other. . . [because] there are in the end no such things as sheer sameness and sheer diversity" [13]. Bradley states also that qualities themselves are in relation. To which we can add that relations themselves are in relation. Each such conjunction of predicates creates a new predicate, $P \ \& \ Q$ say, a new internal relation.

We know opposition when we see it, and we see it everywhere. But without formally exhibiting it systematically we are bound to remain trapped in platitudes. It takes a slight shift in our intellectual attitude, a sustained redirection of our intuition, to dispel the illusions of a spurious simplicity. Then we find that, paraphrasing Whitehead, the complexity of nature is inexhaustible, and so is also the complexity of the mind.

References

1. Asenjo, F.G.: The Logic of Opposition. In: Paraconsistency: The Logical Way to the Inconsistent, pp. 109–140. Marcel Dekker, New York (2002)
2. Bradley, F.H.: Appearance and Reality. Oxford University Press, Oxford
3. Bradley, F.H.: Collected Essays. Oxford University Press, Oxford
4. Buzaglo, M.: The Logic of Concept Expansion. Cambridge University Press, Cambridge
5. Clark, M.: Paradoxes from A to Z. Routledge, London (2002)
6. Asenjo, F.G.: Pregnant Meaning. In: Post-Structuralism and Cultural Theory: The Linguistic Turn and Beyond. Allied Publishers, Delhi (2006)
7. Geckeler, H.: Semántica Estructural y Teoría del Campo Léxico, p. 288. Gredos, Madrid (1976)
8. Grattan-Guinness, I.: The Norton History of the Mathematical Sciences. W.W. Norton, New York (1998)
9. Husserl, E.: Zur Phänomenologie der Intersubjektivität. M. Nijhoff, Hague
10. Geckeler: Ibid, p. 289
11. Ibid, p. 290
12. Wang, H.: Logic, Computers and Sets, p. 411

13. Bradley, F.H.: Collected Essays, p. 664
14. Kant, I.: Critique of Pure Reason. St. Martins Press, New York
15. Klir, G.J., Yuan, B.: Fuzzy Sets and Fuzzy Logic. Prentice Hall, Upper Saddle River
16. McNeill, D., Freiberger, P.: Fuzzy Logic. Simon and Schuster, New York
17. Wiener, N.: A Simplification of the Logic of Relations. In: From Frege to Gödel, p. 224. Harvard University Press, Cambridge (1967)
18. Peirce, C.S.: Writings of Charles S. Peirce, vol. 5, p. 304. Indiana University Press (1993)
19. Proust, M.: In Search of Lost Time. Several editions (1913–1931)
20. Russell, B.: Vagueness. Australasian Journal of Psychology and Philosophy, 84–92 (1923)
21. Wang, H.: Logic, Computers, and Sets. Chelsea Publishing Company, New York (1970)
22. Wolff. Philosophia Prima sive Ontologia (1730)

Part II

Search and Reasoning

5

Collaborative vs. Conflicting Learning, Evolution and Argumentation

Luís Moniz Pereira and Alexandre Miguel Pinto

Centro de Inteligência Artificial (CENTRIA), Universidade Nova de Lisboa, Portugal
{lmp,amp}@di.fct.unl.pt

Summary. We discuss the adoption of a three-valued setting for inductive concept learning. Distinguishing between what is true, what is false and what is unknown can be useful in situations where decisions have to be taken on the basis of scarce, ambiguous, or downright contradictory information. In a three-valued setting, we learn a definition for both the target concept and its opposite, considering positive and negative examples as instances of two disjoint classes. Explicit negation is used to represent the opposite concept, while default negation is used to ensure consistency and to handle exceptions to general rules. Exceptions are represented by examples covered by the definition for a concept that belong to the training set for the opposite concept.

After obtaining the knowledge resulting from this learning process, an agent can then interact with the environment by perceiving it and acting upon it. However, in order to know what is the best course of action to take the agent must know the causes or explanations of the observed phenomena.

Abduction, or abductive reasoning, is the process of reasoning to the best explanations. It is the reasoning process that starts from a set of observations or conclusions and derives their most likely explanations. The term abduction is sometimes used to mean just the generation of hypotheses to explain observations or conclusions, given a theory. Upon observing changes in the environment or in some artifact of which we have a theory, several possible explanations (abductive ones) might come to mind. We say we have several alternative arguments to explain the observations.

One single agent exploring an environment may gather only so much information about it and that may not suffice to find the right explanations. In such case, a collaborative multi-agent strategy, where each agent explores a part of the environment and shares with the others its findings, might provide better results. We describe one such framework based on a distributed genetic algorithm enhanced by a Lamarckian operator for belief revision. The agents communicate their candidate explanations — coded as chromosomes of beliefs — by sharing them in a common pool. Another way of interpreting this communication is in the context of argumentation.

We often encounter situations in which someone is trying to persuade us of a point of view by presenting reasons for it. This is called "arguing a case" or "presenting an argument". We can also argue with ourselves. Sometimes it is easy to see what the issues and conclusions are, and the reasons presented, but sometimes not. In the process of taking all the arguments and trying to find a common ground or consensus we might have to change, or review, some of assumptions of each argument. Belief revision is the process of changing beliefs to take into account a new piece of information. The logical formalization of belief revision is researched in philosophy, in databases, and in artificial intelligence for the design of rational agents.

The resulting framework we present is a collaborative perspective of argumentation where arguments are put together at work in order to find the possible 2-valued consensus of opposing positions of learnt concepts.

H.R. Tizhoosh, M. Ventresca (Eds.): Oppos. Concepts in Comp. Intel., SCI 155, pp. 61–89, 2008.

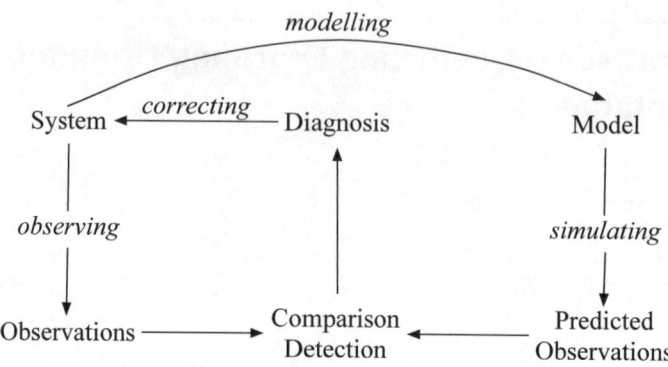

Fig. 5.1. Knowledge Model refinement cycle through Diagnosis

5.1   Introduction

The scientific approach is the most skeptical one towards finding the explanations to natural phenomena. Such skeptical stance leads to a pre-disposition to continuous knowledge revision and refinement based on observations — the solid foundations for any reasonable theory. The endless scientific cycle of theory building and refinement consists of 1) the environment; 2) producing candidate theories that best cover the observations; 3) create and exploit the experiences that will best test and stress the theories; and 4) going back to step 1) by collecting new observations from the environment resulting from the experiences. This cycle is depicted in figure 5.1.

After some iterations along this theory building/refinement cycle the theory built is "good enough" in the sense that the predictions it makes are accurate "enough" concerning the environment observations resulting from experiences. At this point the theory can be used both to provide explanations to observations as well as to produce new predictions.

Throughout the years, scientists from every area of knowledge have relied upon logic to develop and refine theories and to argue about them. Logic has been an indispensable tool to build theories from observations, to use the theories to make predictions, and to revise theories when observational data contradicts the predicted results.

Expressing theories as Logic Programs has become more natural and common as the field of Computational Logic has grown mature and other fields started to use its tools and results. Theories are usually expressed as a set of 'if-then' rules and facts which allow for the derivation, through the use of logic inference, of non-obvious results. When writing such rules and facts, explicit negation, just like explicit affirmation, can be used to formalize sure knowledge which provides inevitable results. Not only has formal argumentation been characterized in terms of Logic Programs, but the various semantics of Logic Programs themselves have been characterized as the result of argumentation between competing program interpretations.

Theories can be further refined by adding special rules taking the form of Integrity Constraints (ICs) . These impose that, whatever the assumptions might be, some conditions must be met. One implicit constraint on every reasonable theory is overall consistency, i.e, it must not be possible to derive one conclusion and its opposition.

Since in the real world the most common situation is one where there is incomplete and updatable information, any system making a serious attempt at dealing with real situations must cope with such complexities. To deal with this issue, the field of Computational Logic has also formalized another form of negation, Default Negation, used to express uncertain knowledge and exceptions, and used to derive results in the absence of complete information. When new information updates the theory, old conclusions might no longer be available (because they were relying on assumptions that become false with the new information), and further new conclusions might now be derived (for an analogous reasons).

The principle we use is thus the *Unknown World Assumption* (UWA) where everything is unknown or undefined until we have some solid evidence of its truthfulness or falseness. This principle differs from the more usual *Closed World Assumption* (CWA) where everything is assumed false until there is solid evidence of its truthfulness. We believe the UWA stance is more skeptical, cautious, and even more realistic than the CWA. We do not choose a fuzzy logic approach due to its necessity of specific threshold values. For such an approach we would need to compute those values *a priori*, possibly recurring to a probabilistic frequency-based calculation. Accordingly, we use a 3-valued logic (with the *undefined* truth value besides the *true* and *false* ones) instead of a more classical 2-valued logic.

We start by presenting the method for theory building from observations we use — a 3-valued logic rule learning method, and in the following section we focus on a method to analyze observations and to provide explanations for them given the learned theory. We show how the possible alternative explanations can be viewed as arguments for and against some hypotheses, and how we can use these arguments in a collaborative way to find better consensual explanations. Conclusions and outlined future work close this chapter.

5.2   Theory Building and Refinement

In real-world problems, complete information about the world is impossible to achieve and it is necessary to reason and act on the basis of the available partial information. In situations of incomplete knowledge, it is important to distinguish between what is true, what is false, and what is unknown or undefined.

Such a situation occurs, for example, when an agent incrementally gathers information from the surrounding world and has to select its own actions on the basis of acquired knowledge. If the agent learns in a two-valued setting, it can encounter the problems that have been highlighted in [22]. When learning in a specific to general way, it will learn a cautious definition for the target concept and it will not be able to distinguish what is false from what is not yet known (see Figure 5.2a) . Supposing the target predicate represents the allowed actions, then the agent will not distinguish forbidden actions from actions with an outcome and this can restrict the agent's acting power. If the agent learns

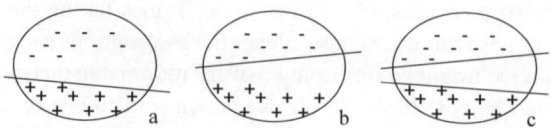

Fig. 5.2. (a,b) two-valued setting, (c) three-valued setting (Taken from [22])

in a general to specific way (i.e., the agent starts with a most general concept and progressively restricts it by adding exceptions as he learns), instead, it will not know the difference between what is true and what is unknown (Figure 5.2b) and, therefore, it can try actions with an unknown outcome. Rather, by learning in a three-valued setting, it will be able to distinguish between allowed actions, forbidden actions, and actions with an unknown outcome (Figure 5.2c) . In this way, the agent will know which part of the domain needs to be further explored and will not try actions with an unknown outcome unless it is trying to expand its knowledge.

In [47] the authors showed that various approaches and strategies can be adopted in Inductive Logic Programming (ILP, henceforth) for learning with Extended Logic Programs (ELP) — including explicit negation — under an extension of well-founded semantics. As in [37, 38], where answer-sets semantics is used, the learning process starts from a set of positive and negative examples plus some background knowledge in the form of an extended logic program. Positive and negative information in the training set are treated equally, by learning a definition for both a positive concept p and its (explicitly) negated concept $\neg p$. Coverage of examples is tested by adopting the *SLX* [3] interpreter for ELP under the Well-Founded Semantics with explicit negation (*WFSX*) defined in [5, 25], and valid for its paraconsistent version [17].

Example 1. Explicit negation

Consider a person who just moved to another city. He has just arrived and so he does not know yet if the neighborhood he is going to live in is dangerous or not.

$$dangerous\_neighborhood \; \leftarrow \; not \; \neg dangerous\_neighborhood$$
$$\neg dangerous\_neighborhood \leftarrow \; not \; dangerous\_neighborhood$$

Suppose now that he learns for sure that the neighborhood is not dangerous at all. In such case the program has another rule (which is actually a fact):

$$\neg dangerous\_neighborhood$$

Default negation is used in the learning process to handle *exceptions* to general rules. Exceptions are examples covered by the definition for the positive concept that belong to the training set for the negative concept or examples covered by the definition for the negative concept that belong to the training set for the positive concept.

In this work, we consider standard ILP techniques to learn a concept and its opposite. Indeed, separately learned positive and negative concepts may conflict and, in order to handle possible *contradiction*, contradictory learned rules are made defeatable by making the learned definition for a positive concept p depend on the default negation of the negative concept $\neg p$, and vice-versa, i.e., each definition is introduced as an exception to the other. This way of coping with contradiction can be even generalized for learning n disjoint classes, or modified in order to take into account preferences among multiple learning agents or information sources (see [45]).

In the learning problem we consider we want to learn an ELP from a background knowledge that is itself an ELP and from a set of positive and a set of negative examples in the form of ground facts for the target predicates.

A learning problem for ELP's was first introduced in [38] where the notion of coverage was defined by means of truth in the answer-set semantics. Here the problem definition is modified to consider coverage as truth in the preferred *WFSX*.

Definition 1 (Learning Extended Logic Programs)
Given:

- *a set $\mathcal{P}$ of possible (extended logic) programs*
- *a set E^+ of positive examples (ground facts)*
- *a set E^- of negative examples (ground facts)*
- *a non-contradictory extended logic program B* (background knowledge [1])

Find:

- *an extended logic program $P \in \mathcal{P}$ such that*
  - $\forall e \in E^+ \cup \neg E^-, \ B \cup P \models_{WFSX} e$ *(completeness)*
  - $\forall e \in \neg E^+ \cup E^-, \ B \cup P \not\models_{WFSX} e$ *(consistency)*
  where $\neg E = \{\neg e | e \in E\}$.

We suppose that the training sets E^+ and E^- are disjoint. However, the system is also able to work with overlapping training sets.

The learned theory will contain rules of the form:

$$p(\mathbf{X}) \leftarrow Body^+(\mathbf{X})$$
$$\neg p(\mathbf{X}) \leftarrow Body^-(\mathbf{X})$$

for every target predicate p, where $\mathbf{X}$ stands for a tuple of arguments. In order to satisfy the completeness requirement, the rules for p will entail all positive examples while the rules for $\neg p$ will entail all (explicitly negated) negative examples. The consistency requirement is satisfied by ensuring that both sets of rules do not entail instances of the opposite element in either of the training sets.

Note that, in the case of extended logic programs, the consistency with respect to the training set is equivalent to the requirement that the program is non-contradictory on the examples. This requirement is enlarged to require that the program be non-contradictory also for unseen atoms, i.e., $B \cup P \not\models L \land \neg L$ for every atom L of the target predicates.

We say that an example e is *covered* by program P if $P \models_{WFSX} e$. Since the *SLX* procedure is correct with respect to *WFSX*, even for contradictory programs, coverage of examples is tested by verifying whether $P \vdash_{SLX} e$.

---

[1] By non-contradictory program we mean a program which admits at least one *WFSX* model.

The approach to learning with extended logic programs considered consists in initially applying conventional ILP techniques to learn a positive definition from E^+ and E^- and a negative definition from E^- and E^+. In these techniques, the *SLX* procedure substitutes the standard Logic Programming proof procedure to test the coverage of examples.

The ILP techniques to be used depend on the level of generality that we want to have for the two definitions: we can look for the Least General Solution (LGS) or the Most General Solution (MGS) of the problem of learning each concept and its complement. In practice, LGS and MGS are not unique and real systems usually learn theories that are not the least nor most general, but closely approximate one of the two. In the following, these concepts will be used to signify approximations to the theoretical concepts.

LGSs can be found by adopting one of the bottom-up methods such as relative least general generalization ($rlgg$) [66] and the GOLEM system [57] [2], inverse resolution [56] or inverse entailment [48]. Conversely, MGSs can be found by adopting a top-down refining method (cf. [49]) and a system such as FOIL [68] or Progol [55].

5.2.1 Strategies for Combining Different Generalizations

The generality of concepts to be learned is an important issue when learning in a three-valued setting. In a two-valued setting, once the generality of the definition is chosen, the extension (i.e., the generality) of the set of false atoms is automatically decided, because it is the complement of the true atoms set. In a three-valued setting, rather, the extension of the set of false atoms depends on the generality of the definition learned for the negative concept. Therefore, the corresponding level of generality may be chosen independently for the two definitions, thus affording four epistemological cases. The adoption of ELP allows case combination to be expressed in a declarative and smooth way.

Furthermore, the generality of the solutions learned for the positive and negative concepts clearly influences the interaction between the definitions. If we learn the MGS for both a concept and its opposite, the probability that their intersection is non-empty is higher than if we learn the LGS for both. Intuitively, this happens because, as explained above, when learning the MGS for a concept we begin with a most permissive definition for that concept and progressively refine it by adding exceptions. It is easy to see that at the very beginning of the learning process, if the MGS is used for both a concept and its opposite, these coincide. As the process of refinement goes on, the intersection of the MGS of the concept and the MGS of its opposite diminishes. Accordingly, the decision as to which type of solution to learn should take into account the possibility of interaction as well: if we want to reduce this possibility, we have to learn two LGS, if we do not care about interaction, we can learn two MGS. In general, we may learn different generalizations and combine them in distinct ways for different strategic purposes within the same application problem.

The choice of the level of generality should be made on the basis of available knowledge about the domain. Two of the criteria that can be taken into account are the damage

---

[2] For a recent implementation see http://www.doc.ic.ac.uk/~shm/Software/golem/

or risk that may arise from an erroneous classification of an unseen object, and the confidence we have in the training set as to its correctness and representativeness.

When classifying an as yet unseen object as belonging to a concept, we may later discover that the object belongs to the opposite concept. The more we generalize a concept, the higher is the number of unseen atoms covered by the definition and the higher is the risk of an erroneous classification. Depending on the damage that may derive from such a mistake, we may decide to take a more cautious or a more confident approach. If the possible damage from an over extensive concept is high, then one should learn the LGS for that concept, if the possible damage is low then one can generalize the most and learn the MGS. The overall risk will depend also on the use of the learned concepts within other rules: we need to take into account the damage that may derive from mistakes made on concepts depending on the target one.

The problem of selecting a solution of an inductive problem according to the cost of misclassifying examples has been studied in a number of works. PREDICTOR [34] is able to select the cautiousness of its learning operators by means of meta-heuristics. These meta-heuristics make the selection based on a user-input penalty for prediction error. In [67] Provost provides a method to select classifiers given the cost of misclassifications and the prior distribution of positive and negative instances. The method is based on the Receiver Operating Characteristic (ROC) [35] graph from signal theory that depicts classifiers as points in a graph with the number of false positives on the X axis and the number of true positive on the Y axis. In [59] it is discussed how the different costs of misclassifying examples can be taken into account into a number of algorithms: decision tree learners, Bayesian classifiers and decision list learners.

As regards the confidence in the training set, we can prefer to learn the MGS for a concept if we are confident that examples for the opposite concept are correct and representative of the concept. In fact, in top-down methods, negative examples are used in order to delimit the generality of the solution. Otherwise, if we think that examples for the opposite concept are not reliable, then we should learn the LGS.

In the following, we present a realistic example of the kind of reasoning that can be used to choose and specify the preferred level of generality, and discuss how to strategically combine the different levels by employing ELP tools to learning.

Example 2. Consider now a person living in a bad neighborhood in Los Angeles. He is an honest man and to survive he needs two concepts, one about who is likely to attack him, on the basis of appearance, gang membership, age, past dealings, etc. Since he wants to take a cautious approach, he maximizes *attacker* and minimizes ¬*attacker*, so that his *attacker*1 concept allows him to avoid dangerous situations.

$$attacker1(X) \leftarrow attacker_{MGS}(X)$$
$$\neg attacker1(X) \leftarrow \neg attacker_{LGS}(X)$$

Another concept he needs is the type of beggars he should give money to (he is a good man) that actually seem to deserve it, on the basis of appearance, health, age, etc. Since he is not rich and does not like to be tricked, he learns a *beggar*1 concept by minimizing *beggar* and maximizing ¬*beggar*, so that his beggar concept allows him to give money strictly to those appearing to need it without faking.

$$beggar1(X) \leftarrow beggar_{LGS}(X)$$
$$\neg beggar1(X) \leftarrow \neg beggar_{MGS}(X)$$

However, rejected beggars, especially malicious ones, may turn into attackers, in this very bad neighborhood. Consequently, if he thinks a beggar might attack him, he had better be more permissive about who is a beggar and placate him with money. In other words, he should maximize $beggar$ and minimize $\neg beggar$ in a $beggar2$ concept.

$$beggar2(X) \leftarrow beggar_{MGS}(X)$$
$$\neg beggar2(X) \leftarrow \neg beggar_{LGS}(X)$$

These concepts can be used in order to minimize his risk taking when he carries, by his standards, a lot of money and meets someone who is likely to be an attacker, with the following kind of reasoning:

$$run(X) \leftarrow lot\_of\_money(X), meets(X,Y), attacker1(Y),$$
$$not\ beggar2(Y)$$
$$\neg run(X) \leftarrow lot\_of\_money(X), give\_money(X,Y)$$
$$give\_money(X,Y) \leftarrow meets(X,Y), beggar1(Y)$$
$$give\_money(X,Y) \leftarrow meets(X,Y), attacker1(Y), beggar2(Y)$$

If he does not have a lot of money on him, he may prefer not to run as he risks being beaten up. In this case he has to relax his attacker concept into $attacker2$, but not relax it so much that he would use $\neg attacker_{MGS}$.

$$\neg run(X) \leftarrow little\_money(X), meets(X,Y), attacker2(Y)$$
$$attacker2(X) \leftarrow attacker_{LGS}(X)$$
$$\neg attacker2(X) \leftarrow \neg attacker_{LGS}(X)$$

The various notions of $attacker$ and $beggar$ are then learned on the basis of previous experience the man has had (see [47]).

5.2.2 Strategies for Eliminating Learned Contradictions

The learned definitions of the positive and negative concepts may overlap. In this case, we have a contradictory classification for the objective literals[3] in the intersection. In order to resolve the conflict, we must distinguish two types of literals in the intersection: those that belong to the training set and those that do not, also dubbed *unseen* atoms (see Figure 5.3).

In the following we discuss how to resolve the conflict in the case of unseen literals and of literals in the training set. We first consider the case in which the training sets are disjoint, and we later extend the scope to the case where there is a non-empty intersection of the training sets, when they are less than perfect. From now onwards, **X** stands for a tuple of arguments.

For unseen literals, the conflict is resolved by classifying them as undefined, since the arguments supporting the two classifications are equally strong. Instead, for literals in the training set, the conflict is resolved by giving priority to the classification stipulated by the training set. In other words, literals in a training set that are covered by the opposite definition are considered as *exceptions* to that definition.

---

[3] An 'objective literal' in a Logic Program is just an atom, possibly explicitly negated. E.g., '$attacker2(X)$' and '$\neg attacker2(X)$' in example 2 are objective literals.

Contradiction on Unseen Literals

For unseen literals in the intersection, the undefined classification is obtained by making opposite rules mutually defeasible, or "non-deterministic" (see [10, 5]). The target theory is consequently expressed in the following way:

$$p(\mathbf{X}) \leftarrow p^+(\mathbf{X}), not \, \neg p(\mathbf{X})$$
$$\neg p(\mathbf{X}) \leftarrow p^-(\mathbf{X}), not \, p(\mathbf{X})$$

where $p^+(\mathbf{X})$ and $p^-(\mathbf{X})$ are, respectively, the definitions learned for the positive and the negative concept, obtained by renaming the positive predicate by p^+ and its explicit negation by p^-. From now onwards, we will indicate with these superscripts the definitions learned separately for the positive and negative concepts.

We want both $p(\mathbf{X})$ and $\neg p(\mathbf{X})$ to act as an exception to the other. In case of contradiction, this will introduce mutual circularity, and hence undefinedness according to *WFSX*. For each literal in the intersection of p^+ and p^-, there are two stable models, one containing the literal, the other containing the opposite literal. According to - *WFSX*, there is a third (partial) stable model where both literals are undefined, i.e., no literal $p(\mathbf{X})$, $\neg p(\mathbf{X})$, $not \, p(\mathbf{X})$ or $not \, \neg p(\mathbf{X})$ belongs to the well-founded (or least partial stable) model. The resulting program contains a recursion through negation (i.e., it is non-stratified) but the top-down *SLX* procedure does not go into a loop because it comprises mechanisms for loop detection and treatment, which are implemented by XSB Prolog through tabling.

Example 3. Let us consider the Example of Section 5.2.1. In order to avoid contradictions on unseen atoms, the learned definitions must be:

$$attacker1(X) \leftarrow attacker^+_{MGS}(X), not \, \neg attacker1(X)$$
$$\neg attacker1(X) \leftarrow attacker^-_{LGS}(X), not \, attacker1(X)$$
$$beggar1(X) \leftarrow beggar^+_{LGS}(X), not \, \neg beggar1(X)$$
$$\neg beggar1(X) \leftarrow beggar^-_{MGS}(X), not \, beggar1(X)$$
$$beggar2(X) \leftarrow beggar^+_{MGS}(X), not \, \neg beggar2(X)$$
$$\neg beggar2(X) \leftarrow beggar^-_{LGS}(X), not \, beggar2(X)$$
$$attacker2(X) \leftarrow attacker^+_{LGS}(X), not \, \neg attacker2(X)$$
$$\neg attacker2(X) \leftarrow attacker^-_{LGS}(X), not \, attacker2(X)$$

Note that $p^+(\mathbf{X})$ and $p^-(\mathbf{X})$ can display as well the undefined truth value, either because the original background is non-stratified or because they rely on some definition learned for another target predicate, which is of the form above and therefore non-stratified. In this case, three-valued semantics can produce literals with the value "undefined", and one or both of $p^+(\mathbf{X})$ and $p^-(\mathbf{X})$ may be undefined. If one is undefined and the other is true, then the rules above make both p and $\neg p$ undefined, since the negation by default of an undefined literal is still undefined. However, this is counterintuitive: a defined value should prevail over an undefined one.

In order to handle this case, we suppose that a system predicate $undefined(X)$ is available[4], that succeeds if and only if the literal X is undefined. So we add the following two rules to the definitions for p and $\neg p$:

---

[4] The *undefined* predicate can be implemented through negation NOT under CWA ($NOT \, P$ means that P is false whereas not means that P is false or undefined), i.e., $undefined(P) \leftarrow NOT \, P, NOT(not \, P)$.

$$p(\mathbf{X}) \leftarrow p^+(\mathbf{X}), undefined(p^-(\mathbf{X}))$$
$$\neg p(\mathbf{X}) \leftarrow p^-(\mathbf{X}), undefined(p^+(\mathbf{X}))$$

According to these clauses, $p(\mathbf{X})$ is true when $p^+(\mathbf{X})$ is true and $p^-(\mathbf{X})$ is undefined, and conversely.

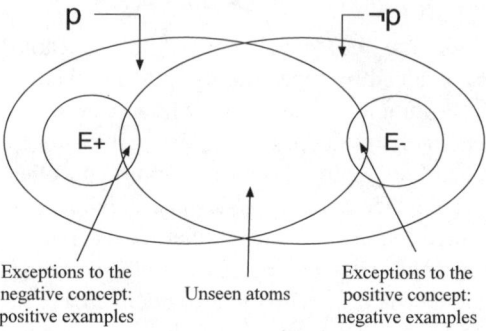

Fig. 5.3. Interaction of the positive and negative definitions on exceptions

Contradiction on Examples

Theories are tested for consistency on all the literals of the training set, so we should not have a conflict on them. However, in some cases, it is useful to relax the consistency requirement and learn clauses that cover a small amount of counterexamples. This is advantageous when it would be otherwise impossible to learn a definition for the concept, because no clause is contained in the language bias that is consistent, or when an overspecific definition would be learned, composed of many specific clauses instead of a few general ones. In such cases, the definitions of the positive and negative concepts may cover examples of the opposite training set. These must then be considered exceptions, which are then due to abnormalities in the opposite concept.

Let us start with the case where some literals covered by a definition belong to the opposite training set. We want of course to classify these according to the classification given by the training set, by making such literals *exceptions*. To handle exceptions to classification rules, we add a negative default literal of the form *not abnorm*$_p(\mathbf{X})$ (resp. *not abnorm*$_{\neg p}(\mathbf{X})$) to the rule for $p(\mathbf{X})$ (resp. $\neg p(\mathbf{X})$), to express possible abnormalities arising from exceptions. Then, for every exception $p(\mathbf{t})$, an individual fact of the form *abnorm*$_p(\mathbf{t})$ (resp. *abnorm*$_{\neg p}(\mathbf{t})$) is asserted so that the rule for $p(\mathbf{X})$ (resp. $\neg p(\mathbf{X})$) does not cover the exception, while the opposite definition still covers it. In this way, exceptions will figure in the model of the theory with the correct truth value. The learned theory thus takes the form:

$$p(\mathbf{X}) \leftarrow p^+(\mathbf{X}), not\ abnorm_p(\mathbf{X}), not\ \neg p(\mathbf{X}) \tag{5.1}$$
$$\neg p(\mathbf{X}) \leftarrow p^-(\mathbf{X}), not\ abnorm_{\neg p}(\mathbf{X}), not\ p(\mathbf{X}) \tag{5.2}$$

$$p(\mathbf{X}) \leftarrow p^+(\mathbf{X}), undefined(p^-(\mathbf{X})) \tag{5.3}$$

$$\neg p(\mathbf{X}) \leftarrow p^-(\mathbf{X}), undefined(p^+(\mathbf{X})) \tag{5.4}$$

Abnormality literals have not been added to the rules for the undefined case because a literal which is an exception is also an example, and so must be covered by its respective definition; therefore it cannot be undefined.

Notice that if E^+ and E^- overlap for some example $p(\mathbf{t})$, then $p(\mathbf{t})$ is classified $false$ by the learned theory. A different behavior would be obtained by slightly changing the form of learned rules, in order to adopt, for atoms of the training set, one classification as default and thus give preference to false (negative training set) or true (positive training set).

Individual facts of the form $abnorm_p(\mathbf{X})$ might be then used as examples for learning a definition for $abnorm_p$ and $abnorm_{\neg p}$, as in [30, 38]. In turn, exceptions to the definitions of $abnorm_p$ and $abnorm_{\neg p}$ might be found and so on, thus leading to a hierarchy of exceptions (for our hierarchical learning of exceptions, see [44, 74]).

Example 4. Consider a domain containing entities a, b, c, d, e, f and suppose the target concept is $flies$. Let the background knowledge be:

$bird(a)$	$has\_wings(a)$	
$jet(b)$	$has\_wings(b)$	
$angel(c)$	$has\_wings(c)$	$has\_limbs(c)$
$penguin(d)$	$has\_wings(d)$	$has\_limbs(d)$
$dog(e)$		$has\_limbs(e)$
$cat(f)$		$has\_limbs(f)$

and let the training set be:

$$E^+ = \{flies(a)\} \qquad E^- = \{flies(d), flies(e)\}$$

A possible learned theory is:

$$flies(X) \leftarrow flies^+(X), not\ abnormal_{flies}(X), not\ \neg flies(X)$$
$$\neg flies(X) \leftarrow flies^-(X), not\ flies(X)$$
$$flies(X) \leftarrow flies^+(X), undefined(flies^-(X))$$
$$\neg flies(X) \leftarrow flies^-(X), undefined(flies^+(X))$$
$$abnormal_{flies}(d) \leftarrow true$$

where $flies^+(X) \leftarrow has\_wings(X)$ and $flies(X)^- \leftarrow has\_limbs(X)$.

The example above and Figure 5.4 show all the possible cases for a literal when learning in a three-valued setting. a and e are examples that are consistently covered by the definitions. b and f are unseen literals on which there is no contradiction. c and d are literals where there is contradiction, but c is classified as undefined whereas d is considered as an exception to the positive definition and is classified as negative.

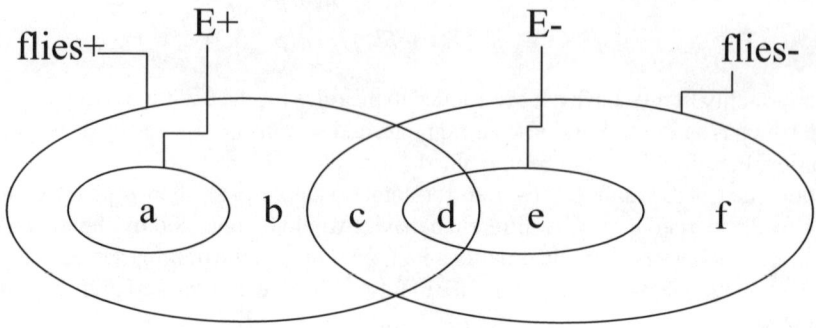

Fig. 5.4. Coverage of definitions for opposite concepts

Identifying contradictions on unseen literals is useful in interactive theory revision, where the system can ask an oracle to classify the literal(s) leading to contradiction, and accordingly revise the least or most general solutions for p and for $\neg p$ using a theory revision system such as REVISE [15] or CLINT [21, 23]. Detecting uncovered literals points to theory extension.

Extended logic programs can be used as well to represent n disjoint classes $p_1, \ldots, p_n$. When one has to learn n disjoint classes, the training set contains a number of facts for a number of predicates $p_1, \ldots, p_n$. Let p_i^+ be a definition learned by using, as positive examples, the literals in the training set classified as belonging to p_i and, as negative examples, all the literals for the other classes. Then the following rules ensure consistency on unseen literals and on exceptions regardless of the algorithm used for learning the p_i^+.

$$p_1(\mathbf{X}) \leftarrow p_1^+(\mathbf{X}), not\ abnormal_{p_1}(\mathbf{X}), not\ p_1(\mathbf{X}), not\ p_2(\mathbf{X}), \ldots, not\ p_n(\mathbf{X})$$
$$p_1(\mathbf{X}) \leftarrow p_1^+(\mathbf{X}), not\ abnormal_{p_1}(\mathbf{X}), not\ p_2(\mathbf{X}), \ldots, not\ p_n(\mathbf{X})$$
$$p_2(\mathbf{X}) \leftarrow p_2^+(\mathbf{X}), not\ abnormal_{p_2}(\mathbf{X}), not\ p_1(\mathbf{X}), not\ p_3(\mathbf{X}), \ldots, not\ p_n(\mathbf{X})$$
$$\cdots \quad \leftarrow \cdots$$
$$p_n(\mathbf{X}) \leftarrow p_n^+(\mathbf{X}), not\ abnormal_{p_n}(\mathbf{X}), not\ p_1(\mathbf{X}), \ldots, not\ p_{n-1}(\mathbf{X})$$

$$p_1(\mathbf{X}) \leftarrow p_1^+(\mathbf{X}), undefined(p_2^+(\mathbf{X})), \ldots, undefined(p_n^+(\mathbf{X}))$$
$$p_2(\mathbf{X}) \leftarrow p_2^+(\mathbf{X}), undefined(p_1^+(\mathbf{X})), undefined(p_3^+(\mathbf{X})), \ldots,$$
$$\qquad undefined(p_n^+(\mathbf{X}))$$
$$\cdots \quad \leftarrow \cdots$$
$$p_n(\mathbf{X}) \leftarrow p_n^+(\mathbf{X}), undefined(p_1^+(\mathbf{X})), \ldots, undefined(p_{n-1}^+(\mathbf{X}))$$

5.3   Observation Analysis and Explanation

After a theory is built it can now be used to analyze observations and to provide explanations for them. Such explanations are sets of abductive hypotheses which, when assumed true under the theory at hand, yield the observations as conclusions. We can also understand each such set of hypotheses as an argument explaining why the observations hold. There can be, of course, many different possible explanations, or arguments. In the end, most of the times, we want to find the single "best" explanation for the observations, and hence we must have some mechanism to identify the "best" solution among the several alternative ones.

5.3.1 Abduction

Deduction and abduction differ in the direction in which a rule like "a entails b" is used for inference. Deduction allows deriving b as a consequence of a ; i.e., deduction is the process of deriving the consequences of what is known. Abduction allows deriving a as a hypothetical explanation of b. Abduction works in reverse to deduction, by allowing the precondition a of "a entails b" to be derived from the consequence b, i.e., abduction is the process of explaining what is known. Charles Saunders Peirce [60] introduced abduction into logic, to mean the use of a rule or hypothetical fact to explain an observation, e.g. "if it rains the grass is wet" is used to explain why the grass is wet, given that it has rained, or vice-versa. In logic, abduction is done from a logical theory T representing a domain and a set of observations O. Abduction is the process of deriving a set of explanations of O according to T. For E to be an explanation of O according to T, it should satisfy two conditions:

- O follows from E and T;
- E is consistent with T.

In formal logic, O and E are assumed to be sets of literals. The two conditions for E being an explanation of O according to theory T are:

- $T \cup E \models O$;
- $T \cup E$ is consistent.

Among the possible explanations E satisfying these two conditions, a condition of minimality is usually imposed to avoid irrelevant facts (i.e. not contributing to the entailment of O) to be included in the explanations. An application of abduction is that of detecting faults in systems: given a theory relating faults with their effects and a set of observed effects, abduction can be used to derive sets of faults that are likely to be the cause of the problem. Belief revision, the process of adapting beliefs in view of new information, is another field in which abduction has been applied. The main problem of belief revision is that the new information may be inconsistent with the corpus of beliefs, while the result of the incorporation must not be inconsistent.

5.3.2 An Argumentation Perspective

When different alternative explanations arise, people argue for and against their theories and others'. For a long time, because of its origins in rhetoric and the law, argumentation has been thought of as a kind of a battle where two (or more) opposing opinions are formalized into arguments, and logic is used for the framing rules for the battle and to decide the outcome. At the end of the battle one of the arguers will be 'right' and the other(s) 'wrong'. The 'winner' is the one whose argument, attacking others' arguments, cannot be counter-attacked. The problem of argumentation becomes more complex when all arguments successfully attack each other's corresponding to different possible opinions, not inevitable conclusions. In this case, argumentation could take a new flavor, one of Collaboration besides Conflict.

5.3.3   Finding Alternative Explanations for Observations

Trying to find explanations for observations can be implemented by simply finding the alternative abductive models that satisfy both the theory's rules and the observations. The latter can be coded as Integrity Constraints (ICs) which are added to the theory thereby imposing the truthfulness of the observations they describe.

Example 5. Running example
    We will use this running example throughout the rest of the chapter.
    Consider the following Logic Program consisting of four rules. According to this program a 'professional' is someone who is a regular employee or someone who is a boss in some company. Also, a non-employee is assumed to be a student as well as all those who are junior (all children should go to school!).

$$professional(X) \leftarrow employee(X) \qquad student(X) \leftarrow not\ employee(X)$$
$$professional(X) \leftarrow boss(X) \qquad\qquad student(X) \leftarrow junior(X)$$

For now keep this example in mind as we will use it to illustrate the concepts and methods we are about to describe. Assume that '$employee/1$', '$boss/1$', and '$junior/1$' are abducible hypotheses.

Adding one single IC to the theory might yield several alternative 2-valued models (sets of abductive hypotheses) satisfying it, let alone adding several ICs.
    In the example above, adding just the Integrity Constraint
'$\perp \leftarrow not\ professional(john)$' — coding the fact that John is a professional — would yield two alternative abductive solutions: $\{employee(john)\}$ and $\{boss(john)\}$.
    When the information from several observations comes in at one single time, several ICs must be added to the theory in order to be possible to obtain the right explanations for the corresponding observations.
    In a fairly complex knowledge domain coded in a complex and lengthy theory, finding each one alternative explanation for a given observation can be quite hard and time consuming, let alone finding the "best" explanation. In general, following Occam's principle, the "best" explanation for any given observation is usually the simplest one, i.e., the one recurring to the fewest number of hypotheses — the minimal set of hypotheses.
    In [46] the authors presented a method for finding the minimal belief revision solution for a set of observations. Therein, each belief corresponds to an abducible hypothesis and the belief revision is the process of finding the set of hypotheses that conforms to the observations by revising their truth value from *true* to *false* or vice-versa. Minimality is required for compliance with the Occam's principle.
    In [46] the authors also code observations as ICs added to the theory, but they recur to finding the support sets for *falsum* ($\perp$)— the special reserved atom for the heads of the rules coding ICs, in order to find the belief revisions necessary to comply to the ICs.

After finding such support sets they can identify the minimal sets of hypotheses that need to be revised in order to prevent *falsum* from being derived.

Here we are also concerned with finding explanations to observations, but the method used is quite different. In a nutshell, we split the set of observations into several smaller subsets; then we create several agents and give each agent the same base theory and a subset of the observations coded as ICs. We then allow each agent to come up with several alternative explanations to its ICs; the explanations need not be minimal sets of hypotheses.

Going back again to our running example, if we also know that John is a student, besides adding the '$\perp \leftarrow not\ professional(john)$' IC we must also add the '$\perp \leftarrow not\ student(john)$' IC.

Finding possible alternative explanations is one problem; finding which one(s) is(are) the "best" is another issue. In the next section we assume "best" means minimal set of hypotheses and we describe the method we use to find such best. Another interpretation of "best" could be "most probable" and in this case the theory inside the agents must contain the adequate probabilistic information. One such possibility would be the one described in [9]. We do not pursue this approach yet, but we consider it for future work, namely following the principles in [9].

5.3.4 Choosing the Best Explanation

Ex contradictione quodlibet. This well-known Latin saying means "Anything follows from contradiction". But contradictory, oppositional ideas and arguments can be combined together in different ways to produce new ideas. Since "anything follows from contradiction" one of the things that might follow from it is a solution to a problem to which several alternative positions contribute.

One well known method for solving complex problems widely used by creative teams is that of 'brainstorming'. In a nutshell, every agent participating in the 'brainstorm' contributes by adding one of his/her ideas to the common idea-pool shared by all the agents. All the ideas, sometimes clashing and oppositional among each other, are then mixed, crossed and mutated. The solution to the problem arises from the pool after a few iterations of this evolutionary process.

The evolution of alternative ideas and arguments in order to find a collaborative solution to a group problem is the underlying inspiration of this work.

Evolutionary Inspiration

Darwin's theory is based on the concept of natural selection: only those individuals that are most fit for their environment survive, and are thus able to generate new individuals by means of reproduction. Moreover, during their lifetime, individuals may be subject to random mutations of their genes that they can transmit to offspring. Lamarck's [42] theory, instead, states that evolution is due to the process of adaptation to the environment that an individual performs in his/her life. The results of this process are then automatically transmitted to his/her offspring, via its genes. In other words, the abilities learned during the life of an individual can modify his/her genes.

Experimental evidence in the biological kingdom has shown Darwin's theory to be correct and Lamarck's to be wrong. However, this does not mean that the process of adaptation (or learning) does not influence evolution. Baldwin [8] showed how learning could influence evolution: if the learned adaptations improve the organism's chance of survival then the chances for reproduction are also improved. Therefore there is selective advantage for genetically determined traits that predisposes the learning of specific behaviors. Baldwin moreover suggests that selective pressure could result in new individuals to be born with the learned behavior already encoded in their genes. This is known as the Baldwin effect. Even if there is still debate about it, it is accepted by most evolutionary biologists.

Lamarckian evolution [43] has recently received a renewed attention because it can model cultural evolution. In this context, the concept of "meme" has been developed. A meme is the cognitive equivalent of a gene and it stores abilities learned by an individual during his lifetime, so that they can be transmitted to his offspring.

In the field of genetic programming [41], Lamarckian evolution has proven to be a powerful concept and various authors have investigated the combination of Darwinian and Lamarckian evolution.

In [46] the authors propose a genetic algorithm for belief revision that includes, besides Darwin's operators of selection, mutation and crossover, a logic based Lamarckian operator as well. This operator differs from Darwinian ones precisely because it modifies a chromosome coding beliefs so that its fitness is improved by experience rather than in a random way. There, the authors showed that the combination of Darwinian and Lamarckian operators are useful not only for standard belief revision problems, but especially for problems where different chromosomes may be exposed to different constraints, as in the case of a multi-agent system. In these cases, the Lamarckian and Darwinian operators play different roles: the Lamarckian one is employed to bring a given chromosome closer to a solution (or even find an exact one) to the current belief revision problem, whereas the Darwinian ones exert the role of randomly producing alternative belief chromosomes so as to deal with unencountered situations, by means of exchanging genes amongst them.

Evolving Beliefs

Belief revision is an important functionality that agents must exhibit: agents should be able to modify their beliefs in order to model the outside world. What's more, as the world may be changing, a pool of separately and jointly evolved chromosomes may code for a variety of distinct belief evolution potentials that can respond to world changes as they occur. This dimension has been explored in [46] with specific experiments to that effect. Mark that it is not our purpose to propose here a competitor to extant classical belief revision methods, in particular as they apply to diagnosis. More ambitiously, we do propose a new and complementary methodology, which can empower belief revision — any assumption based belief revision — to deal with time/space distributed, and possibly intermittent or noisy laws about an albeit varying artifact or environment, possibly by a multiplicity of agents which exchange diversified genetically encoded experience. We consider a definition of the belief revision problem that consists in removing a contradiction from an extended logic program by modifying the truth value of a selected set of literals corresponding to the abducible hypotheses. The program contains as well clauses with *falsum* ($\perp$) in the head, representing ICs. Any model of the program must ensure the body of ICs false for the program to be non-contradictory. Contradiction may also arise in an extended logic program when both a literal L and its opposite $\neg L$ are obtainable in the model of the program. Such a problem has been widely studied in the literature, and various solutions have been proposed that are based on abductive logic proof procedures. The problem can be modeled by means of a genetic algorithm, by assigning to each abducible of a logic program a gene in a chromosome. In the simplest case of a two valued revision, the gene will have the value 1 if the corresponding abducible is *true* and the value 0 if the abducible is *false* . The fitness functions that can be used in this case are based on the percentage of ICs that are satisfied by a chromosome. This is, however, an over-simplistic approach since it assumes every abducible is a predicate with arity 0, otherwise a chromosome would have as many genes as the number of all possible combinations of ground values for variables in all abducibles.

Specific Belief Evolution Method

In multi-agent joint belief revision problems, agents usually take advantage of each other's knowledge and experience by explicitly communicating messages to that effect. In our approach, however, we introduce a new and complementary method (and some variations of it), in which we allow knowledge and experience to be coded as genes in an agent. These genes are exchanged with those of other agents, not by explicit message passing but through the crossover genetic operator. Crucial to this endeavor, a logic-based technique for modifying cultural genes, i.e. memes, on the basis of individual agent experience is used.

The technique amounts to a form of belief revision, where a meme codes for an agent's belief or assumptions about a piece of knowledge, and which is then diversely modified on the basis of how the present beliefs may be contradicted by laws (expressed as ICs). These mutations have the effect of attempting to reduce the number of unsatisfied constraints. Each agent possesses a pool of chromosomes containing such diversely modified memes, or alternative assumptions, which cross-fertilize Darwinianly amongst themselves. Such an experience in genetic evolution mechanism is aptly called Lamarckian.

Since we will subject the sets of beliefs to an evolutionary process (both Darwinian and Lamarckian) we will henceforth refer to this method as "Belief Evolution" (BE) instead of the classical "Belief Revision" (BR).

General Description of the Belief Evolution Method

Each agent keeps a population of chromosomes and finds a solution to the BE problem by means of a genetic algorithm. We consider a formulation of the distributed BE problem where each agent has the same set of abducibles and the same program expressed theory, but is exposed to possibly different constraints. Constraints may vary over time, and can differ because agents may explore different regions of the world. The genetic algorithm we employ allows each agent to cross over its chromosomes with chromosomes from other agents. In this way, each agent can be prepared in advance for situations that it will encounter when moving from one place to another.

The algorithm proposed for BE extends the standard genetic algorithm in two ways:

- crossover is performed among chromosomes belonging to different agents[5],
- a Lamarckian operator called Learn is added in order to bring a chromosome closer to a correct revision by changing the value of abducibles

The Structure of a Chromosome

In BR and BE, each individual hypothesis is described by the truth value of all the abducibles. Several possibilities of increasing complexity and expressive power arise now. Concerning truth values we can have 2-valued and 3-valued revisions — if we are not considering multi-valued logics. Orthogonally to this criterion we can eventualy encode more information in each gene of a chromosome. In particular, some possibilities are:

- each gene encodes a ground literal — all its variables are bound to fixed values
- each gene encodes a literal with non-ground variables plus constraints restricting the possible values for the free variables

Surely there are many other possibilities for the information each gene can encode, but in this work we restrict ourselves to the first one above. In such case, we represent a chromosome as a list of genes and memes, and different chromosomes may contain information about different genes. This implies a major difference to traditional genetic algorithms where every chromosome refers exactly to the same genes and the crossover and mutation operations are somewhat straightforward.

The memes in a chromosome will be just like genes — representing abducibles — but they will have extra information. Each meme has associated with it a counter keeping record of how many times the meme has been confirmed or refuted. Each time a meme is confirmed this value is increased, and each time it is refuted the value decreases. This value provides thus a measure of confidence in the corresponding meme.

Example 6. Running example (cont.)

Continuing with our running example, let us assume that both $professional(john)$ and $student(john)$ have been observed.

We can create two agents, each with the same rule-set theory, and split the observations among them. We would have thus

$$
\begin{array}{ll}
\text{Agent 1:} & \\
& \leftarrow\ not\ professional(john) \\
professional(X) \leftarrow & employee(X) \\
professional(X) \leftarrow & boss(X) \\
student(X) & \leftarrow\ not\ employee(X) \\
student(X) & \leftarrow\ junior(X)
\end{array}
$$

---

[5] Similarly to what happens with island models [75].

Agent 2:

$$\leftarrow not\ student(john)$$
$$professional(X) \leftarrow employee(X)$$
$$professional(X) \leftarrow boss(X)$$
$$student(X) \quad \leftarrow not\ employee(X)$$
$$student(X) \quad \leftarrow junior(X)$$

In the simplest case where a gene encodes an abductive ground literal Agent 1 would come up with two alternative abductive solutions for its IC '$\bot \leftarrow not\ professional$ $(john)$': $\{employee(john)\}$ and $\{boss(john)\}$. Moreover, Agent 2 would come up with two other alternative abductive solutions for its IC '$\bot \leftarrow not\ student(john)$': $\{not\ employee(john)\}$ and $\{junior(john)\}$.

Crossover

When a chromosome is a list of abducible hypotheses (with or without constraints over variables), as it is in the case we present here, the crossover and mutation operations cannot fallback into the well-known versions of standard genetic algorithms. If two different chromosomes, each encoding information about different abducibles, are to be crossed over there are more possibilities other than simply selecting cut points and switching genes between cut points.

As described above, each agent produces several chromosomes which are lists of abducible hypotheses needed to respect the ICs the agent knows. Since each agent knows only some ICs the abductive answer the algorithm seeks should be a combination of the partial answers each agent comes up with. In principle, the overlap on abducibles among two chromosomes coming from different agents should be less than total — after all, each agent is taking care of its own ICs which, in principle, do not refer to the exact same abducibles. Therefore, crossing over such chromosomes can simply turn out to be the merging of the chromosomes, i.e., the concatenation of the lists of abducibles.

If several ICs refer to the exact same abducibles the chromosomes from different agents will contain either the same gene — in which case we can see this as an 'agreement' between the agents as far as the corresponding abducible is concerned — or genes stating contradictory information about the same abducible. In this last case if the resulting concatenated chromosome turns out to be inconsistent in itself the fitness function will filter it out by assigning it a very low value.

Example 7. Running example (cont.)
    Continuing with our running example, recall that Agent 1 would come up with two alternative abductive solutions for its IC
'$\bot \leftarrow not\ professional(john)$': $\{employee(john)\}$ and $\{boss(john)\}$. Moreover, Agent 2 would come up with two other alternative abductive solutions for its IC '$\bot \leftarrow$ $not\ student(john)$': $\{not\ employee(john)\}$ and $\{junior(john)\}$.
    The crossing over of these chromosomes will yield the four combinations $\{employee(john), not\ employee(john)\}$, $\{employee(john), junior(john)\}$, $\{boss(john), not\ employee(john)\}$, and $\{boss(john), junior(john)\}$.

The first resulting chromosome is contradictory so it will be filtered out by the fitness function. The second chromosome correspond to the situation where John is a junior employee who is still studying — a quite common situation, actually. The third chromosome corresponds to the situation where John is a senior member of a company — a 'boss' — who is taking some course (probably a post-graduation study). The last chromosome could correspond to the situation of a young entrepreneur who, besides owning his/hers company, is also a student — this is probably an uncommon situation and, if necessary, the fitness function can reflect that "unprobability".

Mutation

When considering a list of abducible literals the mutation operation resembles the standard mutation of genetic algorithms by changing one gene to its opposite; in this case negating the truth value of the abducted literal.

Example 8. Running example (cont.)

In the example we have been using this could correspond to mutating the chromosome $\{not\ employee(john)\}$ to $\{employee(john)\}$, or to mutating the chromosome $\{junior(john)\}$ to $\{not\ junior(john)\}$.

The Lamarckian Learn operator

The Lamarckian operator Learn can change the values of variables of an abducible in a chromosome c_i so that a bigger number of constraints is satisfied, thus bringing c_i closer to a solution. Learn differs from a normal belief revision operator because it does not assume that all abducibles are false by CWA before the revision but it starts from the truth values that are given by the chromosome c_i. Therefore, it has to revise the values of variables of some abducibles and, in the particular case of an abducible without variables, from *true* to *false* or from *false* to *true* .

In the running example this could correspond, for example, to changing the chromosome $\{junior(john)\}$ to $\{junior(mary)\}$, where 'mary' is another value in the domain range of the variable for abducible $junior/1$.

This Lamarckian *Learn* operator will introduce an extra degree of flexibility allowing for changes to a chromosome to induce the whole belief evolution algorithm to search a solution considering new values for variables.

The Fitness Functions

Various fitness functions can be used in belief revision. The simplest fitness function is the following

$$Fitness(c_i) = \frac{\frac{n_i}{n}}{1 + NC} \tag{5.5}$$

where n_i is the number of integrity constraints satisfied by chromosome c_i, n is the total number of integrity constraints, and NC is the number of contradictions in chromosome c_i. We will call it an accuracy fitness function.

5.4   Argumentation

In [28], the author shows that preferred maximal scenarios (with maximum default negated literals — the hypotheses) are always guaranteed to exist for NLPs; and that when these yield 2-valued complete (total), consistent, admissible scenarios, they coincide with the Stable Models of the program. However, preferred maximal scenarios are, in general, 3-valued. The problem we address now is how to define 2-valued complete models based on preferred maximal scenarios. In [64] the authors took a step further from what was achieved in [28], extending its results. They did so by completing a preferred set of hypotheses rendering it approvable, ensuring whole model consistency and 2-valued completeness.

The resulting semantics thus defined, dubbed Approved Models [64], is a conservative extension to the widely known Stable Models semantics [32] in the sense that every Stable Model is also an Approved Model. The Approved Models are guaranteed to exist for every Normal Logic Program, whereas Stable Models are not. Concrete examples in [64] show how NLPs with no Stable Models can usefully model knowledge, as well as produce additional models. Moreover, this guarantee is crucial in program composition (say, from knowledge originating in divers sources) so that the result has a semantics. It is important too to warrant the existence of semantics after external updating, or in Stable Models based self-updating [1].

For the formal presentation and details of the Approved Models semantics see [64].

5.4.1   Intuition

Most of the ideas and notions of argumentation we are using here come from the Argumentation field — mainly from the foundational work of Phan Minh Dung in [28]. In [64] the *Reductio ad Absurdum* reasoning principle is also considered. This has been studied before in [62], [63], and [65].

Definition 2. *Argument.* *In [28] the author presents an argument as*

> *"an abstract entity whose role is solely determined by its relations to other arguments. No special attention is paid to the internal structure of the arguments."*

In this paper we will pay attention to the internal structure of an argument by considering an *argument* (or set of hypotheses) as a set S of abducible literals of a NLP P.

We have seen before examples of Extended Logic Programs — with explicit negation. In [18] the authors show that a simple syntactical program transformation applied to an ELP produces a Normal Logic Program with Integrity Constraints which has the exact same semantics as the original ELP.

Example 9. Transforming an ELP into a NLP with ICs
Taking the program

$$dangerous\_neighborhood \; \leftarrow \; not \; \neg dangerous\_neighborhood$$
$$\neg dangerous\_neighborhood \; \leftarrow \; not \; dangerous\_neighborhood$$

we just transform the explicitly negated literal $\neg dangerous\_neighborhood$ into the positive literal $dangerous\_neighborhood^n$, and the original $dangerous\_neighborhood$ literal is converted into $dangerous\_neighborhood^p$

Now, in order to ensure consistency, we just need to add the IC
$\perp \leftarrow dangerous\_neighborhood^p, dangerous\_neighborhood^n$. The resulting transformed program is

$$dangerous\_neighborhood^p \leftarrow not\ dangerous\_neighborhood^n$$
$$dangerous\_neighborhood^n \leftarrow not\ dangerous\_neighborhood^p$$
$$\perp \leftarrow dangerous\_neighborhood^p, dangerous\_neighborhood^n$$

Now know that we can just consider NLPs with ICs without loss of generality and so, henceforth, we will assume just that case. NLPs are in fact the kind of programs most Inductive Logic Programming learning systems produce.

5.4.2   Assumptions and Argumentation

Previously, we have seen that assumptions can be coded as abducible literals in Logic Programs and that those abducibles can be packed together in chromosomes. The evolutionary operators of genetic and memetic crossover, mutation and fitness function applied to the chromosomes provide a means to search for a consensus of the initial assumptions since it will be a consistent mixture of these.

Moreover, the 2-valued contradiction removal method presented in subsection 5.2.2 is a very superficial one. That method removes the contradiction between $p(\mathbf{X})$ and $\neg p(\mathbf{X})$ by forcing a 2-valued semantics for the ELP to choose either $p(\mathbf{X})$ or $\neg p(\mathbf{X})$ since they now are exceptions to one another. It is a superficial removal of the contradiction because the method does not look into the reasons why both $p(\mathbf{X})$ and $\neg p(\mathbf{X})$ hold simultaneously. The method does not go back to find the underlying assumptions supporting both $p(\mathbf{X})$ and $\neg p(\mathbf{X})$ to find out which assumptions should be revised in order to restore overall consistency. Any one such method must fall back into the principles of argumentation: to find the arguments supporting one conclusion in order to prevent it if it leads to contradiction.

One such 'deeper' method for contradiction removal is presented in [46]. In this chapter we have presented another alternative method inspired by evolution.

5.4.3   Collaborative Opposition

In [28] the author shows that the Stable Models of a NLP coincide with the 2-valued complete Preferred Extensions which are self-corroborating arguments. However, as it is well known, not all NLPs have Stable Models. In fact, [31] showed that the NLPs with Odd Loops Over Negation (OLONs) [6] are the one who might have no Stable Models. It is always possible to argue that when an ILP system is building the NLP it can detect if there are such OLONs and do something about them. However, in a distributed knowledge environment, e.g. a Semantic Web, several NLPs can be produced by several

---

[6] An OLON is just a loop or cycle in the program's dependency graph for some literal where the number of default negations along the loop is odd.

ILP systems and the NLPs may refer to the same literals. There is a reasonable risk that when merging the NLPs together OLONs might appear which were not present in each NLP separately. The works in [62], [63], [64], and [65] show how to solve OLONs.

The more challenging environment of a Semantic Web is one possible 'place' where the future intelligent systems will live in. Learning in 2-values or in 3-values are open possibilities, but what is most important is that knowledge and reasoning will be shared and distributed. Different opposing concepts and arguments will come from different agents. It is necessary to know how to conciliate those opposing arguments, and how to find 2-valued consensus as much as possible instead of just keeping to the least-commitment 3-valued consensus. In [64] the authors describe another method for finding such 2-valued consensus in an incremental way. In a nutshell, we start by merging together all the opposing arguments into a single one. The conclusions from the theory plus the unique merged argument are drawn and, if there are contradictions against the argument or contradictions inside the argument we non-deterministically choose one contradicted assumption of the argument and revise its truth value. The iterative repetition of this step eventually ends up in a non-contradictory argument (and all possibilities are explored because there is a non-deterministic choice).

In a way, the evolutionary method we presented in subsection 5.3.4 implements a similar mechanism to find the consensus non-contradictory arguments.

5.5 Conclusions

The two-valued setting that has been adopted in most work on ILP and Inductive Concept Learning in general is not sufficient in many cases where we need to represent real world data. This is for example the case of an agent that has to learn the effect of the actions it can perform on the domain by performing experiments. Such an agent needs to learn a definition for allowed actions, forbidden actions and actions with an unknown outcome, and therefore it needs to learn in a richer three-valued setting.

The programs that are learnt will contain a definition for the concept and its opposite, where the opposite concept is expressed by means of explicit negation. Standard ILP techniques can be adopted to separately learn the definitions for the concept and its opposite. Depending on the adopted technique, one can learn the most general or the least general definition.

The two definitions learned may overlap and the inconsistency is resolved in a different way for atoms in the training set and for unseen atoms: atoms in the training set are considered exceptions, while unseen atoms are considered unknown. The different behavior is obtained by employing negation by default in the definitions: default abnormality literals are used in order to consider exceptions to rules, while non-deterministic rules are used in order to obtain an unknown value for unseen atoms.

We have also presented an evolution-inspired algorithm for performing belief revision in a multi-agent environment. The standard genetic algorithm is extended in two ways: first the algorithm combines two different evolution strategies, one based on Darwin's and the other on Lamarck's evolutionary theory and, second, chromosomes from different agents can be crossed over with each other. The Lamarckian evolution strategy is obtained be means of an operator that changes the genes (or, better, the memes) of an agent in order to improve their fitness.

Lamarckian and Darwinian operators have complimentary functions: Lamarckian operators are used to get closer to a solution of a given belief revision problem, while Darwinian operators are used in order to distribute the acquired knowledge amongst agents. The contradictions that may arise between chromosomes of memes from different agents are of a distinct nature from the contradictions arising from the learning process. The former correspond to different alternative explanations to the observations, whereas the latter correspond to uncertainties in the learned concepts.

Moreover, we can also bring the same evolutionary algorithm to a single agent's mind. In such case, we can think of the agent's mind as a pool of several sub-agents, each considering one aspect of the environment, each acting as a specialist in some sub-domain.

We have presented too a new and productive way to deal with oppositional concepts in a collaboration perspective, in different degrees. We use the contradictions arising from opposing arguments as hints for the possible collaborations. In so doing, we extend the classical conflictual argumentation giving a new treatment and new semantics to deal with the contradictions.

The direction of our future work includes three main axis: continuing theory development, implementing prototypes and exploring possible applications.

Most recently, we have been exploring the constructive negation [51] reasoning mechanism. In a nutshell, constructive negation concerns getting answers to queries by imposing inequality constraints on the values of variables (e.g., getting an answer of the form "all birds fly, except for the penguins"). Such mechanism is particularly well suited for the integration of answers coming from different agents, e.g., one agent can learn the general rule that "all birds fly", and another might learn only the exceptional case that "penguins do not fly", and that "penguins are birds". Constructive negation can play a synergistic rôle in this matter by gracefully merging the knowledge of the different agents into a single consistent, integrated, more specialized and refined one.

Our future efforts will therefore engage in bringing together the three branches we described — learning, belief evolution, and argumentation — under the scope of constructive negation. Besides the theoretical research we have been doing, and will continue to do in this area, we have already under way a practical implementation of this constructive negation reasoning mechanism [58] on top of XSB Prolog [71].

Also, one application field for all this work is Ambient Intelligence [14]. In a nutshell, Ambient Intelligence concerns intelligent software agents embedded in real-world environments and that are sensitive and responsive to the presence of people. Such environments are constantly changing as different people come in and out of play, each of which may influence the rest of the environment.

We envisage a framework for Ambient Intelligence where agents interact with users

(i) with the aim of monitoring them for ensuring some degree of consistence and coherence in user behavior and, possibly,

(ii) with the objective of training them in some particular task.

In our view, a system which is a realization of the envisaged framework will bring to a user the following potential advantages: the user is relieved of some of the responsibilities related to her behavior, as directions about the "right thing" to do are constantly and punctually provided. She is assisted in situations where she perceived herself as inadequate, in some respect, to perform her activities or tasks. She is possibly told how to cope with unknown, unwanted or challenging circumstances. She interacts with a "Personal Assistant" that improves in time, both in its "comprehension" of the user needs, cultural level, preferred kinds of explanations, etc. and in its ability to cope with the environment.

References

1. Alferes, J.J., Brogi, A., Leite, J.A., Pereira, L.M.: Evolving logic programs. In: Flesca, S., Greco, S., Leone, N., Ianni, G. (eds.) JELIA 2002. LNCS (LNAI), vol. 2424, pp. 50–61. Springer, Heidelberg (2002)
2. Alferes, J.J., Damásio, C.V., Pereira, L.M.: SLX - A top-down derivation procedure for programs with explicit negation. In: Bruynooghe, M. (ed.) Proc. Int. Symp. on Logic Programming, MIT Press, Cambridge (1994)
3. Alferes, J.J., Pereira, L.M.:
   http://xsb.sourceforge.net/manual2/node179.html
4. Alferes, J.J., Pereira, L.M.: An argumentation theoretic semantics based on non-refutable falsity. In: Dix, J., et al. (eds.) NMELP, pp. 3–22. Springer, Heidelberg (1994)
5. Alferes, J.J., Moniz Pereira, L.: Reasoning with Logic Programming. LNCS, vol. 1111. Springer, Heidelberg (1996)
6. Alferes, J.J., Pereira, L.M., Przymusinski, T.C.: Classical negation in non-monotonic reasoning and logic programming. Journal of Automated Reasoning 20, 107–142 (1998)
7. Bain, M., Muggleton, S.: Non-monotonic learning. In: Muggleton, S. (ed.) Inductive Logic Programming, pp. 145–161. Academic Press, London (1992)
8. http://www.psych.utoronto.ca/museum/baldwin.htm
9. Baral, C., Gelfond, M., Rushton, J., Nelson,: Probabilistic reasoning with answer sets. In: Lifschitz, V., Niemelä, I. (eds.) LPNMR 2004. LNCS (LNAI), vol. 2923, pp. 21–33. Springer, Heidelberg (2003)
10. Baral, C., Gelfond, M.: Logic programming and knowledge representation. Journal of Logic Programming 19(20), 73–148 (1994)
11. Baral, C., Subrahmanian, V.S.: Dualities between alternative semantics for logic programming and nonmonotonic reasoning. J. Autom. Reasoning 10(3), 399–420 (1993)
12. Bondarenko, A., Dung, P.M., Kowalski, R.A., Toni, F.: An abstract, argumentation-theoretic approach to default reasoning. Artif. Intell. 93, 63–101 (1997)
13. Chan, P., Stolfo, S.: Meta-learning for multistrategy and parallel learning. In: Proceedings of the 2nd International Workshop on Multistrategy Learning, pp. 150–165 (1993)
14. Costantini, S., Dell'Acqua, P., Pereira, L.M., Toni, F.: Towards a Model of Evolving Agents for Ambient Intelligence. In: Sadri, F., Stathis, K. (eds.) Procs. Symposium on Artificial Societies for Ambient Intelligence (ASAmI 2007) (2007); Extended version submited to a Journal
15. Damásio, C.V., Nejdl, W., Pereira, L.M.: REVISE: An extended logic programming system for revising knowledge bases. In: Doyle, J., Sandewall, E., Torasso, P. (eds.) Knowledge Representation and Reasoning, pp. 607–618. Morgan Kaufmann, San Francisco (1994)
16. Damásio, C.V., Pereira, L.M.: Abduction on 3-valued extended logic programs. In: Marek, V.W., Truszczyński, M., Nerode, A. (eds.) LPNMR 1995. LNCS, vol. 928, pp. 29–42. Springer, Heidelberg (1995)

17. Damásio, C.V., Pereira, L.M.: A survey on paraconsistent semantics for extended logic programs. In: Gabbay, D., Smets, P. (eds.) Handbook of Defeasible Reasoning and Uncertainty Management Systems, vol. 2, pp. 241–320. Kluwer Academic Publishers, Dordrecht (1998)
18. Damásio, C.V., Pereira, L.M.: Default Negated Conclusions: Why Not? In: Herre, H., Dyckhoff, R., Schroeder-Heister, P. (eds.) ELP 1996. LNCS, vol. 1050, pp. 103–117. Springer, Heidelberg (1996)
19. De Raedt, L.: Interactive Theory Revision: An Inductive Logic Programming Approach. Academic Press, London (1992)
20. De Raedt, L., Bleken, E., Coget, V., Ghil, C., Swennen, B., Bruynooghe, M.: Learning to survive. In: Proceedings of the 2nd International Workshop on Multistrategy Learning, pp. 92–106 (1993)
21. De Raedt, L., Bruynooghe, M.: Towards friendly concept-learners. In: Proceedings of the 11th International Joint Conference on Artificial Intelligence, pp. 849–856. Morgan Kaufmann, San Francisco (1989)
22. De Raedt, L., Bruynooghe, M.: On negation and three-valued logic in interactive concept learning. In: Proceedings of the 9th European Conference on Artificial Intelligence (1990)
23. De Raedt, L., Bruynooghe, M.: Interactive concept learning and constructive induction by analogy. Machine Learning 8(2), 107–150 (1992)
24. Dix, J.: A Classification-Theory of Semantics of Normal Logic Programs: I, II. Fundamenta Informaticae XXII(3), 227–255, 257–288 (1995)
25. Dix, J., Pereira, L.M., Przymusinski, T.: Prolegomena to logic programming and nonmonotonic reasoning. In: Dix, J., Przymusinski, T.C., Moniz Pereira, L. (eds.) NMELP 1996. LNCS, vol. 1216, pp. 1–36. Springer, Heidelberg (1997)
26. Drobnic, M., Gams, M.: Multistrategy learning: An analytical approach. In: Proceedings of the 2nd International Workshop on Multistrategy Learning, pp. 31–41 (1993)
27. Džeroski, S.: Handling noise in inductive logic programming. Master's thesis, Faculty of Electrical Engineering and Computer Science, University of Ljubljana (1991)
28. Dung, P.M.: On the acceptability of arguments and its fundamental role in nonmonotonic reasoning, logic programming and n-person games. Artif. Intell. 77(2), 321–358 (1995)
29. Dung, P.M., Kowalski, R.A., Toni, F.: Dialectic proof procedures for assumption-based, admissible argumentation. Artif. Intell. 170(2), 114–159 (2006)
30. Esposito, F., Ferilli, S., Lamma, E., Mello, P., Milano, M., Riguzzi, F., Semeraro, G.: Cooperation of abduction and induction in logic programming. In: Flach, P.A., Kakas, A.C. (eds.) Abductive and Inductive Reasoning, Pure and Applied Logic. Kluwer, Dordrecht (1998)
31. Fages, F.: Consistency of Clark's completion and existence of stable models. Methods of Logic in Computer Science 1, 51–60 (1994)
32. Gelfond, M., Lifschitz, V.: The stable model semantics for logic programming. In: Kowalski, R., Bowen, K.A. (eds.) Proceedings of the 5th Int. Conf. on Logic Programming, pp. 1070–1080. MIT Press, Cambridge (1988)
33. Gelfond, M., Lifschitz, V.: Logic programs with classical negation. In: Proceedings of the 7th International Conference on Logic Programming ICLP 1990, pp. 579–597. MIT Press, Cambridge (1990)
34. Gordon, D., Perlis, D.: Explicitly biased generalization. Computational Intelligence 5(2), 67–81 (1989)
35. Green, D.M., Swets, J.M.: Signal detection theory and psychophysics. John Wiley and Sons Inc., New York (1966)
36. Greiner, R., Grove, A.J., Roth, D.: Learning active classifiers. In: Proceedings of the Thirteenth International Conference on Machine Learning (ICML 1996) (1996)
37. Inoue, K.: Learning abductive and nonmonotonic logic programs. In: Flach, P.A., Kakas, A.C. (eds.) Abductive and Inductive Reasoning, Pure and Applied Logic. Kluwer, Dordrecht (1998)

38. Inoue, K., Kudoh, Y.: Learning extended logic programs. In: Proceedings of the 15th International Joint Conference on Artificial Intelligence, pp. 176–181. Morgan Kaufmann, San Francisco (1997)
39. Jenkins, W.: Intelog: A framework for multistrategy learning. In: Proceedings of the 2nd International Workshop on Multistrategy Learning, pp. 58–65 (1993)
40. Kakas, A.C., Mancarella, P.: Negation as stable hypotheses. In: LPNMR, pp. 275–288. MIT Press, Cambridge (1991)
41. Koza, J.R.: Genetic Programming: On the Programming of Computers by Means of Natural Selection. MIT Press, Cambridge (1992)
42. Jean Baptiste Lamarck:
    http://www.ucmp.berkeley.edu/history/lamarck.html
43. Grefenstette, J.J.: Lamarckian learning in multi-agent environments (1991)
44. Lamma, E., Riguzzi, F., Pereira, L.M.: Learning in a three-valued setting. In: Proceedings of the Fourth International Workshop on Multistrategy Learning (1988)
45. Lamma, E., Riguzzi, F., Pereira, L.M.: Agents learning in a three-valued setting. Technical report, DEIS - University of Bologna (1999a)
46. Lamma, E., Pereira, L.M., Riguzzi, F.: Belief revision via lamarckian evolution. New Generation Computing 21(3), 247–275 (2003)
47. Lamma, E., Riguzzi, F., Pereira, L.M.: Strategies in combined learning via logic programs. Machine Learning 38(1-2), 63–87 (2000)
48. Lapointe, S., Matwin, S.: Sub-unification: A tool for efficient induction of recursive programs. In: Sleeman, D., Edwards, P. (eds.) Proceedings of the 9th International Workshop on Machine Learning, pp. 273–281. Morgan Kaufmann, San Francisco (1992)
49. Lavrač, N., Džeroski, S.: Inductive Logic Programming: Techniques and Applications. Ellis Horwood (1994)
50. Leite, J.A., Pereira, L.M.: Generalizing updates: from models to programs. In: Dix, J., Moniz Pereira, L., Przymusinski, T.C. (eds.) LPKR 1997. LNCS (LNAI), vol. 1471. Springer, Heidelberg (1998)
51. Liu, J.Y., Adams, L., Chen, W.: Constructive Negation Under the Well-Founded Semantics. J. Log. Program. 38(3), 295–330 (1999)
52. Malý, M.: Complexity of revised stable models. Master's thesis, Comenius University Bratislava (2006)
53. Michalski, R.: Discovery classification rules using variable-valued logic system VL1. In: Proceedings of the Third International Conference on Artificial Intelligence, pp. 162–172. Stanford University (1973)
54. Michalski, R.: A theory and methodology of inductive learning. In: Michalski, R., Carbonell, J., Mitchell, T. (eds.) Machine Learning - An Artificial Intelligence Approach, vol. 1, pp. 83–134. Springer, Heidelberg (1984)
55. Muggleton, S.: Inverse entailment and Progol. New Generation Computing, Special issue on Inductive Logic Programming 13(3-4), 245–286 (1995)
56. Muggleton, S., Buntine, W.: Machine invention of first-order predicates by inverting resolution. In: Muggleton, S. (ed.) Inductive Logic Programming, pp. 261–280. Academic Press, London (1992)
57. Muggleton, S., Feng, C.: Efficient induction of logic programs. In: Proceedings of the 1st Conference on Algorithmic Learning Theory, Ohmsma, Tokyo, Japan, pp. 368–381 (1990)
58. http://centria.di.fct.unl.pt/~lmp/software/contrNeg.rar
59. Pazzani, M.J., Merz, C., Murphy, P., Ali, K., Hume, T., Brunk, C.: Reducing misclassification costs. In: Proceedings of the Eleventh International Conference on Machine Learning (ML 1994), pp. 217–225 (1994)
60. http://www.peirce.org/

61. Pereira, L.M., Alferes, J.J.: Well founded semantics for logic programs with explicit nega-tion. In: Proceedings of the European Conference on Artificial Intelligence ECAI 1992, pp. 102–106. John Wiley and Sons, Chichester (1992)
62. Pereira, L.M., Pinto, A.M.: Revised stable models - a semantics for logic programs. In: Bento, C., Cardoso, A., Dias, G. (eds.) EPIA 2005. LNCS (LNAI), vol. 3808, pp. 29–42. Springer, Heidelberg (2005)
63. Pereira, L.M., Pinto, A.M.: Reductio ad absurdum argumentation in normal logic programs. In: Argumentation and Non-monotonic Reasoning (ArgNMR 2007) workshop at LPNMR 2007, pp. 96–113 (2007)
64. Pereira, L.M., Pinto, A.M.: Approved Models for Normal Logic Programs. In: LPAR, pp. 454–468. Springer, Heidelberg (2007)
65. Pinto, A.M.: Explorations in revised stable models — a new semantics for logic programs. Master's thesis, Universidade Nova de Lisboa (February 2005)
66. Plotkin, G.: A note on inductive generalization. In: Machine Intelligence, vol. 5, pp. 153–163. Edinburgh University Press (1970)
67. Provost, F.J., Fawcett, T.: Analysis and visualization of classifier performance: Comparison under imprecise class and cost distribution. In: Proceedings of the Third International Con-ference on Knowledge Discovery and Data Mining (KDD 1997), AAAI Press, Menlo Park (1997)
68. Quinlan, J.: Learning logical definitions from relations. Machine Learning 5, 239–266 (1990)
69. Quinlan, J.R.: C4.5: Programs for Machine Learning. Morgan Kaufmann, San Mateo (1993)
70. Reiter, R.: On closed-word data bases. In: Gallaire, H., Minker, J. (eds.) Logic and Data Bases, pp. 55–76. Plenum Press (1978)
71. Sagonas, K.F., Swift, T., Warren, D.S., Freire, J., Rao, P.: The XSB Programmer's Manual Version 1.7.1 (1997)
72. Soares, L.: Revising undefinedness in the well-founded semantics of logic programs. Mas-ter's thesis, Universidade Nova de Lisboa (2006)
73. Van Gelder, A., Ross, K.A., Schlipf, J.S.: The well-founded semantics for general logic pro-grams. Journal of the ACM 38(3), 620–650 (1991)
74. Vere, S.A.: Induction of concepts in the predicate calculus. In: Proceedings of the Fourth International Joint Conference on Artificial Intelligence (IJCAI 1975), pp. 281–287 (1975)
75. Whitley, D., Rana, S., Heckendorn, R.B.: The Island Model Genetic Algorithm: On Separa-bility, Population Size and Convergence (1998)

A   Appendix

The definition of WFSX that follows is taken from [2] and is based on the alternating fix points of Gelfond-Lifschitz Γ-like operators.

Definition 3. *The Γ-operator. Let P be an extended logic program and let I be an in-terpretation of P. $\Gamma_P(I)$ is the program obtained from P by performing in the sequence the following four operations:*

- *Remove from P all rules containing a default literal $L = not\ A$ such that $A \in I$.*
- *Remove from P all rules containing in the body an objective literal L such that $\neg L \in I$.*
- *Remove from all remaining rules of P their default literals $L = not\ A$ such that $not\ A \in I$.*
- *Replace all the remaining default literals by proposition u.*

In order to impose the coherence requirement, we need the following definition.

Definition 4. *Seminormal Version of a Program.*
The seminormal version of a program P is the program P_s obtained from P by adding to the (possibly empty) Body of each rule $L \leftarrow$ Body the default literal not $\neg L$, where $\neg L$ is the complement of L with respect to explicit negation.
In the following, we will use the following abbreviations: $\Gamma(S)$ for $\Gamma_P(S)$ and $\Gamma_s(S)$ for $\Gamma_{P_s}(S)$.

Definition 5. *Partial Stable Model.*
An interpretation $T \cup not\ F$ is called a partial stable model *of P iff $T = \Gamma\Gamma_s T$ and $F = H^E(P) - \Gamma_s T$.*
Partial stable models are an extension of stable models *[32] for extended logic programs and a three-valued semantics. Not all programs have a partial stable model (e.g., $P = \{a, \neg a\}$) and programs without a partial stable model are called* contradictory.

Theorem 1. *WFSX Semantics.*
Every non-contradictory program P has a least (with respect to $\subseteq$) partial stable model, the well-founded model of P denoted by $WFM(P)$.

Proof. To obtain an iterative "bottom-up" definition for $WFM(P)$ we define the following transfinite sequence $\{I_\alpha\}$:

$$I_0 = \{\}; \quad I_{\alpha+1} = \Gamma\Gamma_S I_\alpha ; \quad I_\delta = \bigcup\{I_\alpha | \alpha < \delta\}$$

where δ is a limit ordinal. There exists a smallest ordinal λ for the sequence above, such that I_λ is the smallest fix point of $\Gamma\Gamma_S$. Then, $WFM(P) = I_\lambda \cup not\ (H^E(P) - \Gamma_S I_\lambda)$.

$\square$

6

Proof-Number Search and Its Variants

H. Jaap van den Herik[1] and Mark H.M. Winands[2]

[1] Tilburg centre for Creative Computing (TiCC), Faculty of Humanities,
Tilburg University, The Netherlands
[2] Maastricht ICT Competence Centre (MICC), Faculty of Humanities and Sciences,
Maastricht University, The Netherlands
{herik,m.winands}@micc.unimaas.nl

Summary. Proof-Number (PN) search is a best-first adversarial search algorithm especially suited for finding the game-theoretical value in game trees. The strategy of the algorithm may be described as developing the tree into the direction where the opposition characterised by value and branching factor is to expect to be the weakest. In this chapter we start by providing a short description of the original PN-search method, and two main successors of the original PN search, i.e., PN^2 search and the depth-first variant *Proof-number and Disproof-number Search* (PDS). A comparison of the performance between PN, PN^2, PDS, and $\alpha\beta$ is given. It is shown that PN-search algorithms clearly outperform $\alpha\beta$ in solving endgame positions in the game of Lines of Action (LOA). However, memory problems make the plain PN search a weaker solver for harder problems. PDS and PN^2 are able to solve significantly more problems than PN and $\alpha\beta$. But PN^2 is restricted by its working memory, and PDS is considerably slower than PN^2. Next, we present a new proof-number search algorithm, called PDS-PN. It is a two-level search (like PN^2), which performs at the first level a depth-first PDS, and at the second level a best-first PN search. Hence, PDS-PN selectively exploits the power of both PN^2 and PDS. Experiments show that within an acceptable time frame PDS-PN is more effective for really hard endgame positions. Finally, we discuss the depth-first variant df-pn. As a follow up of the comparison of the four PN variants, we compare the algorithms PDS and df-pn. However, the hardware conditions of the comparison were different. Yet, experimental results provide promising prospects for df-pn. We conclude the article by seven observations, three conclusions, and four suggestions for future research.

6.1 Endgame Solvers

Most modern game-playing computer programs use the adversarial search method called the $\alpha\beta$ algorithm [16] for online game-playing [11]. However, the $\alpha\beta$ search even with its enhancements is sometimes not sufficient to play well in the endgame. A variety of factors may cause this lack of effectiveness, for instance the complexity (in Go) and the depth of search (in many endgames). In some games, such as Chess, the latter problem is solved by the use of endgame databases [22]. Due to memory constraints this solution is only feasible for endgames with a relatively small state-space complexity, although nowadays the size may be considerable.

In the last three decades many other search approaches have been proposed, tested and thoroughly investigated (for an overview see [12]). Two lines of research focused

H.R. Tizhoosh, M. Ventresca (Eds.): Oppos. Concepts in Comp. Intel., SCI 155, pp. 91–118, 2008.
springerlink.com   © Springer-Verlag Berlin Heidelberg 2008

on the possibilities of the opponent and the potential threats of the opponent. The development started with the idea of conspiracy-number search as developed by McAllester [17] and worked out by Schaeffer [30]. This idea was heuristical by nature. It inspired Allis [1] to propose PN Search, a specialised binary (win or non-win) search method for solving games and for solving difficult endgame positions [2].

PN search is a best-first adversarial search algorithm especially suited for finding the game-theoretical value in game trees. In many domains PN search outperforms $\alpha\beta$ search in proving the game-theoretic value of endgame positions. The PN-search idea is a heuristic, which prefers expanding slim subtrees over wide ones. PN search or a variant thereof has been successfully applied to the endgame of Awari [2], Chess [6], Checkers [31, 32, 33], Shogi [34], and Go [15]. Since PN search is a best-first search, it has to store the whole search tree in memory. When the memory is full, the search has to end prematurely.

To overcome this problem PN^2 was proposed by Allis [1] as an algorithm to reduce memory requirements in PN search. It is elaborated upon in Breuker [5]. Its implementation and testing for chess positions is extensively described in Breuker, Uiterwijk, and Van den Herik [8]. PN^2 performs two levels of PN search, one at the root and one at the leaves of the first level. As in the B* algorithm [4], a search process is started at the leaves to obtain a more accurate evaluation. Although PN^2 uses far less memory than PN search, it does not fully overcome the memory obstacle.

Therefore, the idea behind the MTD(f) algorithm [25] was applied to PN variants: try to construct a depth-first algorithm that behaves as its corresponding best-first search algorithm. This idea became a success. In 1995, Seo formulated a depth-first iterative-deepening version of PN search, later called PN* [34]. The advantage of this variant is that there is no need to store the whole tree in memory. The disadvantage is that PN* is slower than PN [29].

Other depth-first variants are PDS [18] and df-pn [21]. Although their generation of nodes is even slower than PN*'s, they are building smaller search trees. Hence, they are in general more efficient than PN*.

In this chapter we will investigate several PN-search algorithms, using the game of *Lines of Action (LOA)* [26] as test domain. We will concentrate on the *offline* application of the PN-search algorithms. The number of positions they can solve (i.e., the post-mortem analysis quality) is tested on a set of endgame positions. Moreover, we will investigate to what extent the algorithms are restricted by their working memory *or* by the search speed.

The chapter is organised as follows. In Section 6.1 we discuss the need for special algorithms to solve endgame positions. Section 6.2 describes PN, PN^2, PN*, PDS, and df-pn. In Section 6.3 two enhancements of PN and PN^2 are described. In Section 6.4 we examine the offline solution power and the solution time of three PN variants, in relation to those of $\alpha\beta$. In Section 6.5 we explain the working of PDS-PN by elaborating on PDS and the idea of two-level search algorithms. Then, in Section 6.6, the results of experiments with PDS-PN on a set of endgame positions are given. Subsequently, we briefly discuss df-pn and compare the results by PDS and df-pn in Section 6.7. Finally, in Section 6.8 we present seven observations, three conclusions and four suggestions for future research.

6.2   Five Proof-Number Search Algorithms

In this section we give a short description of PN search (Subsection 6.2.1), PN^2 search (Subsection 6.2.2) and three depth-first variants of PN search. Recently, three depth-first PN variants, PN*, PDS, and df-pn have been proposed, which solved the memory problem of PN-search algorithms. They will be discussed in Subsection 6.2.3, 6.2.4, and 6.2.5.

6.2.1   Proof-Number Search

Proof-Number (PN) search is a best-first search algorithm especially suited for finding the game-theoretical value in game trees [1]. Its aim is to prove the true value of the root of a tree. A tree can have three values: *true*, *false*, or *unknown*. In the case of a forced win, the tree is *proved* and its value is true. In the case of a forced loss or draw, the tree is *disproved* and its value is false. Otherwise the value of the tree is unknown. In contrast to other best-first algorithms PN search does not need a domain-dependent heuristic evaluation function to determine the most-promising node to be expanded next [2]. In PN search this node is usually called the *most-proving* node. PN search selects the most-proving node using two criteria: (1) the shape of the search tree (the branching factor of every internal node) and (2) the values of the leaves. These two criteria enable PN search to treat game trees with a non-uniform branching factor efficiently. The strategy of the algorithm may be described as developing the tree into the direction where the opposition characterised by value and branching factor is to expect to be the weakest.

Below we explain PN search on the basis of the AND/OR tree depicted in Figure 6.1, in which a square denotes an OR node, and a circle denotes an AND node. The numbers

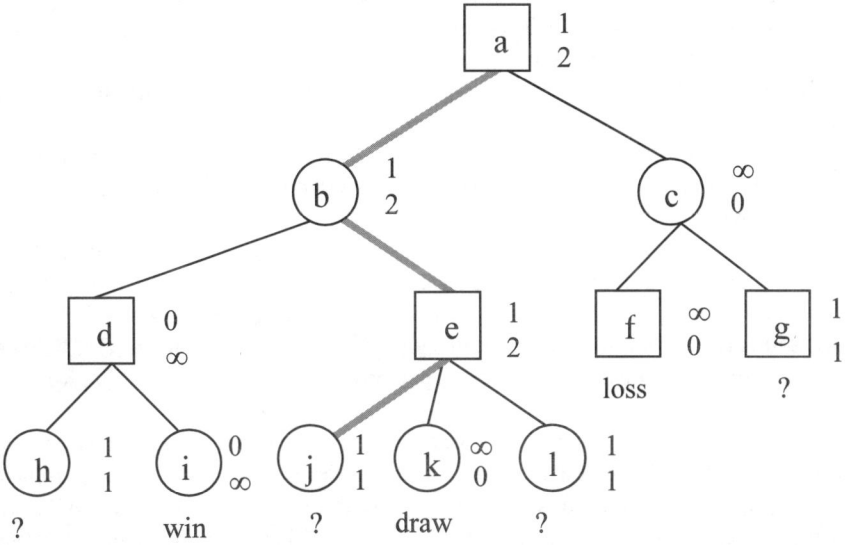

Fig. 6.1. An AND/OR tree with proof and disproof numbers

to the right of a node denote the proof number (upper) and disproof number (lower). A *proof number* (*pn*) represents the minimum number of leaf nodes which have to be proved in order to prove the node. Analogously, a *disproof number* (*dpn*) represents the minimum number of leaf nodes which have to be disproved in order to disprove the node. Because the goal of the tree is to prove a forced win, winning nodes are regarded as proved. So, they have proof number 0 and disproof number ∞ (e.g., node *i*). Lost or drawn nodes are regarded as disproved (e.g., nodes *f* and *k*). They have proof number ∞ and disproof number 0. Unknown leaf nodes have a proof and disproof number of unity (e.g., nodes *g*, *h*, *j*, and *l*). The proof number of an internal AND node is equal to the sum of its children's proof numbers, since to prove an AND node all the children have to be proved. The disproof number of an AND node is equal to the minimum of its children's disproof numbers, since to disprove an AND node it suffices to disprove one child. The proof number of an internal OR node is equal to the minimum of its children's proof numbers, since to prove an OR node it suffices to prove one child. The disproof number of an internal OR node is equal to the sum of its children's disproof numbers, since to disprove an OR node all the children have to be disproved.

The procedure of selecting the most-proving node to expand is as follows. We start at the root. Then, at each OR node the child with the lowest proof number is selected as successor, and at each AND node the child with the lowest disproof number is selected as successor. Finally, when a leaf node is reached, it is expanded (which makes the leaf node an internal node) and the newborn children are evaluated. This is called *immediate evaluation*. The selection of the most-proving node (*j*) in Figure 6.1 is given by the bold path.

The number of node traversals to select the most-proving node can have a negative impact on the execution time. Therefore, Allis [1] proposed the following minor enhancement. The updating process can be terminated when the proof and disproof number of a node do not change. From this node we can start the next most-proving node selection. For an adequate description of implementation details we refer to Allis *et al.* [2], where the essentials for implementation are given.

In the naive implementation, proof and disproof numbers are each initialised to unity in the unknown leaves. In other implementations, the proof number and disproof number are set to 1 and *n*, respectively, for an OR node (and the reverse for an AND node), where *n* is the number of legal moves. In LOA this would mean that we take the *mobility* of the moving player in the position into account, which is an important feature in the evaluation function as well [37]. The effect of this enhancement is tested in Section 6.3. We would like to remark that there are other possibilities to initialise the proof and disproof numbers. Allis [1] applies domain-specific knowledge to set the variables in Awari. Saito *et al.* [27] apply an algorithm called MC-PNS in the game of Go. It gives a value to the proof and disproof number by performing a Monte-Carlo evaluation at the leaf node of the tree.

Here we reiterate that a disadvantage of PN search is that the whole search tree has to be stored in memory. When the memory is full, the search process has to be terminated prematurely. A partial solution is to delete proved or disproved subtrees [1]. In the next subsections we discuss two main variants of PN search that handle the memory problem more adequately.

6.2.2 PN² Search

For an appropriate description we repeat a few sentences of our own. PN^2 is first described by Allis [1], as an algorithm to reduce memory requirements in PN search. It is elaborated upon in Breuker [5]. Its implementation and testing for chess positions is extensively described in Breuker *et al.* [8]. PN^2 consists of two levels of PN search. The first level consists of a PN search (PN_1), which calls a PN search at the second level (PN_2) for an evaluation of the most-proving node of the PN_1-search tree. This PN_2 search is bound by a maximum number of nodes to be stored in memory. The number is a fraction of the size of the PN_1-search tree. The fraction $f(x)$ is given by the logistic-growth function [3], x being the size of the first-level search:

$$f(x) = \frac{1}{1 + e^{\frac{a-x}{b}}} \ , \tag{6.1}$$

with parameters a and b, both strictly positive. The number of nodes y in a PN_2-search tree is restricted to the minimum of this fraction function and the number of nodes which can still be stored. The formula to compute y is:

$$y = min(x \times f(x), N - x), \tag{6.2}$$

with N the maximum number of nodes to be stored in memory.

   The PN_2 search is stopped when the number of nodes stored in memory exceeds y or the subtree is (dis)proved. After completion of the PN_2 search, the children of the root of the PN_2-search tree are preserved, but subtrees are removed from memory. The children of the most-proving node (the root of the PN_2-search tree) are not immediately evaluated by a second-level search; evaluation of such a child node happens only after its selection as most-proving node. This is called *delayed evaluation*. We remark that for PN_2-search trees immediate evaluation is used. The essentials of our implementation are given in [5].

   As we have seen in Subsection 6.2.1, proved or disproved subtrees can be deleted. If we do not delete proved or disproved subtrees in the PN_2 search the number of nodes searched is the same as y, otherwise we can continue the search longer. The effect of *deleting (dis)proved PN_2 subtrees* is tested in Section 6.3.

6.2.3 PN*

In 1995, Seo formulated the first depth-first iterative-deepening version of PN search, later called PN* [34]. PN* uses a method called *multiple-iterative deepening*. Instead of iterating only at the root node such as in the ordinary iterative deepening, it iterates also at AND nodes. To each AND node a threshold is given. The subtree rooted at that node is continued to be searched as long as the proof number is below the assigned threshold. To keep iterative deepening effective, the method is enhanced by storing the expanded nodes in a transposition table.

6.2.4   PDS

The disadvantage of PN* is that it has difficulties to disprove a (sub)tree, which harms its solving performance [29]. Nagai [18, 19] proposed a second depth-first search algorithm, called *Proof-number and Disproof-number Search* (PDS), which is a straight extension of PN*. Instead of using only proof numbers such as in PN*, PDS uses disproof numbers too.[1] Moreover, PDS uses *multiple-iterative deepening* in *every* node. To keep iterative deepening effective, the method is enhanced by storing the expanded nodes in a *TwoBig* transposition table [7]. PDS uses two thresholds in searching, one for the proof numbers and one for the disproof numbers. We note that PDS suffers from the Graph-History Interaction (GHI) problem (cf. [9]).[2] In the present implementation this problem is ignored [19]. In Section 6.5.2 we will describe PDS in detail.

6.2.5   df-pn

Nagai [20, 21] has introduced a third depth-first PN algorithm, called df-pn (depth-first proof-number search). It is mainly a variant of PDS. The algorithm df-pn does not perform iterative deepening at the root node. As with PDS, df-pn uses two thresholds for a node, one as a limit for proof numbers (pnt) and one for disproof numbers (dnt). In contrast to PDS, it sets the thresholds of both proof number and disproof number at the root node to ∞. Once the thresholds are assigned to a node, the subtree rooted in that node is stopped to be searched if the proof number (pn) or disproof number (dpn) is larger than or equal to their corresponding threshold. Obviously, the condition $pn < \infty$ and $dn < \infty$ holds if the tree is not solved. As the search goes more deeply, the threshold values are distributed among the descendant nodes.

At an OR node p we select the child n with lowest pn just like in the regular PN search. Assume that there is a node s with the second lowest pn value. The thresholds of node n are set in the following way:

$$pnt_n = min(pnt_p, pn_s + 1), \tag{6.3}$$
$$dnt_n = dnt_p - dn_p + dn_n. \tag{6.4}$$

Similarly, at an AND node p we select the child n with lowest dn just like in the regular PN search. Assume that there is a node s with the second lowest dpn value. The thresholds of node n are set in the following way:

$$pnt_n = pnt_p - pn_p + pn_n, \tag{6.5}$$
$$dnt_n = min(dnt_p, dpn_s + 1). \tag{6.6}$$

Contrary to PDS, it has been proved that df-pn always selects the most-proving node [20]. Initially, the results of df-pn were mixed [21, 28]. Although df-pn sometimes solves positions faster than PDS, it may solve in practice fewer positions [28]. It turns

---

[1] We recall that PN and PN^2 use disproof numbers too.

[2] In a search graph a node's value may be dependent on the path leading to it. Different paths may lead to different values. Hence, it is difficult to determine the value of any node unambiguously. The problem is known as the Graph-History Interaction (GHI) problem (see [10, 23]).

out that the standard df-pn algorithm suffers more from the GHI problem than PDS. It has a fundamental problem when applied to a domain with repetitions [14]. Nagai proposed an ad-hoc way to solve this problem for Tsume-Shogi [20]. Recently, Kishimoto and Müller proposed a solution, which adequately handles the GHI problem in df-pn [13, 15].

To prevent multiple recreations of a subtree due to multiple-iterative deepening, Pawlewicz and Lew [24] developed the $1 + \epsilon$ *trick* enhancement that might improve df-pn considerably. It transforms formula 3 to formula 6.7 for the child's proof-number threshold in a OR node:

$$pnt_n = min(pnt_p, \lceil pn_s(1 + \epsilon) \rceil), \tag{6.7}$$

where ϵ is a small real number greater than zero.

6.3   Two Enhancements of PN and PN^2

Below, we test the effect of enhancing PN and PN^2 with (1) adding mobility and (2) deleting (dis)proved PN$_2$ subtrees. For PN and PN^2, all nodes evaluated for the termination condition during the search are counted. The node count is equal to the number of nodes generated. The maximum number of nodes searched is 50,000,000. The maximum number of nodes stored in memory is 1,000,000. These numbers roughly corresponds to the tournament conditions of the Computer Olympiad with respect to Pentium III. The parameters (a,b) of the growth function used in PN^2 are set at (1800K, 240K) according to the suggestions in Breuker *et al.* [8].

In the first experiment, we tested PN search and PN^2 with the mobility enhancements on a test set of 116 LOA positions.[3] The results are shown in Table 6.1. In the second column we see that PN search solved 85 positions using mobility; without mobility it solved 53 positions. PN^2 search using mobility solved 109 positions and without it solved only 91 positions. Next, in the third column we see that on a set of 53 positions solved by both PN algorithms, PN search using mobility is roughly 5 times faster in nodes than PN search without mobility. Finally, in the fourth column we see that on a set of 91 positions solved by both PN^2 algorithms, PN^2 search using mobility is more than 6 times faster in nodes than PN^2 search without using mobility. In general we may conclude that mobility speeds up the PN and PN^2 with a factor 5 to 6. The time spent on the mobility extension is estimated at 15% of the total computing time. Owing to mobility PN search can make better search decisions and therefore solve many more positions. The underlying reason is that the memory constraint is violated less frequently.

In the second experiment, we tested the effect of deleting (dis)proved subtrees at the PN$_2$ search of the PN^2. The results are shown in Table 6.2. Both variants (not deleting PN$_2$ subtrees and deleting PN$_2$ subtrees) used mobility in the experiment. On a set of 108 positions that both versions were able to solve, we can see that deleting (dis)proved subtrees improves the search by 10%. It also solves one additional position.

In the remainder of this chapter we will use these two enhancements (i.e., mobility and deleting (dis)proved PN$_2$ subtrees) for PN and PN^2.

---

[3] The test set can be found at www.cs.unimaas.nl/m.winands/loa/tswin116.zip.

Table 6.1. Mobility in PN and PN^2

Algorithm	# of pos. solved	Total nodes (53 pos.)	Total nodes (91 pos.)
PN	53	24,357,832	-
PN + Mob.	85	5,053,630	-
PN^2	91	-	345,986,639
PN^2 + Mob.	109	-	56,809,635

Table 6.2. Deleting (dis)proved subtrees at the second-level search PN^2

Algorithm	# of pos. solved	Total nodes (108 pos.)
PN^2 not deleting PN_2 subtrees	108	463,076,682
PN^2 deleting PN_2 subtrees	109	416,168,419

6.4 PN Search Performance

In this section we test the offline performance of three PN-search variants by comparing PN, PN^2, PDS, and $\alpha\beta$ search with each other. The goal is to investigate the effectiveness of the PN-search variants by experiments. We will look how many endgame positions they can solve and how much effort (in nodes and CPU time) they take. For the $\alpha\beta$ depth-first iterative-deepening search, nodes at depth i are counted only during the first iteration that the level is reached. This is how analogous comparisons are done in Allis [1]. For PN, PN^2, and PDS search, all nodes evaluated for the termination condition during the search are counted. For PDS this node count is equal to the number of expanded nodes (function calls of the recursive PDS algorithm). PN, PN^2, PDS, and $\alpha\beta$ are tested on a set of 488 forced-win LOA positions.[4] Two comparisons are made, which are described in Subsection 6.4.1 and 6.4.2.

6.4.1 A General Comparison of Four Search Techniques

In Table 6.3 we compare PN, PN^2, PDS, and $\alpha\beta$ on a set of 488 LOA positions. The maximum number of nodes searched is again 50,000,000. In the second column of Table 6.3 we see that 470 positions were solved by the PN^2 search, 473 positions by PDS, only 356 positions by PN, and 383 positions by $\alpha\beta$. In the third and fourth column the number of nodes and the time consumed are given for the subset of 314 positions, which all four algorithms were able to solve. If we have a look at the third column, we see that PN search builds the smallest search trees and $\alpha\beta$ by far the largest. PN^2 and PDS build larger trees than PN but can solve significantly more positions. This suggests that both algorithms are better suited for harder problems. PN^2 investigates 1.2 times more nodes than PDS, but PN^2 is (more than) 6 times faster than PDS in CPU time for this subset.

---

[4] The test set can be found at www.cs.unimaas.nl/m.winands/loa/tscg2002a.zip.

Table 6.3. Comparing the search algorithms on 488 test positions

Algorithm	# of positions solved (out of 488)	314 positions	
		Total nodes	Total time (ms.)
$\alpha\beta$	383	1,711,578,143	22,172,320
PN	356	89,863,783	830,367
PN^2	470	139,254,823	1,117,707
PDS	473	118,316,534	6,937,581

From the experiments we may draw the following three conclusions.

1. PN-search algorithms clearly outperform $\alpha\beta$ in solving endgame positions in LOA.
2. The memory problems make the plain PN search a weaker solver for the harder problems.
3. PDS and PN^2 are able to solve significantly more problems than PN and $\alpha\beta$.

6.4.2   A Deep Comparison of PN^2 and PDS

For a better insight into how much faster PN^2 is than PDS in CPU time, we did a second comparison. In Table 6.4 we compare PN^2 and PDS on the subset of 463 test positions, which both algorithms were able to solve. Now, PN^2 searches 2.6 times more nodes than PDS. The reason for the decrease of performance is that for hard problems the PN$_2$-search tree becomes as large as the PN$_1$-search tree. Therefore, the PN$_2$-search tree is causing more overhead. However, if we have a look at the CPU times we see that PN^2 is still three times faster than PDS. The reason is that PDS has a relatively large time overhead because of the delayed evaluation. Consequently, the number of nodes generated is higher than the number of nodes expanded. In our experiments, we observed that PDS generated nodes 7 to 8 times slower than PN^2. Such a figure for the overhead is in agreement with experiments performed in Othello and Tsume-Shogi [29]. We remark that Nagai's [19] Othello results showed that PDS was better than PN search, (i.e., it solved the positions faster than PN). Nagai assigned to both the proof number and the disproof number of unknown nodes a 1 in his PN search and therefore did not use the mobility enhancement. In contrast, we incorporated the mobility in the initialisation of the proof numbers and disproof numbers in our PN search. We believe that comparing PDS with a PN-search algorithm without using the mobility component is not fair. Since PDS does not store unexpanded nodes that have a proof number 1 and disproof number 1, we may state that PDS initialises the (dis)proof number of a node by counting the number of its newborn children [19]. So in the PDS search, the mobility enhancement coincides with the initialisation of the (dis)proof number.

From this second experiment we may conclude that PDS is considerably slower than PN^2 in CPU time. Therefore, PN^2 seems to be a better endgame solver under tournament conditions. Counterbalancing this success, we note that PN^2 is still restricted by its working memory and is not fit for solving really hard problems.

Table 6.4. Comparing PDS and PN^2 on 463 test positions

Algorithm	Total nodes	Total time (ms.)
PN^2	1,462,026,073	11,387,661
PDS	562,436,874	34,379,131

6.5   PDS-PN

In Section 6.4 we have observed two facts: (1) the advantage of PN^2 over PDS is that it is faster and (2) the advantage of PDS over PN^2 is that its tree is constructed as a depth-first tree, which is not restricted by the available working memory. In the next sections we try to overcome fact (1) while preserving fact (2) by presenting a new proof-number search algorithm, called PDS-PN [35, 36]. It is a two-level search (like PN^2), which performs at the first level a depth-first Proof-number and Disproof-number Search (PDS), and at the second level a best-first PN search. Hence, PDS-PN selectively exploits the power of both PN^2 and PDS. In this section we give a description of PDS-PN search, which is a two-level search using PDS at the first level and PN at the second level. In Subsection 6.5.1 we motivate why we developed the method. In Subsection 6.5.2 we describe the first-level PDS, and in Subsection 6.5.3 we provide background information on the second-level technique. Finally, in Subsection 6.5.4 the relevant parts of the pseudo code are given.

6.5.1   Motivation

The development of the PDS-PN algorithm was motivated by the clear advantage that PDS is traversing a depth-first tree instead of a best-first tree. Hence, PDS is not restricted by the available working memory. As against this, PN has the advantage of being fast compared to PDS (see Section 6.4).

The PDS-PN algorithm is designed to combine the two advantages. At the first level, the search is a depth-first search, which implies that PDS-PN is not restricted by memory. At the second level the focus is on fast PN. It is a complex balance, but we expect that PDS-PN will be faster than PDS, and PDS-PN will not be hampered by memory restrictions. Since the expectation on the effectiveness of PDS-PN is difficult to prove we have to rely on experiments (see Section 6.6). In the next two subsections we describe PDS-PN.

6.5.2   First Level: PDS

PDS-PN is a two-level search like PN^2. At the first level a PDS search is performed, denoted PN$_1$. For the expansion of a PN$_1$ leaf node, not stored in the transposition table, a PN search is started, denoted PN$_2$.

Proof-number and Disproof-number Search (PDS) [18] is a straightforward extension of PN*. Instead of using only proof numbers such as in PN*, PDS uses disproof numbers too. PDS exploits a method called *multiple-iterative deepening*. Instead of iterating only in the root such as in ordinary iterative deepening, PDS iterates in *all* interior nodes. The advantage of using the multiple-iterative-deepening method is that in most cases it accomplishes to select the most-proving node (see below), not only in the root, but also in the interior nodes of the search tree. To keep iterative deepening effective, the method is enhanced by storing the expanded nodes in a *TwoBig* transposition table [7].

PDS uses two thresholds for a node, one as a limit for proof numbers and one for disproof numbers. Once the thresholds are assigned to a node, the subtree rooted in that node is stopped to be searched if both the proof number and disproof number are larger than or equal to the thresholds *or* if the node is proved or disproved. The thresholds are set in the following way. At the start of every iteration, the proof-number threshold *pnt* and disproof-number threshold *dnt* of a node are equal to the node's proof number *pn* and disproof number *dn*, respectively. If it seems more likely that the node can be proved than disproved (called *proof-like*), the proof-number threshold is increased. If it seems more likely that the node can be disproved than proved (called *disproof-like*), the disproof-number threshold is increased. In passing we note that it is easier to prove a tree in an OR node, and to disprove a tree in an AND node. Below we repeat Nagai's [18] heuristic to determine proof-like and disproof-like.

In an interior OR node n with parent p (direct ancestor) the solution of n is proof-like, if the following condition holds:

$$pnt_p > pn_p \ AND \ (pn_n \leq dn_n \ OR \ dnt_p \leq dn_p), \qquad (6.8)$$

otherwise, the solution of n is disproof-like.

In an interior AND node n with parent p (direct ancestor) the solution of n is disproof-like, if the following condition holds:

$$dnt_p > dn_p \ AND \ (dn_n \leq pn_n \ OR \ pnt_p \leq pn_p), \qquad (6.9)$$

otherwise, the solution of n is proof-like.

When PDS does not prove or disprove the root given the thresholds, it increases the proof-number threshold if its proof number is smaller than or equal to its disproof number, otherwise it increases the disproof-number threshold. Finally, we remark that only expanded nodes are evaluated. This is called *delayed evaluation* (cf. [1]). The expanded nodes are stored in a transposition table. The proof and disproof number of a node are set to unity when not found in the transposition table. Since PDS does not store unexpanded nodes which have a proof number 1 and disproof number 1, it can be said that PDS initialises the proof and disproof number by using the number of children. The mobility enhancement of PN and PN^2 (see Subsection 6.2.1) is already implicitly incorporated in the PDS search.

PDS is a depth-first search algorithm but behaves like a best-first search algorithm. In most cases PDS selects the same node for expansion as PN search. By using transposition tables PDS suffers from the GHI problem (cf. [9]). Especially the GHI evaluation problem can occur in LOA too. For instance, draws can be agreed upon due to the three-fold-repetition rule. Thus, dependent on its history a node can be a draw or can have a different value. However, in the current PDS algorithm we ignore this problem, since we believe that it is less relevant for the game of LOA than for Chess.

Iteration 1:

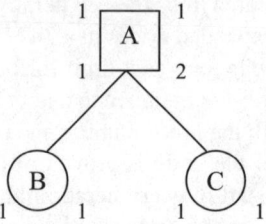

Iteration 2:

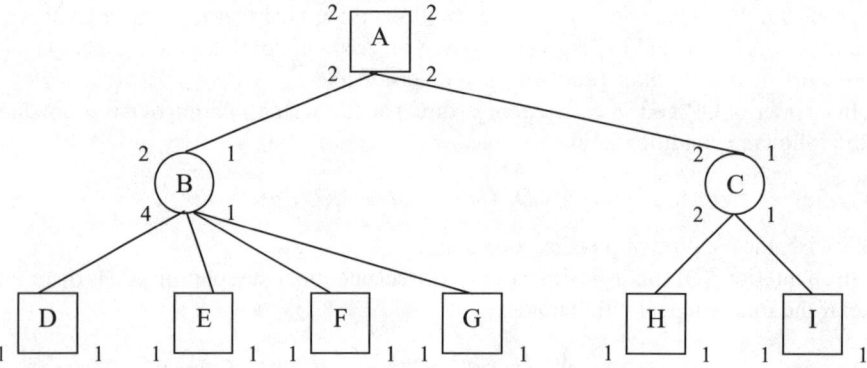

Iteration 3:

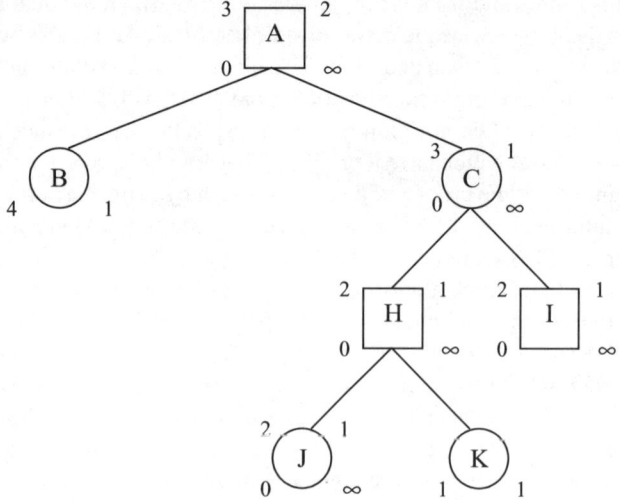

Fig. 6.2. An illustration of PDS

A Detailed Example

A detailed step-by-step example of the working of PDS is given in Figure 6.2. A square denotes an OR node, and a circle denotes an AND node. The numbers at the upper side of a node denote the proof-number threshold (left) and disproof-number threshold (right). The numbers at the lower side of a node denote the proof number (left) and disproof number (right).

In the first iteration (top of Figure 6.2), threshold values of the root A are set to unity. A is expanded, and nodes B and C are generated. The proof number of A becomes 1 and the disproof number becomes 2. Because both numbers are larger than or equal to the threshold values the search stops.

In the second iteration (middle of Figure 6.2), the proof-number threshold is incremented to 2, because the proof number of A (i.e., 1) is the smaller one of both A's proof number and disproof number (i.e., 2). We again expand A and re-generate B and C. The proof number of A is below its proof-number threshold and we continue searching. Now we have to select the child with minimum proof number. Because B and C have the same proof number, the left-most node B is selected. Initially, we set the proof-number and disproof-number threshold of B to its proof and disproof number (both 1). Because B is an AND node we have to look whether the solution of B is disproof-like by checking the appropriate condition (i.e., formula 6.9). The disproof-number threshold of A is not larger than its disproof number (both are 2), therefore the solution of B is not disproof-like but proof-like. Thus, the proof-number threshold of B has to be incremented to 2. Next, node B is expanded and the nodes D, E, F and G are generated. The search in node B is stopped because its proof number (i.e., 4) and disproof number (i.e., 1) are larger than or equal to the thresholds (i.e., 2 and 1, respectively). Node B is stored in the transposition table with proof number 4 and disproof number 1. Then the search backtracks to A. There we have to check whether we still can continue searching A. Since the proof number of A is smaller than its threshold, we continue and subsequently we select C, because this node has now the minimum proof number. The thresholds are set in the same way as in node B. Node C has two children H and I. The search at node C is stopped because its proof number (i.e., 2) and disproof number (i.e., 1) are not below the thresholds. C is stored in the transposition table with proof number 2 and disproof number 1. The search backtracks to A and is stopped because its proof number (i.e., 2) and disproof number (i.e., 2) are larger than or equal to the thresholds. We remark that at this moment B and C are stored because they were expanded.

In the third iteration (bottom of Figure 6.2) the proof-number threshold of A is incremented to 3. Nodes B and C are again generated, but this time we can find their proof and disproof numbers in the transposition table. The node with smallest proof number is selected (C with proof number 2). Initially, we set the proof-number threshold and disproof-number threshold of C to its proof and disproof number (i.e., 2 and 1, respectively). Because C is an AND node we have to look whether the solution is disproof-like by checking condition 6.9. The disproof-number threshold of A is not larger than its disproof number (both are 2), therefore the solution is not disproof-like but proof-like. Thus, the proof-number threshold of C has to be incremented to 3. C has now proof-number threshold 3 and disproof-number threshold 1. Nodes H and I are generated again by expanding C. This time the proof number of C (i.e., 2) is below the proof-number threshold (i.e., 3) and the search continues. The node with minimum disproof number is selected (i.e., H). Initially, we set the proof-number threshold and disproof-number threshold of H to its proof and disproof number (i.e., both 1). Because H is an OR node we have to look whether the solution is proof-like by checking condition 6.8. The proof-number threshold of C (i.e., 3) is larger than its proof number (i.e., 2), therefore the solution is proof-like. Hence, the search expands node H with proof-number threshold 2 and disproof-number threshold 1. Nodes J and K are generated. Because the proof number of H (i.e., 1) is below its threshold (i.e., 2), the node with minimum proof number is selected. Because J is an AND node we have to look whether the solution of J is disproof-like by checking condition 6.9. The disproof-number threshold of H (i.e., 1) is not larger than its disproof number (i.e., 2), therefore the solution of J is not disproof-like but proof-like. J is expanded with proof-number threshold 2 and disproof-number threshold 1. Since node J is a terminal win position its proof number is set to 0 and its disproof number set to ∞. The search backtracks to H. At node H the proof number becomes 0 and the disproof number ∞, which means the node is proved. The search backtracks to node C. The search continues because the proof number of C (i.e., 1) is not larger than or equal to the proof-number threshold (i.e., 3). We select now node I because it has the minimum disproof number. The thresholds of node I are set to 2 and 1, as was done in H. The node I is a terminal win position; therefore its proof number is set to 0 and its disproof number to ∞. At this moment the proof number of C is 0 and the disproof number ∞, which means that the node is proved. The search backtracks to A. The proof number of A becomes 0, which means that the node is proved. The search stops at node A and the tree is proved.

6.5.3   Second Level: PN Search

For an adequate description we reiterate a few sentences from Subsection 6.2.2. At the leaves of the first-level search tree, the second-level search is invoked, similar as in PN^2 search. The PN search of the second-level, denoted PN_2 search, is bounded by the number of nodes that may be stored in memory. The number is a fraction of the size of the PN_1-search tree, for which we take the current number of nodes stored in the transposition table of the PDS search. Preferably, this fraction should start small, and grow larger as the size of the first-level search tree increases. A standard model for this growth is the logistic-growth model [3]. The fraction $f(x)$ is therefore given by the logistic-growth function, x being the size of the first-level search:

$$f(x) = \frac{1}{1 + e^{\frac{a-x}{b}}} , \qquad (6.10)$$

with parameters a and b, both strictly positive. The parameter a determines the transition point of the function: as soon as the size of the first-level search tree reaches a, the second-level search equals half the size of the first-level search. Parameter b determines the S-shape of the function: the larger b, the more stretched the S-shape is. The number of nodes y in a PN_2-search tree is restricted by the minimum of this fraction function and the number of nodes which can still be stored. The formula to compute y is:

$$y = min(x \times f(x), N - x), \qquad (6.11)$$

with N the maximum number of nodes to be stored in memory.

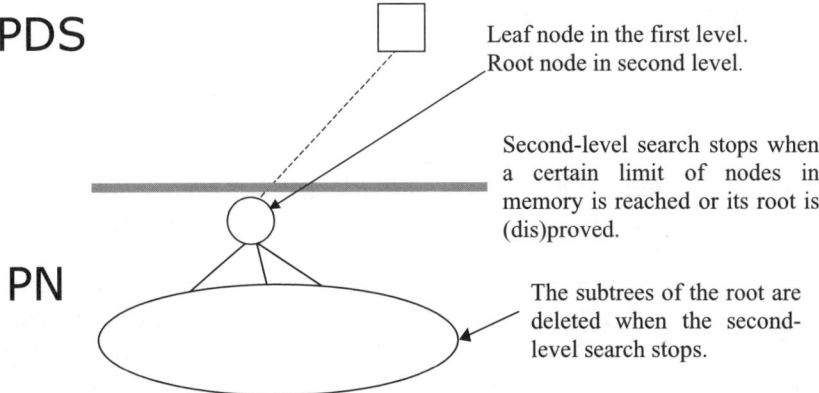

Fig. 6.3. Schematic sketch of PDS-PN

The PN$_2$ search is stopped when the number of nodes stored in memory exceeds y or the subtree is (dis)proved. After completion of the PN$_2$-search tree, only the root of the PN$_2$-search tree is stored in the transposition table of the PDS search. We remark that for PN$_2$-search trees *immediate evaluation* (cf. [1]) is used. This two-level search is schematically sketched in Figure 6.3.

In the second-level search proved or disproved subtrees are deleted. If we do not delete proved or disproved subtrees in the PN$_2$ search, the number of nodes searched becomes the same as y. When we include deletions the second-level search can continue on average considerably longer.

```
//Iterative deepening at root r
procedure NegaPDSPN(r){

  r.proof = 1;
  r.disproof = 1;

  while(true){
    MID(r);
    //Terminate when the root is proved or disproved
    if(r.proof == 0 || r.disproof == 0)
      break;

    if(r.proof <= r.disproof)
      r.proof++;
    else
      r.disproof++;
  }
}
```

Fig. 6.4. PDS-PN: Root node

6.5.4   Pseudo Code for PDS-PN

In this subsection we provide the pseudo code for PDS-PN. For ease of comparison we use similar pseudo code as used by Nagai [18] for the PDS algorithm. The proof number in an OR node and the disproof number in an AND node are equivalent. Analogously, the disproof number in an OR node and the proof number in an AND node are equivalent. As they are dual to each other, an algorithm similar to negamax in the context of minimax searching can be constructed. This algorithm is called NegaPDSPN (see Figure 6.4).

In the following, procedure `MID(n)` performs multiple iterative deepening (see Figure 6.5). The function `proofSum(n)` computes the sum of the proof numbers of all the children. The function `disproofMin(n)` finds the minimum of the disproof numbers of all the children. The procedure `putInTT()` stores information to and `lookUpTT()` retrieves information from the transposition table. `isTerminal(n)` checks whether a node is a win, a loss, or a draw. The procedure `generateChildren (n)` generates the children of the node. By default, the proof number and disproof number of a node are set to unity. The procedure `findChildrenInTT(n)` checks whether the children are already stored in the transposition table. If a hit occurs for a child, its proof number and disproof number are set to the values found in the transposition table. The procedure `PN()` is just the plain PN search. The algorithm is described in Allis [1] and Breuker [5], and is reproduced in the Appendix . The function `computeMaxNodes()` computes the number of nodes, which may be stored for the PN search, according to Equation 6.11.

Finally, the function `selectChild()` selects the child that will be traversed next (see Figure 6.6).

6.6   Experiments

In this section we compare $\alpha\beta$, PN^2, PDS, and PDS-PN search with each other. The goal is to prove experimentally the effectiveness of PDS-PN. We will investigate how many endgame positions it can solve and the effort (in terms of number of nodes and CPU time) it takes compared with $\alpha\beta$, PN^2, PDS. For PDS and PDS-PN we use a *TwoBig* transposition table. In Subsection 6.6.1 we test PDS-PN with different parameters a and b for the growth function. In Subsection 6.6.2 we compare PDS-PN with $\alpha\beta$, PN^2, and PDS on a set of 488 LOA positions in three different ways. In Subsection 6.6.3 we compare PDS-PN with PN^2 on a set of hard LOA problems. Finally, we evaluate the algorithms PDS-PN and PN^2 in solving problems under restricted memory conditions in Subsection 6.6.4.

6.6.1   Parameter Tuning

In the following series of experiments we measured the solving ability with different parameters a and b. For our specific parameter choice we follow Breuker [5], i.e., parameter a takes values of 150K, 450K, 750K, 1050K, and 1350K, and for each value of a parameter b takes values of 60K, 120K, 180K, 240K, 300K, and 360K. The results

```
procedure MID(n){
  //Look up in the transposition table
  lookUpTT(n, &proof, &disproof);
  if(proof == 0 || disproof == 0
  || (proof >= n.proof && disproof >= n.disproof)){
    n.proof = proof; n.disproof = disproof;
    return;
  }

  if(isTerminal(n)){
    if((n.value == true && n.type == AND_NODE)
    ||(n.value == false && n.type == OR_NODE)){
      n.proof = INFINITY; n.disproof = 0;
    }
    else{
      n.proof = 0; n.disproof = INFINITY;
    }
    putInTT(n);
    return;
  }

  generateChildren(n);
  //Avoid cycles
  putInTT(n);

  //Multiple-iterative deepening
  while(true){
    //Check whether the children are already stored in the TT.
     findChildrenInTT(n);
    //Terminate when both pn and dn exceed their thresholds
    if(proofSum(n) == 0 || disproofMin(n) == 0 || (n.proof <=
    disproofMin(n) && n.disproof <= proofSum(n))){
      n.proof = disproofMin(n);
      n.disproof = proofSum(n);
      putInTT(n);
      return;
    }

    proof = max(proof, disproofMin(n));
    n_child = selectChild(n, proof);

    if(n.disproof > proofSum(n) && (proof_child <= disproof_child
      || n.proof <= disproofMin(n)))
      n_child.proof++;
    else
      n_child.disproof++;

    //This is the PDS-PN part
     if(!lookUpTT(n_child)){
      PN(n_child, computeMaxNodes());
      putInTT(n_child);
    }
    else
      MID(n_child);
  }
}
```

Fig. 6.5. PDS-PN: Multiple-Iterative Deepening

```
//Select among children
selectChild(n, proof){

  min_proof = INFINITY;
  min_disproof  = INFINITY;
  for(each child n_child){
    disproof_child = n_child.disproof;
    if(disproof_child != 0)
      disproof_child = max(disproof_child, proof);

    //Select the child with the lowest disproof_child (if there are
    //plural children among them select the child with the lowest
    //n_child.proof)
    if(disproof_child < min_disproof || (disproof_child
        == min_disproof && n_child.proof < min_proof)){
      n_best = n_child;
      min_proof = n_child.proof;
      min_disproof = disproof_child;
    }
  }
  return n_best;
}
```

Fig. 6.6. PDS-PN: Selection mechanism

are given in Table 6.5. For each a holds that the number of solved positions grows with increasing b, when the parameter b is still small. If b is sufficiently large, increasing it will not enlarge the number of solved positions. In the process of parameter tuning we found that PDS-PN solves the most positions with (450K, 300K) (see the bold line in Table 6.5). However, the difference with parameter configurations (150K, 180K), (150K, 240K), (150K, 300K), (150K, 360K), (450K, 360K), and (1350K, 300K) is not significant. On the basis of these results we decided that it is not necessary to perform experiments with a larger a.

6.6.2   Three Comparisons of the Algorithms

In the experiments with PN^2, PDS, and PDS-PN all nodes evaluated during the search are counted; for the $\alpha\beta$ depth-first iterative-deepening searches nodes at depth i are counted only during iteration i. We adopted this method from Allis [1]. It makes a general comparison possible. The maximum number of nodes searched is 50,000,000. The limit corresponds roughly to tournament conditions. The maximum number of nodes stored in memory is 1,000,000. The parameters (a,b) of the growth function used in PN^2 are set at (1800K, 240K) according to the suggestions in Breuker *et al.* [8]. The parameter configuration (450K, 300K) found in the previous subsection will be used for PDS-PN. The smaller value of a corresponds to the smaller PN_1 trees resulting from the use of PDS-PN instead of PN^2. The fact that PDS is much slower than PN is an important factor too.

Table 6.5. Number of solved positions (by PDS-PN) for different values of a and b

a	b	# of solved positions	accuracy (%)	a	b	# of solved positions	accuracy (%)
150,000	60,000	460	94.3	750,000	240,000	463	94.9
150,000	120,000	458	93.9	750,000	300,000	460	94.3
150,000	180,000	466	95.5	750,000	360,000	461	94.5
150,000	240,000	466	95.5	1,050,000	60,000	421	86.3
150,000	300,000	465	95.3	1,050,000	120,000	448	91.8
150,000	360,000	466	95.5	1,050,000	180,000	451	92.4
450,000	60,000	445	91.2	1,050,000	240,000	459	94.1
450,000	120,000	463	94.9	1,050,000	300,000	459	94.1
450,000	180,000	460	94.3	1,050,000	360,000	460	94.3
450,000	240,000	461	94.5	1,350,000	60,000	421	86.3
450,000	**300,000**	**467**	**95.7**	1,350,000	120,000	433	88.7
450,000	360,000	464	95.1	1,350,000	180,000	447	91.6
750,000	60,000	432	88.5	1,350,000	240,000	454	93.0
750,000	120,000	449	92.0	1,350,000	300,000	465	95.3
750,000	180,000	461	94.5	1,350,000	360,000	459	94.1

First Comparison

$\alpha\beta$, PN^2, PDS, and PDS-PN are tested on the same set of 488 forced-win LOA positions as described in Section 6.4. The results are given in Table 6.6. In the first column the four algorithms are mentioned. In the second column we see that 382 positions are solved[5] by $\alpha\beta$, 470 positions by PN^2, 473 positions by PDS, and 467 positions by PDS-PN. The set of 488 positions contains no position that only could be solved by $\alpha\beta$ search. In the third and fourth column the number of nodes and the time consumed are given for the subset of 371 positions, which all four algorithms are able to solve. A look at the third column shows that PDS search builds the smallest search trees and $\alpha\beta$ by far the largest. Like PN^2 and PDS, PDS-PN solves significantly more positions than $\alpha\beta$. This suggests that PDS-PN is a better endgame solver than $\alpha\beta$. As we have seen before, PN^2 and PDS-PN investigate more nodes than PDS, but both are still faster in CPU time than PDS for this subset. Due to the limit of 50,000,000 nodes and the somewhat lower search efficiency, PDS-PN solves three positions fewer than PN^2 (result PDS-PN is 99.4% with respect to PN^2) and six fewer than PDS (result PDS-PN is 98.7% with respect to PDS).

Second Comparison

To investigate whether the memory restrictions are an actual obstacle we increased the limit of nodes searched to 500,000,000 nodes. In this second comparison PN^2 solves now 479 positions and PDS-PN becomes the best solver with a performance of 483 positions. The detailed results are given in Table 6.7.

---

[5] We remark that a slightly less inefficient version of our $\alpha\beta$ implementation could solve 383 positions (see Section 6.4).

Table 6.6. Comparing the search algorithms on 488 test positions with a limit of 50,000,000 nodes

Algorithm	# of positions solved (out of 488)	accuracy (%)	371 positions	
			Total # of nodes	Total time (ms.)
$\alpha\beta$	382	78.3	2,645,022,391	33,878,642
PN^2	470	96.3	505,109,692	3,642,511
PDS	473	96.9	239,896,147	16,960,325
PDS-PN	467	95.7	924,924,336	5,860,908

Table 6.7. Comparing PN^2 and PDS-PN on 488 test positions with a limit of 500,000,000 nodes

Algorithm	# of positions solved (out of 488)	accuracy (%)	479 positions	
			Total # of nodes	Total time (ms.)
PN^2	479	98.2	2,261,482,395	13,295,688
PDS-PN	483	99.0	4,362,282,235	23,398,899

The performance of PDS-PN in Table 6.7 is more effective than that of PN^2, viz. 483 to 479. However, we should thoughtfully take into account the condition for the total number of nodes searched and the time spent. Therefore, we continue our research in the direction of nodes searched and time spent with the 50,000,000 nodes limit. A reason for this decision is that the experimental time constraints are necessary for the PDS experiments.

Third Comparison

For a better insight into the relation between PN^2, PDS, and PDS-PN we performed a third comparison. In Table 6.8 we provide the results of PN^2, PDS, and PDS-PN on a new subset of 457 positions of the principal test set, viz. all positions the three algorithms could solve under the 50,000,000 nodes limit condition. Now, PN^2 searches 2.6 times more nodes than PDS. The reason for the difference of performance is that for hard problems the PN_2-search tree becomes as large as the PN_1-search tree. Therefore, the PN_2-search tree is causing more overhead. However, if we look at the CPU time we see that PN^2 is almost 4 times faster than PDS. PDS has a relatively large time overhead because it performs multiple-iterative deepening at all nodes. PDS-PN searches 3.7 times more nodes than PDS but is still 3 times faster than PDS in CPU time. This is because PDS-PN is focussing more on the fast PN at the second level than on PDS at the first level. PDS-PN searches more nodes than PDS since the PN_2-search tree is repeatedly rebuilt and removed. The overhead is even bigger than PN^2's overhead because the children of the root of the PN_2-search tree are not stored (i.e., this is done to focus more on the fast PN search). It explains why PDS-PN searches 1.4 times more nodes than PN^2. Hence, our provisional conclusions are that on this set of 457 positions and under the 50,000,000 nodes condition: (1) PN^2 outperforms PDS-PN, and (2) PDS-PN is a faster solver than PDS and therefore more effective than PDS.

Table 6.8. Comparing PN^2, PDS and PDS-PN on 457 test positions (all solved) with a limit of 50,000,000 nodes

Algorithm	Total # of nodes	Total time (ms.)
PN^2	1,275,155,583	9,357,663
PDS	498,540,408	36,802,350
PDS-PN	1,845,371,831	11,952,086

Table 6.9. Comparing PN^2 and PDS-PN on 286 hard test positions with a limit of 500,000,000 nodes

| Algorithm | # of positions solved | accuracy | 255 positions | |
	(out of 286)	(%)	Total # of nodes	Total time (ms.)
PN^2	265	92.7	10,061,461,685	57,343,198
PDS-PN	276	96.5	16,685,733,992	84,303,478

6.6.3 Comparing the Algorithms for Hard Problems

Since the impact of the 50,000,000 nodes condition somewhat obscured our provisional conclusions above and since we felt that the 99.4% score by PDS-PN with respect to PN^2 was rather close, we performed a new experiment with a different set of LOA problems in an attempt to find more insights into the intricacies of these complex algorithms. In the new experiment PN^2 and PDS-PN are tested on a set of 286 LOA positions, which were on average harder than the ones in the previous test set.[6] In this context 'harder' means a longer distance to the final position (the solution), i.e., more time is needed. The conditions are the same as in the previous experiments except that the maximum number of nodes searched is set at 500,000,000. The PDS algorithm is not included because it takes too much time given the current node limit. In Table 6.9 we see that PN^2 solves 265 positions and PDS-PN 276. We remark that PN^2 solves 10 positions, which PDS-PN does not solve, but that PDS-PN solves 21 positions that PN^2 does not solve. The ratio in nodes and time between PN^2 and PDS-PN for the positions solved by both (255) is roughly similar to the previous experiments. The reason why PN^2 solves fewer positions than PDS-PN is its being restricted in working memory. We are in a delicate position since new experiments with much more working memory are now on the list to be performed. However, we assume that the nature of PN^2 with respect to using so much memory cannot be overcome. Hence we may conclude that within an acceptable time frame PDS-PN is a more effective endgame solver than PN^2 for hard problems.

6.6.4 Comparing the Algorithms under Reduced Memory

From the experiments in the previous subsection it is clear that PN^2 will not be able to solve very hard problems since it will run out of working memory. To further solidify this statement experimentally, we tested the solving ability of PN^2 and PDS with restricted working memory. In these experiments we started with a memory capacity

---

[6] The test set can be found at www.cs.unimaas.nl/m.winands/loa/tscg2002b.zip.

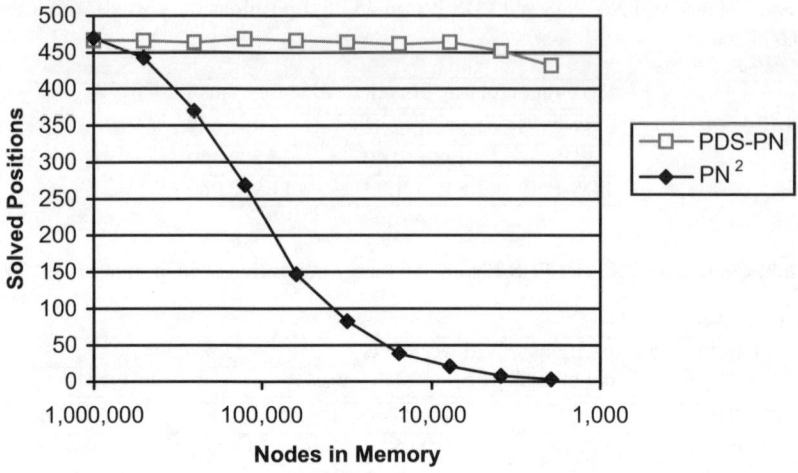

Fig. 6.7. Results with restricted memory

sufficient to store 1,000,000 nodes, subsequently we divided the memory capacity by two at each next step. The parameters a and b were also divided by two. The relation between memory and number of solved positions for both algorithms is given in Figure 6.7. We see that the solving performance rapidly decreases for PN^2. The performance of PDS-PN remains stable for a long time. Only when PDS-PN is restricted to fewer than 10,000 nodes, it begins to solve fewer positions. This experiment suggests that PDS-PN is to be preferred above PN^2 for the very hard problems when the memory capacity is in some way restricted. The reason is that PDS-PN is not suffering from memory constraints. If there are no memory constraints at all, PN^2 is preferred because under those circumstances it is the fastest algorithm.

6.7   Df-pn and PDS Comparison

Pawlewicz and Lew [24] used the set of 286 LOA positions of Subsection 6.6.3 to compare their implementation of df-pn and PDS with each other. Moreover, they enhanced df-pn and PDS with the $1 + \epsilon$ trick. To compare the speed differences in CPU time

Table 6.10. Comparison of df-pn and PDS on 286 hard test positions with a limit of 30 minutes by Pawlewicz and Lew [24]. The rate number r in row A and in column B indicates that algorithm A is r times faster than algorithm B.

Algorithm	df-pn	df-pn $1 + \varepsilon$	PDS	PDS $1 + \varepsilon$
df-pn	1.00	0.63	2.83	2.64
df-pn $1 + \varepsilon$	1.58	1.00	4.46	4.17
PDS	0.35	0.22	1.00	0.93
PDS $1 + \varepsilon$	0.38	0.24	1.07	1.00

for each two methods, they calculated the geometric mean of the ratio of solving times (see Table 6.10). We see that the enhanced df-pn is the most efficient method, plain df-pn is the second best, and both PDS versions are the least efficient. Although the $1 + \epsilon$ trick was not designed for PDS, there is a noticeable difference between enhanced PDS and plain PDS in their experiments. For more details we refer to their paper [24].

Pawlewicz and Lew did neither implement nor tested PN^2 and PDS-PN on their hardware and with their data-structure implementation. Up to now there has not been a direct comparison between df-pn with PDS-PN using the same hardware and the same data-structure implementation. However, even so, we may still conclude that df-pn is an interesting alternative to PDS-PN. In Table 6.10 we see that df-pn is 4 times faster than PDS, while in Table 6.8 (Subsection 6.6.3) we see that PDS-PN is able to search 3 times faster than PDS.

6.8 Conclusions and Future Research

Below we offer seven observations, three conclusions, and four suggestions for future research. Since observations and conclusions are intermingled, we present the conclusions in relation to the observations.

First, we have observed that mobility and deleting (dis)proved PN_2 subtrees speed up PN and PN^2 and increase their ability of solving endgame positions. Second, we have seen that the various PN-search algorithms outperform $\alpha\beta$ in solving endgame positions in LOA. Third, the memory problems make the plain PN search a weak solver for the harder problems. Fourth, PDS and PN^2 are able to solve significantly more problems than PN and $\alpha\beta$.

Our first conclusion is that PN and its variants offer a valuable tool for enhancing programs in endgames. We remark that PN^2 is still restricted by working memory, and that PDS is three times slower than PN^2 (Table 6.4) because of the delayed evaluation.

Our fifth observation is that PDS-PN is able to solve significantly more LOA endgame problems than $\alpha\beta$ search with enhancements. Our sixth observation is that the PDS-PN algorithm is almost as fast as PN^2 when the parameters for its growth function are chosen properly. It turns out that for each a it holds that the number of solved positions grows with increasing b, when the parameter b is still small. If b is sufficiently large, increasing it will not enlarge the number of solved positions. Our seventh observation states that (1) PDS-PN solves more hard positions than PN^2 within an acceptable time frame and (2) PDS-PN is more effective than PN or even PN^2 because it does not run out of memory for hard problems. Moreover, PDS-PN performs quite well under harsh memory conditions. This is especially appropriate for hard problems and for environments with very limited memory such as hand-held computer platforms.

Hence, our second conclusion is that PDS-PN may be a more effective endgame solver for a set of hard problems than PDS and PN^2.

In this chapter we discussed the results of comparing PDS with df-pn as performed by Pawlewicz and Lew [24]. Df-pn was solving the set of hard problems 4 times faster than PDS. In Subsection 6.6.3 we saw that PDS-PN was able to search 3 times faster than PDS, whereas in Section 6.7 df-pn was 4 times faster than PDS. Although the conditions were different, our third conclusion is that df-pn may be an interesting alternative to PDS-PN.

Finally, we believe that there are four suggestions for the near future. The first challenge is testing PDS-PN in other domains with difficult endgames. An example of a game notoriously known for its complicated endgames is the game of Tsume-Shogi (a variant of Shogi). Several hard problems including solutions over a few hundred ply are solved by PN* [34] and PDS [20, 29]. It would be interesting to test PDS-PN on these problems. Second, it would be interesting to have a direct comparison between df-pn with PDS-PN using the same hardware and the same data-structure implementation. The third challenge would be to construct a two-level Proof-Number search variant using df-pn at the first level search, and a plain best-first PN search at the second level. Fourth, one problem clearly remains, viz. that there is no dynamic strategy available that determines when to use PN search instead of $\alpha\beta$ in a real game. This will be subject of future research as well.

Acknowledgements

The authors would like to thank the members of the MICC Games and AI Group for their comments.

References

1. Allis, L.V.: Searching for Solutions in Games and Artificial Intelligence. PhD thesis, Rijksuniversiteit Limburg, Maastricht, The Netherlands (1994)
2. Allis, L.V., van der Meulen, M., van den Herik, H.J.: Proof-number search. Artificial Intelligence 66(1), 91–123 (1994)
3. Berkey, D.D.: Calculus. Saunders College Publishing, New York (1988)
4. Berliner, H.J.: The B*-tree search algorithm: A best-first proof procedure. Artificial Intelligence 12(1), 23–40 (1979)
5. Breuker, D.M.: Memory versus Search in Games. PhD thesis, Universiteit Maastricht, Maastricht, The Netherlands (1998)
6. Breuker, D.M., Allis, L.V., van den Herik, H.J.: How to mate: Applying proof-number search. In: van den Herik, H.J., Herschberg, I.S., Uiterwijk, J.W.H.M. (eds.) Advances in Computer Chess, University of Limburg, Maastricht, The Netherlands, vol. 7, pp. 251–272 (1994)
7. Breuker, D.M., Uiterwijk, J.W.H.M., van den Herik, H.J.: Replacement schemes and two-level tables. ICCA Journal 19(3), 175–180 (1996)
8. Breuker, D.M., Uiterwijk, J.W.H.M., van den Herik, H.J.: The PN^2-search algorithm. In: van den Herik, H.J., Monien, B. (eds.) Advances in Computer Games, IKAT, Universiteit Maastricht, Maastricht, The Netherlands, vol. 9, pp. 115–132 (2001)
9. Breuker, D.M., van den Herik, H.J., Uiterwijk, J.W.H.M., Allis, L.V.: A solution to the GHI problem for best-first search. Theoretical Computer Science 252(1-2), 121–149 (2001)
10. Campbell, M.: The graph-history interaction: On ignoring position history. In: 1985 Association for Computing Machinery Annual Conference, pp. 278–280 (1985)
11. Campbell, M., Hoane Jr., A.J., Hsu, F.-h.: Deep Blue. Artificial Intelligence 134(1-2), 57–83 (2002)

12. Junghanns, A.: Are there practical alternatives to alpha-beta? ICCA Journal 21(1), 14–32 (1998)
13. Kishimoto, A.: Correct and Efficient Search Algorithms in the Presence of Repetitions. PhD thesis, University of Alberta, Edmonton, Canada (2005)
14. Kishimoto, A., Müller, M.: Df-pn in Go: An application to the one-eye problem. In: van den Herik, H.J., Iida, H., Heinz, E.A. (eds.) Advances in Computer Games, USA,, vol. 10, pp. 125–141. Kluwer Academic Publishers, Boston (2003)
15. Kishimoto, A., Müller, M.: A solution to the GHI problem for depth-first proof-number search. Information Sciences 175(4), 296–314 (2005)
16. Knuth, D.E., Moore, R.W.: An analysis of alpha-beta pruning. Artificial Intelligence 6(4), 293–326 (1975)
17. McAllester, D.A.: Conspiracy numbers for min-max search. Artificial Intelligence 35(1), 278–310 (1988)
18. Nagai, A.: A new AND/OR tree search algorithm using proof number and disproof number. In: Proceedings of Complex Games Lab Workshop, pp. 40–45. ETL, Tsukuba, Japan (1998)
19. Nagai, A.: A new depth-first-search algorithm for AND/OR trees. Master's thesis, The University of Tokyo, Tokyo, Japan (1999)
20. Nagai, A.: Df-pn Algorithm for Searching AND/OR Trees and its Applications. PhD thesis, The University of Tokyo, Tokyo, Japan (2002)
21. Nagai, A., Imai, H.: Application of df-pn+ to Othello endgames. In: Proceedings of Game Programming Workshop in Japan 1999, Hakone, Japan, pp. 16–23 (1999)
22. Nalimov, E.V., Haworth, G.M.C., Heinz, E.A.: Space-efficient indexing of chess endgame tables. ICGA Journal 23(3), 148–162 (2000)
23. Palay, A.J.: Searching with Probabilities. PhD thesis, Carnegie Mellon University, Pittsburgh, PA, USA, 1983. Also published by Pitman, Boston (1985)
24. Pawlewicz, J., Lew., Ł.: Improving depth-first pn-search: 1 + ϵ trick. In: van den Herik, H.J., Ciancarini, P., Donkers, H.H.L.M(J.) (eds.) CG 2006. LNCS, vol. 4630, pp. 160–171. Springer, Heidelberg (2007)
25. Plaat, A., Schaeffer, J., Pijls, W., de Bruin, A.: Best-first fixed-depth minimax algorithms. Artificial Intelligence 87(2), 255–293 (1996)
26. Sackson, S.: A Gamut of Games. In: Random House, New York, NY, USA (1969)
27. Saito, J.-T., Chaslot, G., Uiterwijk, J.W.H.M., van den Herik, H.J.: Monte-carlo proof-number search for computer Go. In: van den Herik, H.J., Ciancarini, P., Donkers, H.H.L.M(J.) (eds.) CG 2006. LNCS, vol. 4630, pp. 50–61. Springer, Heidelberg (2007)
28. Sakuta, M.: Deterministic Solving of Problems with Uncertainty. PhD thesis, Shizuoka University, Hamamatsu, Japan (2001)
29. Sakuta, M., Iida, H.: The performance of PN*, PDS and PN search on 6×6 Othello and Tsume-Shogi. In: van den Herik, H.J., Monien, B. (eds.) Advances in Computer Games, Universiteit Maastricht, Maastricht, The Netherlands, vol. 9, pp. 203–222 (2001)
30. Schaeffer, J.: Conspiracy numbers. Artificial Intelligence 43(1), 67–84 (1990)
31. Schaeffer, J.: Game over: Black to play and draw in Checkers. ICGA Journal 30(4), 187–197 (2007)
32. Schaeffer, J., Burch, N., Björnsson, Y., Kishimoto, A., Müller, M., Lake, R., Lu, P., Sutphen, S.: Checkers is solved. Science 317(5844), 1518–1522 (2007)
33. Schaeffer, J., Lake, R.: Solving the game of Checkers. In: Nowakowski, R.J. (ed.) Games of No Chance, pp. 119–133. Cambridge University Press, Cambridge (1996)
34. Seo, M., Iida, H., Uiterwijk, J.W.H.M.: The PN*-search algorithm: Application to Tsume-Shogi. Artificial Intelligence 129(1-2), 253–277 (2001)
35. Winands, M.H.M., Uiterwijk, J.W.H.M., van den Herik, H.J.: PDS-PN: A new proof-number search algorithm: Application to Lines of Action. In: Schaeffer, J., Müller, M., Björnsson, Y. (eds.) CG 2002. LNCS, vol. 2883, pp. 170–185. Springer, Heidelberg (2003)

36. Winands, M.H.M., Uiterwijk, J.W.H.M., van den Herik, H.J.: An effective two-level proof-number search algorithm. Theoretical Computer Science 313(3), 511–525 (2004)

37. Winands, M.H.M., van den Herik, H.J., Uiterwijk, J.W.H.M.: An evaluation function for Lines of Action. In: van den Herik, H.J., Iida, H., Heinz, E.A. (eds.) Advances in Computer Games, USA,, vol. 10, pp. 249–260. Kluwer Academic Publishers, Boston (2003)

Appendix: Pseudo Code for PN Search

Below we give the pseudo code for PN search, which was discussed in Subsection 6.2.1. For ease of comparison we use similar pseudo code as given in Breuker [5]. PN(root, maxnodes) is the main procedure of the algorithm. The procedure evaluate(node) evaluates a position, and assigns one of the following three values to a node: FALSE, TRUE, and UNKNOWN. The proof and disproof numbers of a node are initialised by setProofAndDisproofNumbers(node). The function selectMostProvingNode(node) finds the most-proving node. Expanding the most-proving node is done by expandNode(node). After the expansion of the most-proving node, the new information is backed up by updateAncestors(node, root). The function countNodes() gives the number of nodes currently stored in memory.

```
//The PN-search algorithm
PN(root, maxnodes){

    evaluate(root);
    setProofAndDisproofNumbers(root);

    while(root.proof != 0 && root.disproof != 0
          && countNodes() <= maxnodes){
      //Second Part of the algorithm
      mostProvingNode = selectMostProvingNode(currentNode);
      expandNode(mostProvingNode);
      currentNode = updateAncestors(mostProvingNode, root);
    }
}


//Calculating proof and disproof numbers
setProofAndDisproofNumbers(node){
    if(node.expanded) //Internal node;
      if(node.type == AND_NODE){
        node.proof = 0;
        node.disproof = INFINITY;
        for(each child n){
          node.proof = node.proof + n.proof;
          if(n.disproof < node.disproof)
            node.disproof = n.disproof;
        }
      }
      else{ //OR node
```

```
      node.proof = ProofNode.INFINITY;
      node.disproof = 0;
      for(each child n){
        node.disproof = node.disproof + n.disproof;
        if(n.proof < node.proof)
          node.proof = n.proof;
      }
    }
  else //Leaf
    switch(node.value){
      case FALSE:
          node.proof = INFINITY;
          node.disproof = 0;
      case TRUE:
          node.proof = 0;
          node.disproof = INFINITY;
      case UNKNOWN:
          node.proof = 1;
          node.disproof = 1;
    }
}

//Select the most-proving node
SelectMostProvingNode(node){
  while(node.expanded){
    n = node.children;

    if(node.type == OR_NODE) //OR Node
      while(n.proof != node.proof)
        n = n.sibling;
    else //AND Node
      while(n.disproof != node.disproof)
        n = n.sibling;

    node = n;
  }
  return node;
}

//Expand node
expandNode(node){
    generateAllChildren(node);
    for(each child n){
        evaluate(n);
        setProofAndDisproofNumbers(n);
        //Addition to original code
        if((node.type == OR_NODE && n.proof == 0) ||
           (node.type == AND_NODE && n.disproof == 0))
             break;
    }
```

```
      node.expanded = true;
}

   //Update ancestors
   updateAncestors(node, root){

      do{
         oldProof = node.proof;
         oldDisProof = node.disproof;

         setProofAndDisproofNumbers(node);
         //No change on the path
         if(node.proof == oldProof &&
            node.disproof == oldDisProof)
           return node;
         //Delete (dis)proved trees
         if(node.proof == 0 || node.disproof == 0)
           node.deleteSubtree();

         if(node == root)
           return node;

         node = node.parent;
      }while(true)
}
```

Part III

Optimization

7

Improving the Exploration Ability of Ant-Based Algorithms

Alice Ralickas Malisia

Department of Systems Design Engineering, Univeristy of Waterloo, Canada
amalisia@alumni.uwaterloo.ca

Summary. The chapter discusses the application of Opposition-Based Optimization (OBO) to ant algorithms. Ant Colony Optimization (ACO) is a powerful optimization technique that has been used to solve many complex problems. Despite its successes, ACO is not a perfect algorithm: it can remain trapped in local optima, miss a portion of the solution space or, in some cases, it can be slow to converge. Thus, we were motivated to improve the accuracy and convergence of the current algorithm by extending it with the concept of OBO. In the case of ACO, the application of opposition can be challenging because ACO usually optimizes using a graph representation of problems, where the opposite of solutions and partial components of the solutions are not clearly defined.

The chapter presents two types of opposition-based extensions to the ant algorithm. The first type, called Opposite Pheromone per Node (OPN), involves a modification to the construction phase of the algorithm which affects the decisions of the ants by altering the pheromone values used in the decision. Basically, there is an opposite rate that determines the frequency at which opposite pheromone will be used in the construction step. The second method, Opposite Pheromone Update (OPU), involves an extension to the update phase of the algorithm that performs additional updates to the pheromone content of opposite decisions. The opposition-based approaches were tested using the Travelling Salesman Problem (TSP) and the Grid World Problem (GWP).

Overall, the application of some fundamental opposition concepts led to encouraging results in the TSP and the GWP. OPN led to some accuracy improvements and OPU demonstrated significant speed-ups. However, further work is necessary to fully evaluate the benefits of opposition. Theoretical work involving the application of opposition to graphs is necessary, specifically in establishing the 'opposite graph'.

7.1 Introduction

Ant Colony Optimization (ACO) is classified under the general class of algorithms known as Swarm Intelligence (SI). SI reflects the emergence of collective intelligence from a swarm of simple agents. It is generally defined as a structured collection of interacting organisms which cooperate to achieve a greater goal [1, 14]. It is possible to have genetic cooperation, as it is the case with genetic algorithms, but in SI, it is more of a social interaction. The framework is based on the repeated sampling of solutions to the problem at hand, where each member of the population provides a potential solution. In the case of ACO, the algorithm mimics the social interaction of ants, thus,

H.R. Tizhoosh, M. Ventresca (Eds.): Oppos. Concepts in Comp. Intel., SCI 155, pp. 121–142, 2008.

the population is a colony of ants. Social behaviour increases the ability of individuals to adapt, as they can cooperate and learn from each other. The main idea of SI algorithms is that organisms of a swarm behave in a distributed manner while exchanging information directly or indirectly.

In addition, ACO is a population-based metaheuristic. Metaheuristics are procedures that use heuristics to seek a near-optimal solution with reasonable computation time [8, 9]. The general idea behind a metaheuristic is to create a balance between local improvements and a high-level strategy. They optimize problems through guided search of the solution space [9, 29]. In brief, metaheuristics seek optimality while attempting to reduce computation time.

ACO is a powerful technique that has been used to solve many complex optimization problems, such as the travelling salesman problem [6, 7, 10], quadratic assignment problem [18], vehicle routing [2, 3, 11], and many others [9]. Given the complexity of these problems and their real-world applications, any improvement to the ant algorithm performance is encouraged. Specifically, an increase in accuracy and faster convergence are strongly welcomed. Ant algorithms can become trapped in a local optimum, miss a portion of the solution space or simply be slow to converge because the ants might take a long time to discover and learn to use the best paths. Thus, it is interesting to study and develop more complex behaviour for ant algorithms.

We attempt to improve the accuracy and convergence of ACO by extending the current algorithm with the concept of opposition-based optimization (OBO), which is a subclass opposition-based computing. This chapter provides a general overview how OBO can be applied to ACO. We will discuss two types of opposition-based extensions to the ant algorithm. The first type involves a modification to the construction phase of the algorithm which affects the decisions of the ants by altering the pheromone values used in the decision. This modification is called Opposite Pheromone per Node. The second type is an extension to the update phase of the algorithm that performs additional updates to the pheromone content of opposite decisions. The second extension is called Opposite Pheromone Update. The opposition-based approaches were tested on two different problems: the Travelling Salesman Problem (TSP) and the Grid World Problem (GWP).

The remaining of this chapter is organized as follows. Section 7.2 provides background information, including an overview of ant colony optimization and a review of the work that has been conducted to improve the performance of ant algorithms. Section 7.3 presents the two opposition-based approaches. Section 7.4 includes the experimental results of the opposition-based ant algorithms for the TSP and the GWP. Conclusions and future work are discussed in Sect. 7.5.

7.2 Background Information

This section will describe ACO, followed by a discussion on some of the pitfalls of the algorithm that we address using opposition. It also provides an overview of relevant work that has been conducted to improve the performance of the ant algorithm.

7.2.1 Ant Colony Optimization

The ant algorithm was introduced by M. Dorigo in 1992 [5]. It was developed to solve complex discrete combinatorial optimization problems. The first ACO algorithm was the Ant System (AS) [6], which was designed to solve the TSP.

Since the introduction of ACO, researchers have developed multiple versions to improve the performance of the AS. The extensions tend to focus more on the best solutions found to attempt to guide the ants search more effectively. The Ant Colony System (ACS) is a popular revised version of ACO [9] that achieved considerable accuracy improvements [7, 9]. Other extensions include the Best-Worst Ant System [4], the Max-Min Ant System [30], Ant-Q (extends ACO with reinforcement learning), AntNet (dynamic version of the algorithm designed for the vehicle routing problem), and the ACS combined with local search [9].

The ACO algorithm is inspired from the natural behaviour of trail laying and following by ants [1, 9]. When exploring a region, ants are able to find the shortest path between their nest and a food source. This is possible because the ants communicate with each other indirectly via pheromone deposits they leave behind as they travel. The pheromone deposited by one ant influences the selection of the path by the other ants. A high pheromone concentration increases the probability that the path will be selected. The pheromone deposits work as a form of positive feedback, reinforcing good path choices and guiding the ants to better paths.

Ant System

When applied to an optimization problem, the ACO metaheuristic usually involves solution construction on a graph. The solutions are a path along the graph. The AS algorithm is summarized in Alg. 1. In line 3, the ants are distributed randomly among the nodes. Then, they move between nodes, sequentially adding edges to their current path until they have visited all nodes. The selection of an edge depends on the pheromone content of the edge, represented by values in a $n \times n$ matrix where n is the number of nodes, and the value of the heuristic function of each edge. At each step of construction, ant k selects the next node using a probabilistic action choice rule which dictates the probability with which the ant will choose to go from current node i to next node j (see line 6):

$$p_{ij}^k = \frac{[\tau_{ij}]^\alpha [\eta_{ij}]^\beta}{\sum_{l \in N_i^k} [\tau_{il}]^\alpha [\eta_{il}]^\beta} \quad \text{if } j \in N_i^k, \tag{7.1}$$

where τ_{ij} represents the pheromone content on the edge. Node j is included in N_i^k, the neighbourhood for ant k given its current location i. The neighbourhood only includes nodes that have not been visited by ant k and are connected to node i. The parameter η_{ij} represents the heuristic information. The heuristic value of an edge is a measure of the cost of extending the current partial solution with that edge (typically the inverse of the weight of the edge). The constants α and β represent the influence of pheromone content and heuristic information respectively. The stochastic component of the algorithm, namely probabilistically selecting a component, leads to exploration of a higher number of solutions.

Algorithm 1. Pseudocode of the Ant System

1: Initialize pheromone matrix ($\tau = \tau_o$)
2: **while** Termination condition is not satisfied **do**
3:     Place m ants on random nodes
4:     **for** $k = 1$ to m **do**
5:         **if** Solution construction for ant k is NOT complete **then**
6:             Pick next node j (see (7.1))
7:         **end if**
8:     **end for**
9:     Update pheromone matrix (evaporation and trail update) using (7.2) and (7.3)
10: **end while**

When all the ants have completed their paths the pheromone content is updated (see line 9). First, the pheromone content of the arcs, τ_{ij}, is evaporated based on the evaporation rate, ρ, following the relation

$$\tau_{ij}^{new} = (1 - \rho)\tau_{ij}^{current} \quad 0 < \rho < 1. \tag{7.2}$$

After the evaporation step, the solution of each ant is evaluated and pheromone is deposited on the ant's path relative to the quality of its solution. The ants deposit pheromone on the arcs they visited as follows:

$$\tau_{ij}^{new} = \tau_{ij}^{current} + \sum_{k=1}^{m} \Delta\tau_{ij}^{k}, \tag{7.3}$$

where $\Delta\tau_{ij}^{k}$ is the amount of pheromone ant k contributes to the arc going from node i to node j and m is the total number of ants. The additional pheromone is based on the overall quality of the total path and is defined by

$$\Delta\tau_{ij}^{k} = \begin{cases} \frac{1}{C_k} & \text{if arc is in the path of ant } k, \\ 0 & \text{otherwise,} \end{cases} \tag{7.4}$$

where C_k is the total cost of the solution for ant k. All arcs of one path will receive the same amount of pheromone (i.e. each ant deposits a constant amount of pheromone per edge).

Ant Colony System

Another commonly used version of ACO is the ACS. This version differs from the AS algorithm in three aspects [7, 9]: 1) different selection rule for path construction, 2) trail update only occurs for the best-so-far solution, and 3) local pheromone removal occurs each time an ant visits a node.

The general steps of the ACS algorithm are summarized in Alg. 2. When ants construct their paths in ACS, they use a selection rule that has a strong emphasis on exploitation of previous experience. An ant k located on node i chooses the next node j using a pseudorandom proportional rule described by

$$j = \begin{cases} \underset{l \in N_i^k}{\operatorname{argmax}} \left\{ \tau_{il} [\eta_{il}]^\beta \right\} & \text{if } q < q_o, \\ J & \text{otherwise.} \end{cases} \tag{7.5}$$

The parameter q is a uniform random number between 0 and 1 and q_o is the probability that an ant will use learned knowledge. If $q < q_o$, the ant will select the node with the highest product of pheromone content and heuristic function value. Otherwise, it will use J, which is the node selected by the probabilistic action rule used in the AS (see (7.1)). This pseudorandom rule is a greedy selection approach that will tend to favour edges with a higher heuristic value and a high pheromone content.

The ACS has a pheromone update approach that exploits the best solutions found. The solutions found by the ants during the iteration are compared to the best solution found so far (best-so-far), and if one of the solutions is better, then the best-so-far solution is revised (see line 10). In line 11 the evaporation and deposit of pheromone is only applied to the arcs contained in the current best solution. The update is implemented by

$$\tau_{ij}^{new} = (1 - \rho)\tau_{ij}^{current} + \rho(\Delta\tau_{ij}^{bs}) \quad \forall (i,j) \in T^{bs}, \tag{7.6}$$

where $\Delta\tau_{ij}^{bs}$ is additional pheromone, ρ is the global evaporation rate, and T^{bs} is the best-so-far path. Sometimes the best-iteration path is used for smaller problems [9]. The additional pheromone is calculated using the cost of the best-so-far path.

The ACS includes a local pheromone update to reduce emphasis on exploitation of existing solutions (see line 7). Immediately after an ant adds an arc to its current path the amount of pheromone on the arc is decreased as follows:

$$\tau_{ij}^{new} = (1 - \xi)\tau_{ij}^{current} + \xi\tau_o \quad 0 < \xi < 1, \tag{7.7}$$

where τ_o is the initial amount of pheromone. The parameter ξ is the local evaporation rate, which controls the amount of pheromone that is removed. In the case of the TSP, research indicates that for the ACS, this value should be set to $\frac{1}{nC_{nn}}$, where n represents the number of cities and C_{nn} is cost of the nearest neighbour solution [9]. This local update works to counterbalance the greedy construction rule by reducing the pheromone on the selected edge, thus making it less desirable to the next ant.

7.2.2   Challenges and Drawbacks

Despite being a powerful algorithm, ACO can benefit from performance improvements. Like other optimization techniques, ACO can remain trapped in a local optimum, miss a portion of the solution space or, in some cases, it can be slow to converge. ACO has many applications and deals with complex optimization problems, such as the travelling salesman problem [6, 7, 10], the quadratic assignment problem [18], vehicle routing [2, 3, 11] and many more [9]. Thus, any increase in speed of convergence and accuracy is beneficial.

Algorithm 2. Ant Colony System

1: Initialize pheromone matrix ($\tau = \tau_o$)
2: **while** Termination condition is not satisfied **do**
3:    Place m ants on random nodes
4:    **for** $k = 1$ to m **do**
5:       **if** Solution construction for ant k is NOT complete **then**
6:          Pick next node j (see (7.5))
7:          Apply local pheromone update using (7.7)
8:       **end if**
9:    **end for**
10:   If necessary, revise the overall best (best-so-far) solution
11:   Update pheromone matrix (trail update) according to (7.6)
12: **end while**

Given the fundamental structure of ACO, which involves reinforcement of good solutions, the algorithm can sometimes remain trapped in local optima resulting in reduced accuracy. This situation can occur when a certain component is very desirable on its own, but leads to a sub-optimal solution when combined with other components. Consequently, modifications that can help the algorithm move away from a local optimum will likely lead to an increase in accuracy, are also welcome. Moreover, by moving away from a local optimum, one will increase exploration, which may also lead to improved solution quality.

Another interesting aspect of ACO is that it generates new solutions in every iteration using pheromone information. It is the progress and quality of the pheromone information that will affect the quality of the solutions. ACO can be slow to converge because sometimes it will take quite a number of iterations before the pheromone content of edges start having a strong impact on the ants decisions. Thus, one may improve the ant algorithm by making additional pheromone updates, which will help achieve accurate pheromone information faster.

7.2.3   Related Works

Since the introduction of ACO, researchers have developed multiple versions to improve the performance of the algorithm. The Ant Colony System (ACS) is a popular extension of the original ant algorithm that was developed to improve upon the performance of the AS [9]. The ACS has a greedy selection rule, but provides regular pheromone reduction as a measure to decrease desirability of edges once they are visited [7]. This attempts to prevent all the ants in the colony from generating the same solution. Another successful version of the ant algorithm is the Max-Min Ant System (MMAS) [9, 30]. The MMAS strongly exploits the best tours found, but also limits the range of pheromone content values and initializes the pheromone contents at the upper limit. These modifications led to performance improvements.

In addition, work has been conducted to establish more complex pheromone mechanisms, such as multiple pheromone matrices, and complex pheromone updates. These modifications were implemented so ant algorithms could solve more complex problems and to improve the performance of the ACS. For instance, one particular variant of the ant algorithm, known as the Best-Worst Ant System (BWAS) [4], substracts pheromone content based on the results of the worst ant of the colony. The BWAS also uses a form of pheromone mutation based on concepts from evolutionary computation. To solve bi-criterion optimization problems, Iredi et al. proposed a version of the ACS where two different pheromone trail matrices and two heuristic functions are considered simultaneously [13]. Randall and Montgomery proposed the Accumulated Experience Ant Colony as a method to determine the effect of each component on the overall solution quality [24]. In their approach, the pheromone and heuristic values of an edge are weighted.

Schoonderwoerd and his colleagues were one of the first to elude to the concept of an 'anti-pheromone', where ants would decrease pheromone contents rather than reinforce them [26]. Montgomery and Randall developed three methods based on the concept of anti-pheromone as an attempt to capture complex pheromone behaviour [19]. In the first method, the pheromone content of the elements composing the worst solutions is reduced. Their second alternative combines a pheromone content for the best solution and pheromone content for the worst solution. The ants select edges based on a weighted combination of pheromone and anti-pheromone and the heuristic. Finally, their third approach involves the use of a small number of *explorer-ants* that have a reversed preference for the pheromone. On average, their methods produced better solutions on the smaller TSP problems (less than 200 cities).

Given these existing extensions, their results, and the potential for performance improvement, there was motivation to investigate the application of oppositional concepts to ant colony optimization. Moreover, opposition-based computing has already been successfully applied to reinforcement learning [27, 28, 31], evolutionary algorithms [20, 21, 22, 23] and neural networks [33, 34]. Opposition can potentially lead to a new way of developing more complex pheromone and path selection behaviour for ACO. The use of opposition is very interesting because it provides a structured way to investigate how to modify the ant algorithm.

7.3   Opposition and Ant Colony Optimization

The main idea of opposition-based optimization (OBO) is that by considering "opposites" one can increase the coverage of the solution space leading to increased accuracy and/or faster convergence [17, 20, 21, 22, 23, 31] (see also Chapter 2). OBO provides a general strategy that can be tailored to the technique of interest. Increasing the speed of convergence of the ant algorithm can be seen as a type-I opposition problem (Definitions 1 and 2 on page 14 in Chapter 2), because we can attempt to reach a better solution faster by using opposite guesses. It could also be addressed as a type-II opposition problem (Definitions 5 and 6 on page 17 in Chapter 2), because we can attempt to look for solutions that have opposite fitness, which would guarantee that we are moving to a better location. However, it is not obvious how to determine the solution that has the opposite fitness. Thus, we can use the type-I opposition as an approximation of type-II. An increase in accuracy can be achieved by using opposition to move away from a local optimum. Remaining trapped in a local optimum can be characterized as a type-I opposition problem since picking a guess that is opposite to the current guess will likely lead to a new area of the solution space. The following sections will dicuss the different ways that opposition can be applied to ACO and it will introduce two main opposition-based extensions to the ant algorithm: Opposition per Node and Opposite Pheromone Update.

7.3.1   Motivating Principles

In the case of ACO, the application of opposition is not as straightforward as mapping between an estimate and its opposite. Due to the graph structure of the problems, the opposite of a solution or of the partial components of the solutions are not clearly defined. Given the combinatorial aspect of solutions on a graph, even if only one element of a solution is changed, it leads to a whole set of new solutions. Moreover, simply taking the opposite of every component of the solution might not necessarily lead to a plausible solution because it might not lead to a connected path. It is not clear how one can generate an opposite solution, mainly because of the combinatorial aspect of the applications associated with ACO. To fit in the OBO scheme, the term *opposite* must be related to ant algorithms.

The concept of opposition serves as a starting point for the proposed extensions. The main idea is to think of opposition as a way of increasing the coverage of the solution space, which may lead to greater accuracy and faster convergence.

Ant algorithms do not work by modifying the existing solutions at each iteration; instead, new solutions are created based on the pheromone matrix. It is the pheromone matrix that changes as the algorithm progresses. In algorithms that work with complete solutions, such as genetic algorithms, one can generate an opposite candidate solution and replace the current candidate solution. Then, the algorithm proceeds with the opposite candidate solution. In contrast, in the ant algorithm, even if the opposite solution is generated, one needs to find a way to alter the pheromone content since that is what affects solution creation. To move in the solution space, the algorithm has to move in the pheromone space. Thus, instead of looking at a solution candidate and its opposite, the concept of opposition has to involve the pheromone matrix.

However, it is not easy to define an opposite pheromone matrix because a pheromone matrix is not a point in the solution space. The pheromone matrix is indirectly related to the final solutions. Given the probabilistic nature of the path selection in ACO, a particular pheromone matrix can lead to an array of solutions. This leads to an array of paths and, hence, there is no one-to-one relationship between a particular matrix and a single path. Thus, instead of focusing on an opposite pheromone matrix, another idea is to find a way to use opposition to move in the pheromone matrix solution space.

Consequently, it was determined that opposition could be applied to directly or indirectly affect the pheromone matrix. There are many different ways in which this can be achieved. For example, instead of initializing the pheromone content of all the edges to the same value, one could use opposition to determine a better initial value. However, while there can be many different places where opposition can be applied, we decided to concentrate on two main parts of the ACO algorithm: 1) the construction phase and/or 2) the update phase. These two phases were selected because, unlike initialization, they were present in every iteration of the algorithm. Thus, modifications and extensions to these phases will likely have a greater impact on the performance of the algorithm.

The construction phase can be modified by affecting the ant's decision. This can be done by altering the parameters used by the decision, namely the pheromone content. The modification to the update phase involves changing the way the pheromone is updated. It can be done by making additional updates using other ants. A form of this idea was implemented in the Best-Worst Ant System [4], which uses the worst-ant to remove pheromone. However, there are other ways to affect the update phase. One way is to use opposite components of the current solutions without necessarily creating an opposite solution, which can be seen as identifying the opposite actions of the ants.

With these oppositional modifications, the algorithm is able to move to a new region of pheromone space. By changing the decision rule of the ants or changing the pheromone content used in the decision, one simulates the creation of another pheromone matrix without directly changing the current matrix, which is useful to help the algorithm escape a local optimum. Moreover, this leads to exploration of a higher number of solutions because components with lower pheromone content can be selected. In contrast, in the case of opposition-based pheromone updates, the algorithm is actually moving to a new pheromone matrix, which may eventually lead to an area closer to the optimal solution.

The discussed modifications provide a general framework as to how opposition can extend ACO. The ideas were used to design specific opposition-based algorithms which were tested with travelling salesman and the grid world problems. The next section will describe two successful extensions to the ant algorithm: Opposite Pheromone per Node and Opposite Pheromone Update. Detailed descriptions of these extensions, and of other less successful implementations can be found in [16].

7.3.2 Opposite Pheromone Per Node

The Opposite Pheromone per Node (OPN) is a direct modification of the pheromone value used by the ants to make their selection. It was designed to help the ants try different paths, and addresses the problem where the ant algorithm remains trapped in a local optimum. OPN attempts to move the ants out of their current paths by altering the pheromone they use in their decisions.

Algorithm 3 describes the OPN extension to ACS. The local pheromone update in line 12 and the best-trail update (line 16 are identical to the ones of the normal ACS algorithm. During the construction phase from line 4 to 14, the ants will move from node to node creating a solution until they have visited all nodes, achieved a specific number of steps or reached a goal.

Every time an ant k has to select the next node, the pheromone content used for its decision will depend on the value of a random number λ and the *opposite-rate*, $\check{\lambda}$. The *opposite-rate*, λ_o, determines the rate at which opposite pheromone will be used in the construction step of the algorithm. If $\lambda < \check{\lambda}$ (line 6), then the algorithm calculates the opposite pheromone content for the edges in line 7 and the ant will use the opposite pheromone content, $\check{\tau}$, to pick its next city (line 8). Otherwise, the ant will simply select the next city in line 10 using the original pheromone content.

The opposite pheromone content, $\check{\tau}_{ij}$, for the edge connecting the current node i to an available node j is calculated as follows:

$$\check{\tau}_{ij} = \tau_{min} + \tau_{max} - \tau_{ij}. \tag{7.8}$$

Algorithm 3. Opposite Pheromone per Node Algorithm

1: Initialize pheromone matrix ($\tau = \tau_o$)
2: **while** Termination condition is NOT satisfied **do**
3:     Place m ants on random nodes
4:     **for** $k = 1$ to m **do**
5:         **if** Solution construction for ant k is NOT complete **then**
6:             **if** $\lambda < \check{\lambda}$ **then**
7:                 Calculate opposite pheromone values, $\check{\tau} = \tau_{min} + \tau_{max} - \tau$
8:                 Pick next node j using pseudorandom rule (see (7.5)) with $\check{\tau}$
9:             **else**
10:                Pick next node j using pseudorandom rule (see (7.5)) with τ
11:            **end if**
12:            Apply local pheromone update using (7.7)
13:        **end if**
14:    **end for**
15:    If necessary, revise the overall best (best-so-far) solution
16:    Update pheromone matrix (trail update) according to (7.6)
17: **end while**

The parameters τ_{min} and τ_{max} represent the minimum and maximum available pheromone contents, respectively. In the case of the ACS algorithm, τ_{min} can be the initial pheromone deposit and τ_{max} can be the inverse of the length of the best-so-far path, L_{bs}. These values can be used to determine the opposite pheromone content because, given the pheromone update equations of ACS (see (7.6) and (7.7)), the pheromone content is bounded by the initial pheromone deposit and the global optimal value [9]. With the AS, the maximum and minimum pheromone contents are not bounded. Thus, the opposite can be calculated using the maximum and minimum pheromone contents of the available edges or of the entire pheromone matrix.

7.3.3   Opposite Pheromone Update

The Opposite Pheromone Update (OPU) extends the pheromone update phase of the ant algorithm. This extension focuses more on the convergence issues existing in the ant algorithm. By adding or removing pheromone from opposite edges, the algorithm will modify the pheromone content faster than the ants normally would, which will speed up their learning. OPU performs additional updates using opposition information. When an ant completes its path, pheromone is added to every decision along the path, or in the case of ACS the edges of the best-so-far path. With the OPU extension, pheromone can also be added or removed from opposite edges.

OPU has a standard framework, where opposite edges receive additional updates. However, the specific way to define an opposite edge and how to update the pheromone varies depeding on 1) the version of the ant algorithm that is used, and 2) the problem that is being solved. In the case of the ACS, where pheromone is only added to the best-so-far path, the pheromone levels will not be very high on the other edges. Thus, OPU might work better if pheromone is added to the opposite edges. In constrast, in

---

Algorithm 4. Opposite Pheromone Update Algorithm

1: Initialize pheromone matrix ($\tau = \tau_o$)
2: **while** Termination condition is NOT satisfied **do**
3:    Place m ants on random nodes
4:    **for** $k = 1$ to m **do**
5:       **if** Solution construction for ant k is NOT complete **then**
6:          Pick next node j using pseudorandom rule (see (7.5)) with τ
7:       **end if**
8:    **end for**
9:    Update pheromone matrix (trail update) according to (7.3)
10:   **if** $\check{\lambda} < \check{\lambda}_o$ **then**
11:      Calculate opposition-rating $\breve{o}$ for all edges using (7.9)
12:      Apply opposite pheromone update (see 7.10)
13:   **end if**
14: **end while**

---

the case of the AS, it is probably best to remove pheromone because the AS algorithm deposits pheromone on all generated paths.

Algorithm 4 describes the OPU extension to the AS. The initialization, the termination conditions, and the pheromone trails update (line 9) are identical to the ones of the normal AS algorithm. Also, like in the AS framework, the ants will construct their solutions (line 4 to 8) until they have a complete solution (i.e. visited all nodes, visited a specific number of nodes or found a goal). The selection of the next nodes in line 6 also follows the AS framework.

After the pheromone trails update (line 9), the OPU algorithm will potentially perform an opposite update. The rate at which the opposite update occurs depends on the value of a random number λ and the *opposite-rate*, $\check{\lambda}$. If $\lambda < \check{\lambda}$ (line 10), the *opposition-rating* is calculated in line 11 and the OPU algorithm performs the opposite update in line 12. If $\check{\lambda} = 1$, the opposite update is done in every iteration.

The *opposition-rating* in line 11 of the OPU algorithm is a way to evaluate the degree of opposition of other edges in the graph relative to the current solution. In the case of AS, there are multiple current solutions: the solution found by each ant. In the case of ACS, the current solution is the best-so-far solution. For a given current solution, at every node of that solution, one outgoing edge will receive the trail pheromone update. Then, the *opposition-rating*, $\breve{o}$, is calculated for all the other outgoing edges relative to the winning edge. This rating is used to determine the amount of pheromone to add to the other edges.

There are different ways to evaluate the degree of opposition of an edge. For example, $\breve{o}$ can be calculated using the heuristic function values:

$$\breve{o}_{ij} = \frac{\left| \eta_{ij} - \eta_i^{bs} \right|}{\eta^{\max} - \eta^{\min}}, \tag{7.9}$$

where η_{ij} represents the heuristic function value for the edge going from node i to node j, and η_i^{bs} is the value for the edge outgoing from node i included present in the best path. The values $\eta^{\max}$ and $\eta^{\min}$ are the maximum and minimum heuristic values of the graph. They are used to normalize the *opposition-rating* of the edges. This calculation method was used in OPU experiments for the TSP which can be found in [16].

In problems where each edge has a clearly defined opposite, the *opposition-rating* is straightforward: $\breve{o} = 1$ for the opposite edge and $\breve{o} = 0$ for the other outgoing edges. For example, in a path finding problem where the ants must pick a direction to take at every node, the selected direction (or edge) and opposite-direction pairs are clear. If, at node i, the ant chooses to move "up", then for that particular node, the "down" direction is the opposite edge. Thus, the "down" direction will have an *opposition-rating* of 1, and the "left" and "right" directions will have a rating of 0.

Once the *opposition-rating* is determined in line 11, the pheromone content of the "opposite-rated" edges is updated in line 12. The additional pheromone can be added or removed depending on the type of ant algorithm. The opposite update is based on the equation used in the regular pheromone trail update. In the case of AS, the opposite update equation involves pheromone removal:

$$\tau_{ij}^{new} = \tau_{ij}^{current} - \sum_{k=1}^{m} \breve{o}_{ij} \Delta \tau^k, \tag{7.10}$$

where $\breve{o}_{ij}$ is the *opposition-rating* for the edge, $\tau_{ij}^{current}$ is the current pheromone level on the edge going from node i to node j, $\Delta \tau^k$ is the pheromone added to the path found by ant k (see 7.4). This equation can be modified depending on the ant algorithm used and the application. In some cases, the opposite pheromone value can be divided by a weight to reduce its impact and, instead of a removal, the opposite pheromone can be added. In OPU experiments for the TSP [16], the opposite pheromone was added and it was divided by a weight.

Finally, in the OPU implementation for the AS, where the opposite update involves a removal of pheromone, the opposite update can replace evaporation. Evaporation is used as a way to "forget" bad decisions and thus, removing pheromone from opposite edges is achieving the same goal as evaporation. Keeping both the evaporation and the opposite pheromone update is not necessary, especially if the amount of pheromone being removed is as high as $\Delta \tau^k$ (the pheromone added to edges of the path of ant k). In summary, OPU provides an additional opposite update that can be interpreted as an intelligent evaporation.

7.4 Experimental Evidence

This section will outline some of the experimental results of applying opposition to ACO. The OPN algorithm was tested with the Travelling Salesman Problem (TSP) and the OPU algorithm was tested using the Grid World Problem (GWP). The Wilcoxon rank sum (or Mann-Whitney) test was used to compare the results [12]. If the result of the test comparing the two samples is significant ($p < 0.05$), one can accept the alternative hypothesis that there is a difference between the median of the two samples. Other experimental investigations can be found in [16].

7.4.1 OPN Experiment

The OPN algorithm was compared to the ACS algorithm on 9 symmetric TSP instances, namely att48, eil51, eil76, kroA100, pr124, ch150, d198, lin318 and pcb442 [25]. Table 7.1 provides more details about each instance.

Table 7.1. Overview of the TSP Instances

Instance	#Cities	Optimal Tour
att48	48	10628
eil51	51	426
eil76	76	538
kroA100	100	21282
pr124	124	59030
ch150	150	6528
d198	198	15780
lin318	318	42029
pcb442	442	50778

The TSP is an optimization problem based on the problem faced by a travelling salesman who, given a starting city, wants to take the shortest trip through a set of customer cities, visiting each city only once before returning to the starting point. Mathematically, the TSP involves finding the minimum cost path in a weighted graph, which is an NP-hard problem [15]. A particular TSP instance has a specific number of cities (nodes) and arc weights (typically the distance between the cities).

Experimental setup

The parameters of the ant algorithms were all set to the same values, namely $\beta = 2$, $\rho = 0.1$, $\xi = 0.1$, $m = 10$, and $q_o = 0.9$. These values were selected based on other research done using ACS and TSP [7, 19]. The algorithms completed 100 trials and each trial was terminated after 5000 iterations or if the optimal solution was found. For the OPN algorithm, the *opposite-rate*, λ_o, was set to a fixed rate of 0.0005, 0.001, 0.05, and 0.1.

Experimental results

The accuracy of each algorithm was evaluated in terms of the median final path length, the median accuracy difference with the ACS, the mean and standard deviation of the final path length and the number of times the optimal solution was found. The Wilcoxon rank sum (or Mann-Whitney) test was used to compare the medians of the results [12].

The median accuracy difference between ACS and the OBO algorithms was quantified as follows:

$$\bar{A}_{diff}(\%) = \left(\frac{\bar{A}^{OPN}}{\bar{A}^{ACS}} - 1 \right) \times 100\%. \tag{7.11}$$

where $\bar{A}^{OPN}$ and $\bar{A}^{ACS}$ are the median accuracies of the best path found for the OPN and ACS algorithms. The accuracy of final path found by each algorithm is determined by

$$A = 2 - (\frac{L^{bs}}{L^{opt}}) \times 100\%. \tag{7.12}$$

where L^{bs} is the length of the best path found by the algorithm and L^{opt} is the length of the optimal solution. The accuracy results are reported in Table 7.2.

The results show that the *opposite-rate*, $\check{\lambda}_o$, is an important factor in the success of the OPN strategy. In the smaller instances, $\check{\lambda}_o = 0.1$ provided the best results, but as the number of cities increased the *opposite-rate* had to decrease to achieve good results. A high *opposite-rate* becomes detrimental to a problem with higher number of cities because the use of opposite pheromone becomes too frequent during the path construction. OPN is meant to be a strategy to affect a small number of the decisions in the hopes of helping the ants move from a local optimum. The frequent use of opposite pheromone does not let the ants benefit from their learning.

OPN performed very well in instances with 150 cities and less. Given the appropriate $\check{\lambda}_o$, OPN achieved statistically significant results in all the smaller instances. Moreover, for the instances where optimal solutions were found, the OPN was able to achieve a higher number of optimal solutions. The statistically significant improvements in accuracy ranged from 0.187% to 0.757%. Even if these improvements are below 1%, they are important because the accuracy achieved by the different algorithms is already very high. On problems d198 and lin318, OPN achieved a lower median path length than ACS, but the difference was not statistically significant. In general, the standard deviation from the mean final path length was lower for the OPN algorithm. OPN produced worse results than the ACS for the pcb442 problem. It can be seen that OPN helps improve the accuracy of the ACS, which suggests it is addressing the local optimum trap issue faced by the ant algorithm.

To evaluate the convergence rate of the algorithms, a desired accuracy of 85% was set for the two larger instances (lin318, pcb442) and 95% level of accuracy was set for the other instances. The number of iterations needed to reach the accuracy was used as the convergence measure. The total computational time in seconds is also reported. A speed-up factor, S, was also defined to compare the median number of iterations of ACS relative to the median number of iterations of the OBO algorithm:

$$S = \left(1 - \frac{\bar{n}_I^{OBL}}{\bar{n}_I^{ACS}}\right) \times 100\% \tag{7.13}$$

Table 7.3 summarizes the convergence results for OPN algorithms. Like the accuracy results, the convergence results support the idea that the performance of OPN depends on the *opposite-rate*. The OPN algorithm was able to achieve an increase in convergence rate in the instances with less than 200 cities. The increases ranged from 4.6% to 22.7%. In three of the problem instances, namely kroA100, pr125 and ch150 with speed-up factors of 22.1%, 22.7% and 19.3% respectively, the difference was statistically significant. The standard deviations of the means for the OPN algorithm were generally lower than for the ACS. Typically, the OPN that achieved the best convergence results had the lower standard deviations. OPN did not perform as well for the

Table 7.2. Median ($\bar{A}$), Mean (μ_A), and Standard Deviation (σ_A) of the Accuracy, Accuracy Difference ($\bar{A}_{diff}(\%)$) and Number of Optimal Solutions Found Comparing the AS and OPN for the TSP

Instance	Algorithm	Median	$\bar{A}_{diff}(\%)$	$\mu_A \pm \sigma_A$	#Opt
att48	ACS	10653	–	10682 ± 46	6
	OPN 0.005	10653	0	10687 ± 54	7
	OPN 0.01	10653	0	10674 ± 45	9
	OPN 0.05	10653	0	10670 ± 42	11
	OPN 0.1	10653	0	**10663 ± 36**	**15**
eil51	ACS	428	–	429.3 ± 2.9	7
	OPN 0.005	429.5	-0.354	430.3 ± 3.4	11
	OPN 0.01	428	0	429.6 ± 3.3	9
	OPN 0.05	428	0	428.5 ± 3.5	10
	OPN 0.1	**427** †	**0.236**	428.2 ± **2.3**	**16**
eil76	ACS	548	–	547.7 ± 5.2	1
	OPN 0.005	547	0.189	547.4 ± 5.0	2
	OPN 0.01	548	0	547.2 ± 5.2	5
	OPN 0.05	545 †	0.568	545.4 ± **4.9**	8
	OPN 0.1	**544** †	**0.757**	**545.0** ± 5.1	**11**
kroA100	ACS	21423	–	21530 ± 238	3
	OPN 0.005	21460	-0.177	21542 ± 242	2
	OPN 0.01	21389 †	0.161	21471 ± 235	**14**
	OPN 0.05	**21383** †	**0.187**	**21435 ± 179**	12
	OPN 0.1	21393 †	0.142	21474 ± 219	6
pr124	ACS	59385	–	59475 ± 439	4
	OPN 0.005	59185	0.342	59401 ± **425**	5
	OPN 0.01	59242	0.243	59512 ± 494	6
	OPN 0.05	**59087** †	**0.508**	**59352 ± 425**	**12**
	OPN 0.1	59159	0.385	59431 ± 431	9
ch150	ACS	6641	–	6654 ± 67.8	0
	OPN 0.005	6643	-0.031	6656 ± 74	0
	OPN 0.01	6621 †	0.32	6636 ± 59.7	0
	OPN 0.05	**6601** †	**0.631**	**6612 ± 49.7**	0
	OPN 0.1	6623	0.281	6638 ± 64.1	0
d198	ACS	16093	–	16113 ± 116.9	0
	OPN 0.005	**16076**	**0.11**	**16109** ± 155	0
	OPN 0.01	16105	-0.08	**16109 ± 103**	0
	OPN 0.05	16275 †	-1.18	16275 ± 142	0
	OPN 0.1	16690 †	-3.86	16695 ± 180	0
lin318	ACS	44426	–	44372 ± **495.3**	0
	OPN 0.005	**44230**	**0.496**	44353 ± 519	0
	OPN 0.01	44305	0.305	**44318** ± 519	0
	OPN 0.05	46296 †	-4.72	46402 ± 891	0
	OPN 0.1	49274 †	-12.2	49233 ± 1012	0
pcb442	ACS	**55684**	–	**55610** ± 982	0
	OPN 0.005	55955 †	-0.59	56100 ± 1233	0
	OPN 0.01	57519 †	-0.305	57656 ± 1467	0
	OPN 0.05	63253 †	-4.72	63301 ± 1249	0
	OPN 0.1	64833 †	-12.2	64848 ± **965**	0

Bold values indicate the best results.

† Difference with the ACS median is significant ($p < 0.05$).

Table 7.3. Median ($\bar{I}$), Mean (μ_I), and Standard Deviation (σ_I) of the Number of Iterations, Speed-up Factor (S), and Median Time ($t(s)$) to Reach Desired Accuracy Comparing the AS and OPN for the TSP

Instance	Algorithm	$\bar{I}$		$S(\%)$	$\mu_I \pm \sigma_I$	$t(s)$
att48	ACS	40.5		–	51.6 ± 45.4	0.031
A=95%	OPN 0.005	44.5		-9.88	56.4 ± 40	0.031
	OPN 0.01	40.0		1.23	49.9 ± 33.0	0.031
	OPN 0.05	32.5		19.8	42.0 ± 34.5	0.031
	OPN 0.1	**32.0**		**21**	$\mathbf{39.5 \pm 29.5}$	0.031
eil51	ACS	76.0		–	109.2 ± 103	0.054
A=95%	OPN 0.005	72.0		5.26	123.1 ± 150.5	**0.047**
	OPN 0.01	**66.0**		**13.2**	91.3 ± 81.6	**0.047**
	OPN 0.05	68.5		9.87	$\mathbf{80 \pm 52.9}$	0.062
	OPN 0.1	74.5		1.97	99.1 ± 95	0.078
eil76	ACS	173		–	310 ± 390	**0.203**
A=95%	OPN 0.005	204		-17.63	305 ± 369	0.235
	OPN 0.01	169		2.60	270 ± 492	0.204
	OPN 0.05	**165**		**4.62**	$\mathbf{219 \pm 199}$	0.265
	OPN 0.1	186		-7.51	266 ± 222	0.360
kroA100	ACS	238		–	453 ± 707	0.422
A=95%	OPN 0.005	231		2.94	363 ± 583	0.438
	OPN 0.01	**186**	†	**22.1**	324 ± 614	**0.360**
	OPN 0.05	200		16	$\mathbf{304 \pm 371}$	0.492
	OPN 0.1	284		-19.3	394 ± 380	0.891
pr124	ACS	64.0		–	74.4 ± 46.1	**0.172**
A=95%	OPN 0.005	66.0		-3.13	80.6 ± 62.3	0.187
	OPN 0.01	62.5		2.34	86.2 ± 81.1	0.180
	OPN 0.05	**49.5**	†	**22.7**	$\mathbf{61.7 \pm 46}$	0.188
	OPN 0.1	55.5		13.3	76.5 ± 65.9	0.266
ch150	ACS	468		–	791 ± 915	1.66
A=95%	OPN 0.005	471		-0.749	709 ± 872	1.77
	OPN 0.01	**378**	†	**19.3**	$\mathbf{490 \pm 484}$	**1.48**
	OPN 0.05	379		19.0	$542 \pm \mathbf{443}$	1.95
	OPN 0.1	705	†	-50.7	960 ± 840	4.69
d198	ACS	979		–	1129 ± 725	5.42
A=95%	OPN 0.005	**916**		**6.44**	1095 ± 796	5.36
	OPN 0.01	1050		-7.25	1137 ± 599	6.45
	OPN 0.05	1925	†	-96.6	2145 ± 1011	15.8
	OPN 0.1	5000	†	-410.7	4650 ± 799	54.3
lin318	ACS	**672**		–	$\mathbf{706 \pm 331}$	**9.34**
A=85%	OPN 0.005	700		-4.09	721 ± 329	10.3
	OPN 0.01	752		-11.9	$802 \pm \mathbf{303}$	11.6
	OPN 0.05	2107	†	-213.5	2297 ± 1044	44.2
	OPN 0.1	5000	†	-644	4770 ± 720	140.1
pcb442	ACS	**2399**		–	$\mathbf{2406 \pm 762}$	**67.9**
A=85%	OPN 0.005	2946	†	-22.8	2883 ± 971	88.5
	OPN 0.01	4081	†	-70.1	3977 ± 945	128.6
	OPN 0.05	–	†	–	–	–
	OPN 0.1	–	†	–	–	–

Bold values indicate the best result.

† Difference with the ACS median is significant ($p < 0.05$).

two larger instances. It is also important to note that the computational time (in seconds) for OPN are comparable to AS, and even below ACS in most of the smaller instances. This indicates that the speed-ups achieved do not have a high computational cost. Overall, the results indicate that fixed rate OPN has a faster convergence rate than the normal ACS.

7.4.2   OPU Experiment

The OPU algorithm was compared to the Ant System (AS) using the Grid World Problem (GWP) on three different grid sizes, namely 20×20, 50×50 and 100×100. The GWP involves a $n \times n$ grid where one square is randomly selected as the goal. This means that a direction is assigned to each square of the grid so that when an agent moves using this grid, it will reach the goal in the smallest number of steps. The GWP was selected because it has been previously used as a benchmark problem for studies involving opposition-based Reinforcement Learning (RL) [27, 28, 31].

We adapted the AS algorithm to solve the GWP. In our implementation, the location of the goal is unknown to the ants until they reach it. The ants start in a random square in the grid and travel until they reach the goal, which means that in one iteration they may not travel on every square of the grid. Each square is associated with four pheromone contents, one for each available direction. Also, since ants do not visit every square in every iteration, the evaporation was only applied to the squares visited by the ants. Complete details of the actual implementation can be found in [16].

Experimental Setup

The algorithms were terminated after 10000 iterations. Each algorithm completed 100 trials on each grid set. The parameters of the ant algorithms were set to the following values: $\alpha = 1$, $\rho = 0.001$, $\tau_o = 1$, and $m = 10$. These parameters were selected based on some general experimentation. The initial pheromone value (τ_o) was set to the high value of 1 to encourage more exploration in the early stages of the algorithm, so that the ants do not focus too fast on a single direction.

In the case of the OPU extension the removal of pheromone was done on every iteration ($\check{\lambda}_o = 1$). Also, since the GWP has clearly defined opposites, the edges that are true opposites have an *opposition-rating* of 1 and the other have a rating of 0. For example, if, at square (node) i, the ant chose to move up, then for that particular square (node) the up direction is part of the current solution, the down direction will have an *opposition-rating* of 1, and the left and right directions will have a rating of 0.

Experimental Results

The perfomance of each algorithm was evaluated based on the accuracy of the final policy and the convergence rate of the algorithm. The Wilcoxon rank sum (or Mann-Whitney) test was used to compare the medians of the results [12]. The accuracy or quality of the policies is determined by comparing them to an optimal policy. This

Table 7.4. Median ($\bar{A}$), Mean (μ_A), and Standard Deviation (σ_A) of the Accuracy Comparing the AS and OPU for the GWP

Instance	Algorithm	$\bar{A}$	$\mu_A \pm \sigma_A$
20×20	AS	98.00	98.05 ± 0.818
	OPU	98.25 †	98.26 ± 0.801
50×50	AS	97.10	97.23 ± 0.612
	OPU	97.84 †	97.87 ± 0.487
100×100	AS	93.21	93.4 ± 0.7235
	OPU	96.67 †	96.74 ± 0.465

† Difference with the AS median is significant ($p < 0.05$).

Table 7.5. Median ($\bar{I}$), Mean (μ_I), and Standard Deviation (σ_I) of the Number of Iterations, Speed-up Factor (S), and Median Time ($t(s)$) to Reach a 90% Accuracy Comparing the AS and OPU for the GWP

Instance	Algorithm	$\bar{I}$	$S(\%)$	$\mu_I \pm \sigma_I$	$t(s)$
20×20	AS	321	–	343.6 ± 113.9	0.312
	OPU	108.5 †	66.2	114.8 ± 25.4	0.109
50×50	AS	1885.5	–	1885.1 ± 212.1	10.45
	OPU	755.5 †	59.9	752.4 ± 62.7	4.28
100×100	AS	6155	–	6018.2 ± 563.4	135.3
	OPU	3048.5 †	50.5	2995.4 ± 197.3	72.1

† Difference with the AS median is significant ($p < 0.05$).

accuracy calculation, which was used in other work with GWP experiments [31], is defined as follows:

$$A_{\pi^*} = \frac{\|(\pi^* \cap \pi_1) \cup (\pi^* \cap \pi_2)\|}{n \times n}, \qquad (7.14)$$

where π^* is the policy being evaluated and π_1 and π_2 represent the two optimal possibilities for each square given a goal. Table 7.4 reports the overall accuracy results including the median, mean and standard deviation of the accuracy.

The OPU extension performed very well. OPU improved the accuracy for all grid sizes. The difference of the medians is statistically significant for the smaller size ($p < 0.05$) and very significant ($p < 0.01$) for the 50×50 and 100×100 grids. In the 100×100 grid case, the median accuracy was improved by 3.7%, which is good considering that the base accuracy is already above 90%. Moreover, comparing to the AS, the OPU mean accuracies are all higher and their standard deviations are all lower.

In order to evaluate the convergence rate of the algorithms, a desired accuracy of 90% was set. The median, mean and standard deviation of the number of iterations to reach the desired accuracy were used as comparative measures. The Wilcoxon test was used to statistically compare the median number of iterations. The speed-up factor (see (7.13)) and the computational time to reach the accuracy are also reported. Table 7.5 summarizes the convergence results for the AS and OPU algorithms.

The OPU algorithm was significantly faster than the AS. It achieved a speed-up factor of 66%, 60% and 50% for the 20×20, 50×50, and 100×100 grids, respectively. The mean number of iterations and their standard deviation were also lower. The lower standard deviations indicate that the OPU will reliably reach the 90% accuracy with fewer number of iterations that the AS. It is also important to note that the computational time (in seconds) for the OPU are also below the AS, which shows that the speed-up achieved does not have a high computational cost.

7.4.3   Discussion

While the application of opposition to ACO can be challenging, in general, results indicate that the use of opposition can be beneficial. Specifically, the OPN approach, using opposite pheromone for some decisions was beneficial for improving the accuracy for the TSP. The OPU extension, which involved performing additional updates during the best trail update phase, led to excellent results for the GWP. The OPU method applied to the GWP led to accuracy improvements in all grid sizes and convergence speed-ups reaching 66%. It was interesting to see that the performance improvements were relatively similar for all grid sizes.

One fundamental difference between the TSP and the GWP is that, in the GWP, the "opposite" is clearly defined. For each square in a grid, there are two sets of opposite pairs: up/down and left/right. Each direction has a unique opposite. In the TSP, a choice made by the ant at a certain node does not have a clearly defined opposite. Also, a straight mathematical opposite might not even be defined. Simply defining opposites with respect to the length of the edge might not make sense because, in some solutions, you need to take a longer edge to get an overall shorter path. In the GWP, the partial components of the solution are all the perfect components, which may be a reason why OPU, by removing pheromone in rejected directions, is very advantageous for the GWP. In the TSP, the algorithm makes local sacrifices for global success, which may explain why OPN is helpful for the TSP.

Moreover, in the GWP, the path travelled by the ants from their starting point to the goal is unidirectional. Thus, it is possible to define an "opposite" path that makes sense. This opposite path would include all the decisions that would bring the ants away from the goal. In the TSP, the solutions are *bidirectional*: going in the opposite direction of the path makes no difference in the final solution. Therefore, defining the "opposite" path is not as straightforward and so problem-type dependent. The combinatorial aspect of the TSP complicates the definition of an opposite path. Changing a single component in the solution brings a new array of possibilities. The partial components of a solution are all dependent.

The speed-ups achieved with the use of opposite pheromone updates can be explained by the fact that the algorithm is rapidly moving toward the final optimal pheromone matrix. With usual pheromone updates, the algorithm takes very small steps moving towards the final pheromone matrix. In contrast, the opposite pheromone updates allow the algorithm to take very large guided jumps toward the optimal solution by removing or adding more pheromone in the appropriate regions.

7.5   Conclusions and Future Work

The work of investigating the application of opposition to ACO is just beginning. The use of some fundamental opposition concepts, such as the use of opposite pheromone and performing opposite updates, led to encouraging results in the TSP and the GWP. Thus, opposition is a way that can provide benefits to ant algorithms, but more work is needed to fully develop the OBO framework for ACO.

While the OPN extension proved successful for the smaller TSP instances, more work is required to determine all the benefits of this extension. The results show that *opposite-rate*, $\check{\lambda}_o$, is a key element in the success of the OPN algorithm. Thus, selecting the rate wisely can lead to a better accuracy at a faster rate. Additional investigations in new ways to vary the pheromone rate are necessary.

Further work is needed to explore the application of opposition to different versions of the ant algorithm, namely the Max-Min Ant System and the Best-Worst Ant System. Continuing the investigation with the ACS and the AS is also necessary so that performance differences can be clearly understood. It is also possible that applying opposition to ant algorithms will eventually generate a new form of the algorithm, which will be separate from the existing ACO frameworks. There should also be some experiments with the concept of opposition in combination with local search. It would be important to determine if the benefits of opposition complement those achieved through local search.

While it is true that the GWP is not a typical ACO problem, it helped reinforce some of the good results achieved with the TSP. Some of the differences might be attributed to the implementations, the use of different ACO versions and different opposition algorithms. However, the problem is what defines the algorithm that is used. Thus, future work should include more applications of ACO.

Another potential issue is that, in the TSP, pheromone matrices lead to an array of possible solutions. There is no one-to-one relation between the pheromone matrix and a solution. Therefore, it might be important to establish rules on how to generate an actual opposite solution in a graph, so that there can be an exact fitness value. Additionally, it is important to establish how to compute the opposite pheromone matrix. The GWP is a little different from the TSP, in that the pheromone matrix was directly related to a solution, which may be one reason why OPU performed well with the GWP. This work explored opposite pheromone values and opposite updates; however, it did not create a direct relation between two pheromone matrices.

The most important work that needs to be developed is fundamental theoretical work with opposition and graph theory. While the GWP was an application that worked well with opposition, the true nature of ant algorithms are graphs like in the TSP. Thus, it is crucial to establish a strong theoretical base regarding opposition and graphs. As it has already been mentioned, opposition is not clearly defined in TSP, which springs from that fact that opposition is not clearly defined in graphs. Research has established opposite actions [27, 28, 31], opposite estimates [20, 21, 22], and opposite transfers functions [33, 34]. Perhaps, the next step is to establish the "opposite graph".

References

1. Bonabeau, E., Dorigo, M., Theraulaz, G.: Swarm Intelligence: From Natural to Artificial Systems. Oxford University Press, New York (1999)
2. Bullnheimer, B., Hartl, R.F., Strauss, C.: Applying the Ant System to the Vehicle Routing Problem. In: Osman, I.H., Voβ, S., Martello, S., Roucairol, C. (eds.) Meta-Heuristics: Advances and Trends in Local Search Paradigms for Optimization, pp. 109–120. Kluwer Academic Publishers, Dordrecht (1998)
3. Bullnheimer, B., Hartl, R.F., Strauss, C.: An Improved Ant System Algorithm for the Vehicle Routing Problem. Ann. Oper. Res. 89, 319–328 (1999)
4. Cordón, O., de Viana, I.F., Herrera, F., Moreno, L.: A New ACO Model Integrating Evolutionary Computation Concepts: The Best-Worst Ant System. In: Proc. of the 2nd Int. Workshop on Ant Algorithms (ANTS 2000), Brussels, Belgium, pp. 22–29 (2000)
5. Dorigo, M.: Optimization, Learning and Natural Algorithms (in Italian). PhD dissertation, Dipartimento di Elettronica, Politecnico di Milano, Italy (1992)
6. Dorigo, M., Maniezzo, V., Colorni, A.: The Ant System: Optimization by a Colony of Cooperating Agents. SMC 26, 29–41 (1996)
7. Dorigo, M., Gambardella, L.M.: Ant Colony System: A Cooperative Learning Approach to the Traveling Salesman Problem. IEEE Transactions On Evolutionary Computation 1(1), 53–66 (1997)
8. Dorigo, M., Stützle, T.: The Ant Colony Optimization Metaheuristic: Algorithm, Applications, and Advances. In: Glover, F., Kochenberger, G.A. (eds.) Handbook of Metaheuristics, pp. 55–82. Kluwer Academic Publishers, Boston (2003)
9. Dorigo, M., Stützle, T.: Ant Colony Optimization. MIT Press, Cambridge (2004)
10. Gambardella, L.M., Dorigo, M.: Solving Symmetric and Asymmetric TSPs by Ant Colonies. In: Proc. IEEE Int. Conf. on Evolutionary Computation (ICEC 1996), Nagoya, Japan, pp. 622–627 (1996)
11. Gambardella, L.M., Taillard, E., Agazzi, G.: MACS-VRPTW: A multiple Ant Colony System for Vehicle Routing Problems with Time Windows. In: Corne, D., Dorigo, M., Glover, F. (eds.) New Ideas in Optimization, pp. 63–76. McGraw-Hill, New York (1999)
12. Hollander, M., Wolfe, D.A.: Nonparametric Statistical Methods. Wiley, Chichester (1973)
13. Iredi, S., Merkle, D., Midderndorf, M.: Bi-Criterion Optimization with Multi Colony Ant Algorithms. In: Zitzler, E., Deb, K., Thiele, L., Coello Coello, C.A., Corne, D.W. (eds.) EMO 2001. LNCS, vol. 1993, pp. 359–372. Springer, Heidelberg (2001)
14. Kennedy, J., Eberhart, R.C., Shi, Y.: Swarm Intelligence. Morgan Kaufmann, San Mateo (2001)
15. Lawler, E.L., Lenstra, J.K., Rinnooy-Kan, A.H.G., Shmoys, D.B.: The Travelling Salesman Problem. Wiley, New York (1985)
16. Malisia, A.R.: Investigating the Application of Opposition-Based Ideas to Ant Algorithm. MASc. thesis, University of Waterloo, ON, Canada (2007), http://hdl.handle.net/10012/3233
17. Malisia, A.R., Tizhoosh, H.R.: Applying Opposition-Based Ideas to the Ant Colony System. In: Proc. of IEEE Swarm Intelligence Symposium, Honolulu, HI, April 1-5, pp. 182–189 (2007)
18. Maniezzo, V., Colorni, A.: The Ant System Applied to the Quadratic Assignment Problem. IEEE Trans. Knowl. Data Eng. 11(5), 769–778 (1999)
19. Montgomery, J., Randall, M.: Anti-Pheromone as a Tool for Better Exploration of Search Spaces. In: Proc. 3rd Int. Workshop on Ant Algorithms (ANTS 2002), Brussels, Belgium, pp. 100–110 (September 2002)

20. Rahnamayan, S., Tizhoosh, H.R., Salama, M.M.: Opposition-Based Differential Evolution Algorithms. In: Proc. IEEE Congress on Evolutionary Computation, Vancouver, July 16-21, pp. 7363–7370 (2006)
21. Rahnamayan, S., Tizhoosh, H.R., Salama, M.M.: A Novel Population Initialization Method for Accelerating Evolutionary Algorithms. Computers and Mathematics with Applications 53(10), 1605–1614 (2007)
22. Rahnamayan, S., Tizhoosh, H.R., Salama, M.M.: Opposition-Based Differential Evolution (ODE) With Variable Jumping Rate. In: Proc. of IEEE Symposium on Foundations of Computational Intelligence (FOCI 2007), Hawaii, April 1-5, pp. 81–88 (2007)
23. Rahnamayan, S., Tizhoosh, H.R., Salama, M.M.A.: Opposition-Based Differential Evolution. IEEE Transactions of Evolutionary Computation (in press, 2008)
24. Randall, M., Montgomery, J.: The Accumulated Experience Ant Colony for the Travelling Salesman Problem. In: Proc. of Inaugural Workshop on Artificial Life, Adelaide, Australia, pp. 79–87 (2001)
25. Reinelt, G.: TSPLIB - A traveling salesman problem library. ORSA J. Comput. 3, 376–384 (1991)
26. Schoonderwoerd, R., Holland, O.E., Bruten, J.L., Rothkrantz, L.J.M.: Ant-Based Load Balancing in Telecommunications Networks. Adaptive Behavior 2, 169–207 (1996)
27. Shokri, M., Tizhoosh, H.R., Kamel, M.S.: Opposition-Based Q(λ) Algorithm. In: Proc. IEEE International Joint Conf. on Neural Networks (IJCNN), Vancouver, July 16-21, pp. 646–653 (2006)
28. Shokri, M., Tizhoosh, H.R., Kamel, M.S.: Opposition-Based Q(λ) with Non-Markovian Update. In: Proc. IEEE Symposium on Approximate Dynamic Programming and Reinforcement Learning (ADPRL 2007), Hawaii, April 1-5, pp. 288–295 (2007)
29. Song, Y., Irving, M.R.: Optimisation techniques for electrical power systems. II. Heuristic optimisation methods. Power Engineering Journal 15(1), 151–160 (2001)
30. Stützle, T., Hoos, H.H.: MAX-MIN Ant System. Future Generation Computer Systems 16(8), 889–914 (2000)
31. Tizhoosh, H.R.: Opposition-Based Learning: A New Scheme for Machine Intelligence. In: Proc. Int. Conf. on Computational Intelligence for Modelling Control and Automation - CIMCA 2005, Vienna, Austria, vol. I, pp. 695–701 (2005)
32. Tizhoosh, H.R.: Opposition-Based Reinforcement Learning. Journal of Advanced Computational Intelligence and Intelligence Informatics 10(4), 578–585 (2006)
33. Ventresca, M., Tizhoosh, H.R.: Improving the Convergence of Backpropagation by Opposite Transfer Functions. In: Proc. IEEE International Joint Conf. on Neural Networks (IJCNN), Vancouver, July 16-21, pp. 9527–9534 (2006)
34. Ventresca, M., Tizhoosh, H.R.: Opposite Transfer Functions and Backpropagation Through Time. In: Proc. IEEE Symposium on Foundations of Computational Intelligence (FOCI 2007), Hawaii, April 1-5, pp. 570–577 (2007)

8

Differential Evolution Via Exploiting Opposite Populations

Shahryar Rahnamayan[1] and H.R. Tizhoosh[2]

[1] Faculty of Engineering and Applied Science, University of Ontario Institute of Technology
(UOIT), Canada
Shahryar.Rahnamayan@uoit.ca
[2] Department of Systems Design Engineering, University of Waterloo, Canada
tizhoosh@uwaterloo.ca

Summary. The concept of opposition can contribute to improve the performance of population-based algorithms. This chapter presents an overview of a novel opposition-based scheme to accelerate an evolutionary algorithm, differential evolution (DE). The proposed opposition-based DE (ODE) employs opposition-based computation (OBC) for population initialization and also for generation jumping. Opposite numbers, representing anti-chromosomes, have been utilized to improve the convergence rate of the classical DE. A test suite with 15 well-known benchmark functions is employed for experimental verification. Descriptions for the DE and ODE algorithms, and a comparison strategy are provided. Results are promising and confirm that the ODE outperforms its parent algorithm DE. This work can be regarded as an initial study to exploit oppositional concepts to expedite the optimization process for any population-based approach.

8.1 Introduction

Evolutionary algorithms (EAs) are well-established techniques to approach problems with mixed-type variables, many local optima, and with undifferentiable or non-analytical functions [1]. Among various kinds of evolutionary algorithms, differential evolution (DE) is well known for its effectiveness and robustness. Many comparative studies confirm that the DE outperforms many other optimizers [5]. Finding more accurate solution(s) in a shorter period of time for complex black-box problems is still a crucial target of research on evolutionary algorithms.

In this chapter, opposition-based schemes including opposition-based population initialization and generation jumping, will be described. The differential evolution (DE) is selected as a parent algorithm to verify the acceleration effect of the proposed schemes. A set of well-known complex benchmark functions is employed to experimentally compare and analyze the algorithms. Results confirm that Opposition-Based Differential Evolution (ODE) performs better than DE in terms of convergence speed and solution accuracy.

The main purpose of this and previous works has been to introduce a new notion into nonlinear continuous optimization via innovative metaheuristics, namely *the notion of opposition*. Although, all conducted experiments utilize DE as a parent algorithm, the proposed schemes are defined at the population level and, hence, have an inherent potential to be utilized for acceleration of other population-based algorithms.

H.R. Tizhoosh, M. Ventresca (Eds.): Oppos. Concepts in Comp. Intel., SCI 155, pp. 143–160, 2008.
springerlink.com

The organization of this chapter is as follows: A short review of differential evolution is given in section 8.2. The main reasons to select DE as a parent algorithm are explained in section 8.3. Opposition-based differential evolution is described in section 8.4. Experimental verifications are elaborated in section 8.5. Finally, the chapter is concluded in section 8.6.

8.2   Differential Evolution (DE)

Differential evolution (DE) is a population-based optimization algorithm based on the idea of genetic annealing which was used to solve the Chebyshev polynomial fitting problem [1]. In order to solve the Chebyshev problem in continuous space, a modified genetic annealing algorithm from bit-string to floating-point encoding and a consequent switch from logical operators to arithmetic ones were proposed [2, 3, 4]. During these experiments, the differential mutation operator to perturb the population of vectors was discovered. Additionally, by using differential mutation, discrete recombination, and pair-wise selection, it was recognized that an annealing mechanism is not needed; it was removed completely and DE was born.

Let us assume that $X_{i,G}(i = 1, 2, ..., N_p)$ are candidate solution vectors in generation G (N_p : population size). Like other evolutionary algorithms, DE starts with an initial population, which is usually generated in a random manner. A typical vector of the initial population can be generated as follows [5]:

$$X_{i,j} = l_j + \text{RAND}_j(0, 1) \times (l_j - u_j) \quad \text{with} \quad j = 1, 2, ..., D, \qquad (8.1)$$

where D is the problem dimensionality; l_j and u_j are the lower and the upper boundaries of the jth variable, respectively, and $\text{RAND}(0, 1)$ is a uniformly generated random number in $[0, 1]$.

Successive populations are generated by adding the weighted difference of two randomly selected vectors to a third randomly selected vector. For classical DE (see Algorithm 1), the mutation, crossover, and selection operators are straightforwardly defined.

8.2.1   Mutation

For each vector $X_{i,G}$ in generation G a mutant vector $V_{i,G}$ (see line 9 of Algorithm 1) is defined by

$$V_{i,G} = X_{a,G} + F(X_{c,G} - X_{b,G}), \qquad (8.2)$$

where $i = \{1, 2, ..., N_p\}$, and a, b, and c are mutually different random integer indices selected from $\{1, 2, ..., N_p\}$. Further, the variables i, a, b, and c are different so that $N_p \geq 4$ is necessary. The factor $F \in [0, 2]$ is a real constant which determines the amplification of the added differential variation of $(X_{c,G} - X_{b,G})$ [5]. Larger values for F result in higher diversity in the generated population and lower values cause faster convergence.

8.2.2 Crossover

DE utilizes the crossover operation to generate new solutions by shuffling competing vectors and also to increase the population diversity. For the classical DE (lines $10 - 16$ of Algorithm 1), the binary crossover is utilized. It defines the following trial vector:

$$U_{i,G} = (U_{1i,G}, U_{2i,G}, ..., U_{Di,G}), \tag{8.3}$$

where

$$U_{ji,G} = \begin{cases} V_{ji,G} & \text{if RAND}_j(0,1) \le C_r \vee j = k, \\ X_{ji,G} & \text{otherwise.} \end{cases} \tag{8.4}$$

$C_r \in (0,1)$ is the predefined crossover rate, and $\text{RAND}_j(0,1)$ is the jth evaluation of a uniform random number generator. The parameter $k \in \{1, 2, ..., D\}$ is a random index chosen once for each i to make sure that at least one parameter is always selected from the mutated vector $V_{ji,G}$. The most common values for C_r are in the range of $(0.4, 1)$ [17]. Figure 8.1 illustrates a pictorial example for the binary crossover.

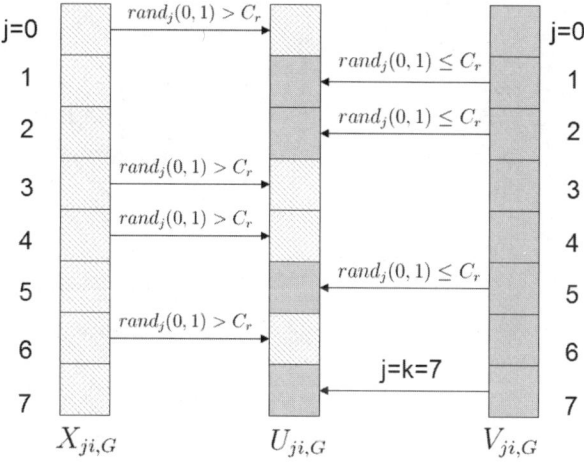

Fig. 8.1. A pictorial example for the binary crossover in DE ($k = 7$) [5]. When $\text{RAND}_j(0,1) \le C_r \vee j = k$, then the variable is copied from $V_{ji,G}$, otherwise copied from $X_{ji,G}$ to $U_{ji,G}$.

8.2.3 Selection

Selection is a mechanism to decide which vector ($U_{i,G}$ or $X_{i,G}$) should be a member of next (new) generation $G + 1$. For a minimization problem, the vector with the lower objective function value is chosen (greedy selection). If $f(U_i) \le f(X_i)$, then U_i is selected; otherwise X_i will be chosen (lines $17 - 23$ of Algorithm 1).

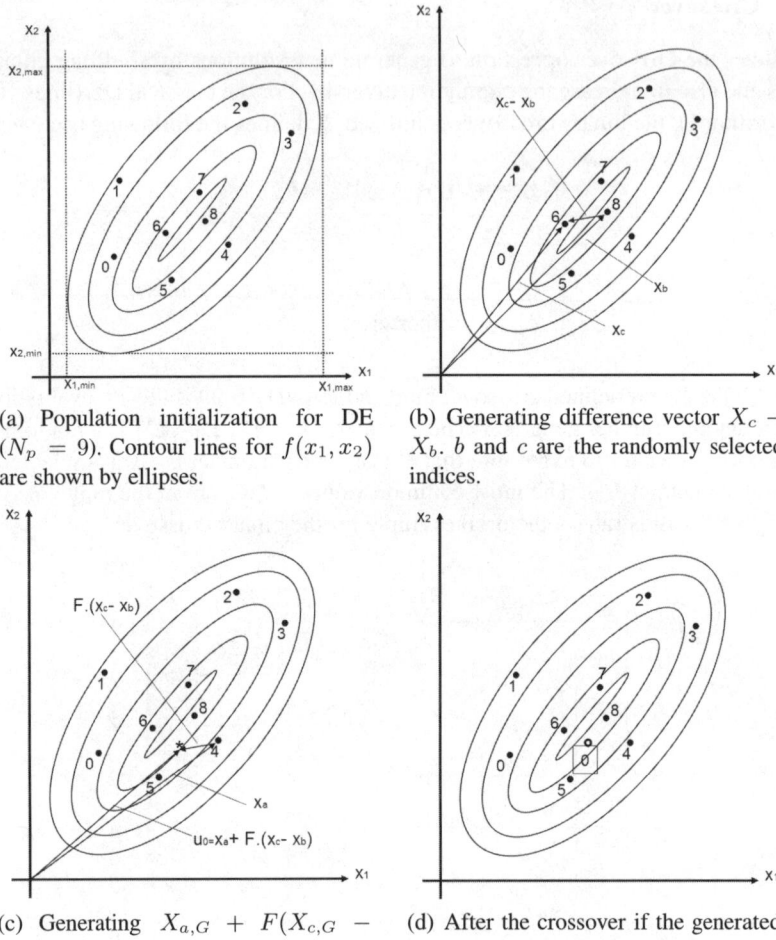

(a) Population initialization for DE ($N_p = 9$). Contour lines for $f(x_1, x_2)$ are shown by ellipses.

(b) Generating difference vector $X_c - X_b$. b and c are the randomly selected indices.

(c) Generating $X_{a,G} + F(X_{c,G} - X_{b,G})$. a is the third randomly selected index.

(d) After the crossover if the generated vector has lower objective value, then it will be replaced with the vector 0.

Fig. 8.2. Illustration of one generate-and-test cycle for DE (starting from vector 0) [5]

This evolutionary cycle (i.e., mutation, crossover, and selection) is repeated N_p times to generate new populations. These successive generations are produced until the termination conditions are satisfied. One generate-and-test cycle for DE is presented in Figure 8.2.

The starting point for the mutation, crossover and selection is indicated by the comments in the algorithm. The algorithm terminates (line 5) when the best achieved fitness value (BFV) is smaller than the value-to-reach (VTR), or the number of function calls (NFC) exceeds the predefined maximum number of function calls (MAX$_{NFC}$). The termination strategy can be defined differently based on the application or the purpose of the experiment. The number of generations, the execution time, or some population

Algorithm 1. Differential Evolution (DE). P_0: Initial population, N_p: Population size, V: Noise vector, U: Trial vector, D: Problem dimension, BFV: Best achieved fitness value, VTR: Value-to-reach, NFC: Number of function calls, MAX_{NFC}: Maximum number of function calls, F: Mutation constant, $\text{RAND}(0, 1)$: Uniformly generated random number, C_r: Crossover, $f(\cdot)$: Objective function, P': Population of the next generation.

1: Generate uniformly distributed random population P_0
2: NFC $\leftarrow 0$
3: Evaluate individuals of P_0
4: NFC $\leftarrow$ NFC $+ N_p$
5: **while** (BFV $>$ VTR **and** NFC $<$ MAX_{NFC}) **do**
6:     {Generate-and-Test-Loop}
7:     **for** $i = 0$ to N_p **do**
8:         Select three parents X_a, X_b, and X_c randomly from current population where $i \neq a \neq b \neq c$
            {Mutation}
9:         $V_i \leftarrow X_a + F \times (X_c - X_b)$
            {Crossover}
10:        **for** $j = 0$ to D **do**
11:            **if** $\text{RAND}(0, 1) < C_r$ **then**
12:                $U_{i,j} \leftarrow V_{i,j}$
13:            **else**
14:                $U_{i,j} \leftarrow X_{i,j}$
15:            **end if**
16:        **end for**
            {Selection}
17:        Evaluate U_i
18:        NFC $\leftarrow$ NFC $+ 1$
19:        **if** $(f(U_i) \leq f(X_i))$ **then**
20:            $X'_i \leftarrow U_i$
21:        **else**
22:            $X'_i \leftarrow X_i$
23:        **end if**
24:    **end for**
25:    $X \leftarrow X'$
26: **end while**

statistics (e.g., diversity or the improvement rate) are some commonly used termination criteria.

8.2.4 DE in Optimization Field

A summary classification of optimization methods can be seen in Figure 8.3. According to the proposed classification scheme for optimization methods, DE is a population-based, nonlinear, continuous and global optimization algorithm [1].

Studies have been conducted to enhance the performance of the classical DE algorithm by adaptive determination of DE control parameters. For instance, the fuzzy adaptive differential evolution algorithm (FADE) was introduced by Liu and Lampinen

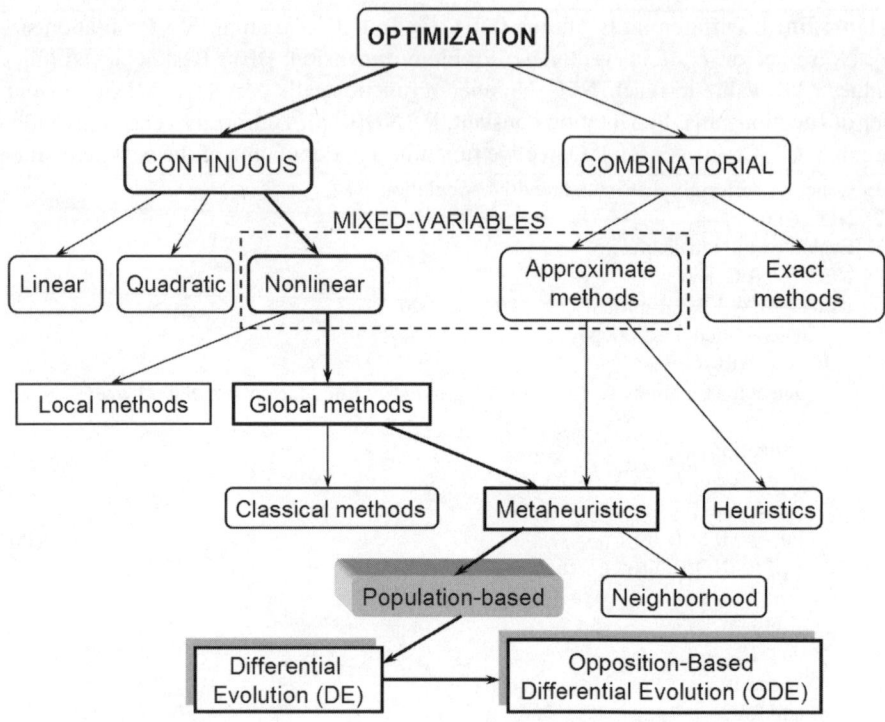

Fig. 8.3. A simple classification scheme of optimization methods [1]

[24]. They employed a fuzzy logic controller to set the mutation and crossover rates. In the same direction, Brest et al. [20] proposed self-adaptive DE. Teo [30] proposed a dynamic population sizing strategy based on self-adaptation, and Ali and Törn [25] introduced auxiliary population and automatic calculating of the amplification factor F for the difference vector.

Other researchers have experimented with multi-population ideas. Tasoulis et al. [31] proposed parallel DE where they assign each subpopulation to a different processor node. Shi et al. [32] partitioned high-dimensional search spaces into smaller spaces and used multiple cooperating subpopulations to find the solution. They called this method cooperative co-evolutionary differential evolution.

Hybridization with different algorithms is another direction for improvement of DE. Sun et al. [26] proposed a new hybrid algorithm based on a combination of DE and estimation of distribution algorithm. This technique uses a probability model to determine promising regions in order to focus the search process on those areas. Noman and Iba [35] incorporated local search into the classical DE. They employed fittest-individual refinement which is a crossover-based local search. Fan and Lampinen [33] introduced a new local search operation, trigonometric mutation, in order to obtain a better trade-off between convergence speed and robustness. Kaelo and Ali [34] employed reinforcement learning and some other schemes for generating fitter trial points.

All mentioned approaches are proposed to increase the convergence rate and/or the accuracy of DE. For more details about DE extensions, the reader is referred to literature [1, 5, 27].

8.3 Why Differential Evolution?

Differential evolution is a simple and compact metaheuristic which directly operates on continuous variables (arithmetic operators instead of logical operators). Unlike many binary versions of the genetic algorithms, DE works with the floating-point numbers. This removes encoding and decoding of the variables which is a source of inaccuracy. Consequently, this feature makes DE scalable for high-dimensional problems and also time and memory efficient.

Another reason for choosing DE is that it does not need a probability density function to adapt the control parameters (unlike most evolutionary strategies) or any probability distribution pattern for the mutation (unlike genetic algorithms or evolutionary programming). DE's different mutation and crossover schemes distinguish it from other evolutionary algorithms [5].

Additionally, handling mixed integers, discrete and continuous variables makes DE more applicable for a wider range of real-world applications. The main advantage of DE while working with integer variables is that it internally works on a continuous space and only switches to the integer space during the evaluation of the objective function. This characteristic supports higher accuracy compared to some other well-known algorithms (e.g., GAs) which perform in the reverse manner [27]. Extensions of classical DE are capable of handling boundary constraints and also nonlinear function constraints which both are commonly required in the real-world problems [5].

Many comparative studies report higher robustness, convergence speed, and solution quality of the DE when compared to other evolutionary algorithms for both benchmark functions and real-world applications. A comprehensive performance study is provided in [5]. The authors first compare DE to 16 other optimizers against five well-known thirty-dimensional test functions (namely, Rosenbrock, Ackley, Griewangk, Rastrigin, and Schwefel). Consequently, they explored eight function-based comparative studies (e.g., unconstrained optimization, multi-constraints nonlinear optimization, and multi-objective optimization) and also eleven application-oriented performance comparison studies (e.g., multi-sensor fusion, earthquake relocation, image registration, and optimization of neural networks). Finally, they conclude [5] "[...] DE may not always be the fastest method, it is usually the one that produces the best results, although the number of cases in which it is also the faster is significant. DE also proves itself to be robust, both in how control parameters are chosen and in the regularity with which it finds the true optimum. [...] As these researchers have found, DE is a good first choice when approaching a new and difficult global optimization problem is defined with continuous and/or discrete parameters."

8.4 Opposition-Based Differential Evolution

In his primary paper on opposition-based learning (OBL), Tizhoosh proposed to use *anti-chromosmes* for GAs [12]. For every selected chromosome a corresponding

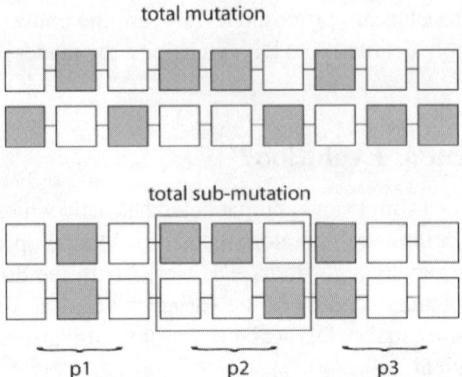

Fig. 8.4. Generation of anti-chromosomes [12]

anti-chromosome can be generated. The initial chromosomes are generally generated randomly meaning that they can possess high or low fitness. However, in a complex problem it is usually very likely that the initial populations do not contain optimal solutions. Hence, in lack of any a-priori knowledge, it is reasonable to look at anti-chromosomes simultaneously. Considering the search direction and its opposite at the same time will bear more likelihood to reach the best population in a shorter time (for more motivation on OBL see also Chapters 1-2). Tizhoosh also suggested to use total or sub-total-mutation to generate opposite candidate solutions (Figure 8.4).

Similar to all population-based optimization algorithms, two main steps are distinguishable for DE, namely population initialization and producing new generations by evolutionary operations such as mutation, crossover, and selection. The opposition-based differential evolution (ODE) [6, 9, 13] will enhance these two steps by considering opposite solutions. For black-box optimization – which is a general assumption for optimization methods – there is no information about the shape of the problem landscape such that type II opposition can only be approximated via type I opposition (see Definitions 2 and 6 in Chapter 2). The pseudo-code of ODE is presented in Algorithm 2 [13].

8.4.1 Opposition-Based Population Initialization

In absence of domain knowledge, uniform random number generation is generally the only choice to create an initial population. But as mentioned before, by utilizing type-I opposition it is possible to obtain fitter starting candidates. Block (1) from Figure 8.5 shows the implementation of opposition-based population initialization (lines $5 - 12$ of Algorithm 2). The following steps explain that procedure [6]:

Step 1. Initialize the population $P(N_p)$ randomly,
Step 2. Create the opposite population OP by

$$OP_{i,j} = a_j + b_j - P_{i,j}, \quad \text{with} \quad i = 1, 2, ..., N_p; j = 1, 2, ..., D, \quad (8.5)$$

Algorithm 2. Pseudo-code of Opposition-Based Differential Evolution (ODE) in order to solve a minimization problem [Adopted from [13]]. P_0: Initial population, OP_0: Opposite of initial population, P: Current population, OP: Opposite of current population, D: Problem dimension, $[l_j, u_j]$: Range of the j-th variable, J_r: Jumping rate, C_r: Crossover rate, $\min_j^p / \max_j^p$: Minimum/maximum value of the j-th variable in the current population. Lines **1-12** and **33-42** are implementations of opposition-based population initialization and opposition-based generation jumping, respectively.

```
 1: Generate uniformly distributed random population P₀
 2: NFC ← 0
 3: Evaluate individuals of P₀
 4: NFC ← NFC + Nₚ
    {**Begin of Opposition-Based Population Initialization**}
 5: for i = 0 to Nₚ do
 6:    for j = 0 to D do
 7:       OP₀ᵢ,ⱼ ← lⱼ + uⱼ − P₀ᵢ,ⱼ
 8:    end for
 9: end for
10: Evaluate individuals of OP₀
11: NFC ← NFC + Nₚ
12: Select Nₚ fittest (best) individuals from P₀ and OP₀ as initial population P₀
    {Begin of DE's Evolution Steps}
13: while ( BFV > VTR and NFC < MAX_NFC ) do
14:    for i = 0 to Nₚ do
15:       Select three parents Pᵢ₁, Pᵢ₂, and Pᵢ₃ randomly from current population where
          i ≠ i₁ ≠ i₂ ≠ i₃
16:       Vᵢ ← Pᵢ₁ + F × (Pᵢ₂ − Pᵢ₃)
17:       for j = 0 to D do
18:          if RAND(0, 1) < Cᵣ then
19:             Uᵢ,ⱼ ← Vᵢ,ⱼ
20:          else
21:             Uᵢ,ⱼ ← Pᵢ,ⱼ
22:          end if
23:       end for
24:       Evaluate Uᵢ
25:       NFC ← NFC + 1
26:       if (f(Uᵢ) ≤ f(Pᵢ)) then
27:          P′ᵢ ← Uᵢ
28:       else
29:          P′ᵢ ← Pᵢ
30:       end if
31:    end for
32:    P ← P′
       {**Begin of Opposition-Based Generation Jumping**}
33:    if RAND(0, 1) < Jᵣ then
34:       for i = 0 to Nₚ do
35:          for j = 0 to D do
36:             OPᵢ,ⱼ ← MINⱼᵖ + MAXⱼᵖ − Pᵢ,ⱼ
37:          end for
38:       end for
39:       Evaluate individuals of OP₀
40:       NFC ← NFC + Nₚ
41:       Select Nₚ fittest (best) individuals from P and OP as current population P
42:    end if
43: end while
```

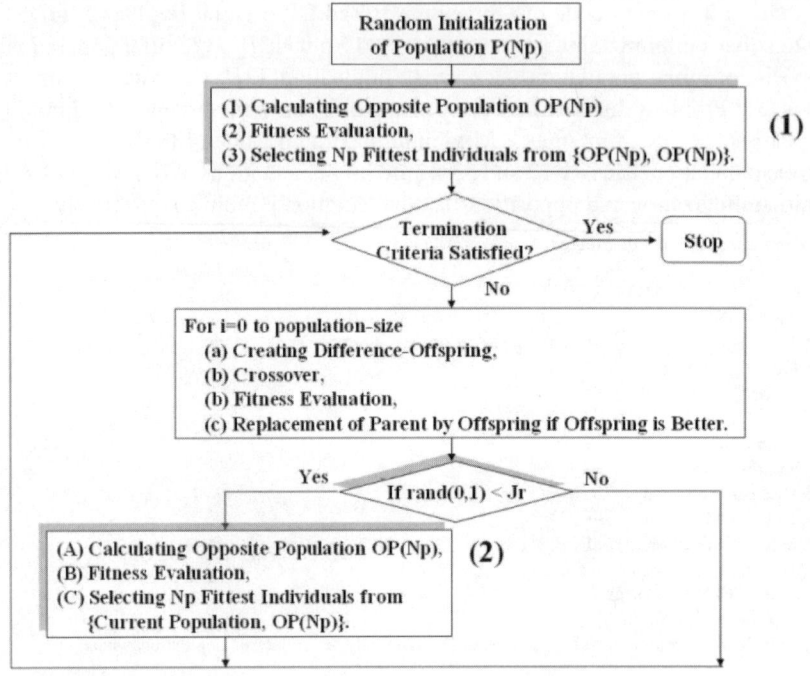

Fig. 8.5. Gray boxes extend DE to ODE. Block (1): Opposition-based initialization, Block (2): Opposition-based generation jumping (J_r: jumping rate, RAND(0, 1): uniformly generated random number, N_p: population size).

where P_i and OP_i denote the ith individual of the current population and its corresponding opposite, respectively, and $[l_j, u_j]$ is the range of the jth variable.

Step 3. Select the N_p fittest (best) individuals from $P \cup OP$ as the initial population.

According to the above procedure, $2N_p$ function evaluations are required instead of N_p for the regular random population initialization. But, by the opposition-based initialization, the parent algorithm can start with the fitter initial individuals instead.

8.4.2   Opposition-Based Generation Jumping

By applying a similar approach mentioned in Sec. 8.4.1 to the current population, which means selecting N_p best individuals from the current and corresponding opposite populations, the evolutionary process can be forced to jump to a fitter generation (the generation with fitter individuals). After generating new populations, the opposite population is calculated and the N_p fittest (best) individuals are selected from the union of the current and opposite population based on a jumping rate $J_r \in (0, 0.4)$ [13, 15]. In order to calculate the opposite population for generation jumping, the opposite of each variable is calculated dynamically; that is, the maximum and minimum values of each variable

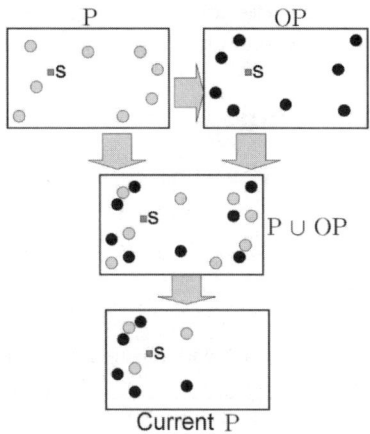

Fig. 8.6. Example to visualize the opposition-based generation jumping in 2D space ($N_p = 8$)

in the current population ($[\text{MIN}_j^p, \text{MAX}_j^p]$) are used to calculate opposite points instead of using variables' predefined interval boundaries ($[l_j, u_j]$):

$$\text{OP}_{i,j} = \text{MIN}_j^p + \text{MAX}_j^p - \text{P}_{i,j}, i = 1, 2, ..., N_p; j = 1, 2, ..., D. \qquad (8.6)$$

If the opposites are calculated within variables' static boundaries, it is possible to jump outside of the already shrunken search space and lose the knowledge of the already reduced space. Hence, we calculate opposite points by using variables' current interval in the population ($[\text{MIN}_j^p, \text{MAX}_j^p]$) which is, as the search does progress, increasingly smaller than the corresponding initial range $[l_j, u_j]$. Block (2) from Figure 8.5 illustrates the implementation of opposition-based generation jumping (lines $33 - 42$ of Algorithm 2).

A pictorial example for opposition-based generation jumping procedure in 2D space is illustrated in Figure 8.6. The letter 'S' indicates the location of the optimal solution. Dark and light circles represent the points and the opposite points, respectively.

In [14], we established mathematical proofs and experimental evidence to verify the advantage of opposite points compared to additional random points when dealing with high-dimensional problems (see also Chapter 2 for more discussions on the formalism of opposition). Both experimental and mathematical results confirmed that opposite points are more beneficial than additional independent random points. We can conclude that the opposition-based learning can be utilized to accelerate optimization methods since considering the pair x and its opposite $\breve{x}$ has apparently a higher fitness probability than pure randomness.

8.5 Experimental Verifications

In this section, the convergence metrics are defined and DE and ODE are compared experimentally over well-known benchmark functions (section 8.5.1). Also, the contribution of opposite points to the achieved acceleration rate is investigated by replacing them with random points (section 8.5.2).

8.5.1   Comparison of DE and ODE

A set of 15 well-known benchmark functions [6, 13, 15], which contains 7 unimodal $(f_1, f_2, f_3, f_6, f_{10}, f_{11}, f_{14})$ and 8 multimodal functions $(f_4, f_5, f_7, f_8, f_9, f_{12}, f_{13}, f_{15})$, has been selected for performance verification of ODE. The definition of the benchmark functions is given in Table 8.1.

Table 8.1. List of employed benchmark functions (unimodal and multimodal)

Function	*Search Space*
$f_1(X) = \sum\limits_{i=1}^{D} x_i^2$	$[-5.12, 5.12]^D$
$f_2(X) = \sum\limits_{i=1}^{D} ix_i^2$	$[-5.12, 5.12]^D$
$f_3(X) = \sum\limits_{i=1}^{D} \left(\sum\limits_{j=1}^{i} x_j \right)^2$	$[-65, 65]^D$
$f_4(X) = 10D + \sum\limits_{i=1}^{D} (x_i^2 - 10\cos(2\pi x_i))$	$[-5.12, 5.12]^D$
$f_5(X) = \sum\limits_{i=1}^{D} \frac{x_i^2}{4000} - \prod\limits_{i=1}^{D} \cos\left(\frac{x_i}{\sqrt{i}} \right) + 1$	$[-600, 600]^D$
$f_6(X) = \sum\limits_{i=1}^{D} |x_i|^{(i+1)}$	$[-1, 1]^D$
$f_7(X) = -20\exp\left(-0.2\sqrt{\frac{\sum\limits_{i=1}^{D} x_i^2}{D}} \right) - \exp\left(\frac{\sum\limits_{i=1}^{D} \cos(2\pi x_i)}{D} \right) + 20 + e$	$[-32, 32]^D$
$f_8(X) = \sin^2(3\pi x_1) + \sum\limits_{i=1}^{D-1} (x_i - 1)^2(1 + \sin^2(3\pi x_{i+1})) + (x_D - 1)(1 + \sin^2(2\pi x_D))$	$[-10, 10]^D$
$f_9(X) = -\sum\limits_{i=1}^{D} \sin(x_i)(sin(ix_i^2/\pi))^{2m}, \, (m = 10)$	$[0, \pi]^D$
$f_{10}(X) = \sum\limits_{i=1}^{D} x_i^2 + \left(\sum\limits_{i=1}^{D} 0.5ix_i \right)^2 + \left(\sum\limits_{i=1}^{D} 0.5ix_i \right)^4$	$[-5, 10]^D$
$f_{11}(X) = \sum\limits_{i=1}^{D} |x_i| + \prod\limits_{i=1}^{D} |x_i|$	$[-10, 10]^D$
$f_{12}(X) = \sum\limits_{i=1}^{D} (\lfloor x_i + 0.5 \rfloor)^2$	$[-100, 100]^D$
$f_{13}(X) = \sum\limits_{i=1}^{D} |x_i \sin(x_i) + 0.1x_i|$	$[-10, 10]^D$
$f_{14}(X) = \exp\left(-0.5 \sum\limits_{i=1}^{D} x_i^2 \right)$	$[-1, 1]^D$
$f_{15}(X) = 1 - \cos(2\pi \| x \|) + 0.1 \| x \|, \text{ where } \| x \| = \sqrt{\sum\limits_{i=1}^{D} x_i^2}$	$[-100, 100]^D$

We compare the convergence speed of DE and ODE by measuring the number of function calls (NFC) which is the most commonly used metric in literature [5, 7, 8, 9, 10, 11, 19]; a smaller NFC means higher convergence speed. The termination criterion is to find a value smaller than the value-to-reach (VTR) before reaching the maximum number of function calls MAX$_{\text{NFC}}$. In order to minimize the effect of the stochastic nature of the algorithms on the metric, the reported number of function calls (NFC) for

Table 8.2. Parameter settings for conducted experiments

Parameter name	Setting	Reference
population size (N_p)	100	[20, 21, 22]
differential amplification factor (F)	0.5	[11, 20, 23, 24, 25]
crossover probability constant (C_r)	0.9	[11, 20, 23, 24, 25]
jumping rate constant (J_r)	0.3	[9, 13, 18]
maximum number of function calls (MAX$_{NFC}$) 10^6		[9, 13, 18]
value to reach (VTR)	10^{-8}	[10]
mutation strategy	$DE/rand/1/bin$	[5, 20, 23, 26, 27]

Table 8.3. Comparison of DE, ODE, and RDE. The best result for each case is highlighted in boldface. Results for RDE have been discussed in section 8.5.2 (corresponding results for replacing the opposite points with random points).

		DE			ODE			RDE		
F	D	NFC	SR	SP	NFC	SR	SP	NFC	SR	SP
f_1	30	87748	1	87748	47716	1	**47716**	115096	1	115096
f_2	30	96488	1	96488	53304	1	**53304**	126780	1	126780
f_3	20	177880	1	177880	168680	1	**168680**	231152	1	231152
f_4	10	328844	1	328844	70389	0.76	**92617**	501875	0.96	522786
f_5	30	113428	1	113428	69342	0.96	**72231**	149744	1	149744
f_6	30	25140	1	25140	8328	1	**8328**	29096	1	29096
f_7	30	169152	1	169152	98296	1	**98296**	222784	1	222784
f_8	30	101460	1	101460	70408	1	**70408**	138308	1	138308
f_9	10	191340	0.76	**251763**	213330	0.56	380946	306900	0.60	511500
f_{10}	30	385192	1	385192	369104	1	**369104**	498200	1	498200
f_{11}	30	187300	1	187300	155636	1	**155636**	244396	1	244396
f_{12}	30	41588	1	41588	23124	1	**23124**	54316	1	54316
f_{13}	30	411164	1	411164	337532	1	**337532**	927230	0.24	3863458
f_{14}	10	19528	1	19528	15704	1	**15704**	23156	1	23156
f_{15}	10	37824	1	37824	24260	1	**24260**	46800	1	46800
	SR$_{ave}$		0.98			0.95			0.92	

each function is the average over 50 different trials. The number of times, for which the algorithm successfully reaches the VTR for each test function is measured as the success rate SR:

$$SR = \frac{\text{number of times reached VTR}}{\text{total number of trials}}. \tag{8.7}$$

In order to combine these two measures (NFC and SR), a new measure, called success performance has been introduced as follows [10]:

$$SP = \frac{\text{average of NFC over successful runs)}}{SR}. \tag{8.8}$$

The parameter setting for all conducted experiments is summarized in Table 8.2.

In order to maintain a reliable and fair comparison, the parameter settings are kept the same for all conducted experiments, unless we mention new settings. Besides, for all experiments, the reported values are the average of the results for 50 independent runs. In addition, and most importantly, extra fitness evaluations required for the

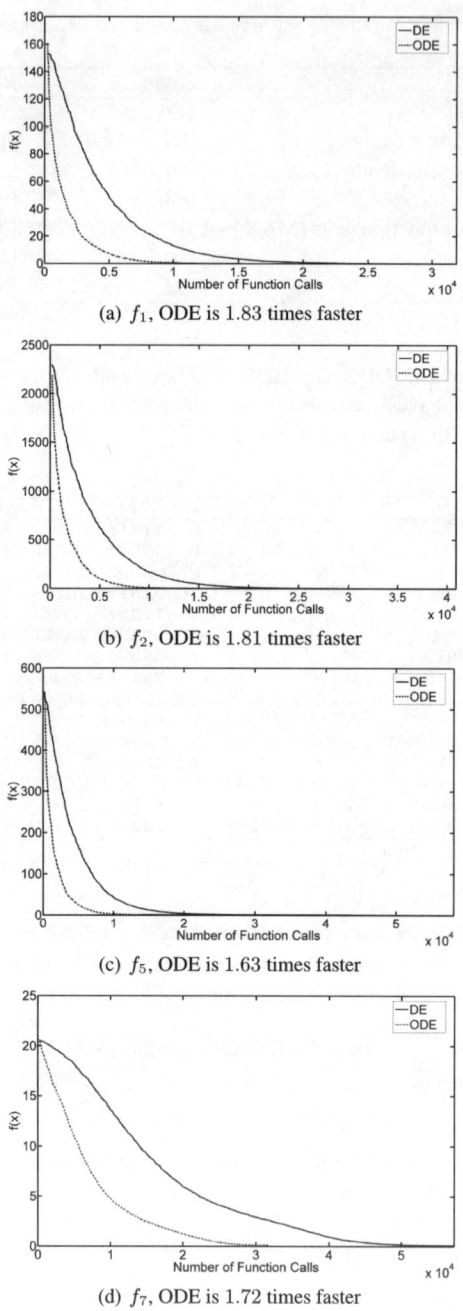

(a) f_1, ODE is 1.83 times faster

(b) f_2, ODE is 1.81 times faster

(c) f_5, ODE is 1.63 times faster

(d) f_7, ODE is 1.72 times faster

Fig. 8.7. Sample convergence graphs (best solution vs. number of function calls). As seen, ODE (dotted curve) shows better convergence speed than DE (solid curve) because it needs small amount of function calls to find the solution. is calculated by $\frac{\text{NFC}_{\text{DE}}}{\text{NFC}_{\text{ODE}}}$.

opposite points (both in population initialization and also generation jumping phases) are counted as well to accurately measure the benefit in spite of the additional overhead for computing the opposites.

The results for DE and ODE to solve the test problems are given in Table 8.3 (the results in the last column will be discussed in section 8.5.2). ODE outperforms DE on 14 benchmark functions with respect to the success performance. Some sample performance comparison graphs are presented in Figure 8.7. ODE (dotted curve) shows better convergence speed than DE (solid curve) because it needs smaller number of function calls to find the solution. With the same parameter settings for both algorithms and fixing the jumping rate for the ODE ($J_r = 0.3$), their success rates are comparable while ODE shows better convergence speed than DE. DE has a better success rate (SR) than ODE on 3 functions (f_4, f_5, and f_9). The jumping rate is an important control parameter which, if optimally set, can achieve even better results. Detailed discussions about this parameter can be found in [13].

On 7 multimodal functions (out of 8), ODE performs better than DE. This means that the opposition-based extension performs well even when the function contains many optima.

8.5.2 Contribution of Opposite Points

In this section, we verify whether the achieved acceleration rate for DE is really due to utilizing opposite points. For this purpose, all parts of the proposed algorithm remain unchanged and instead of using opposite points for the population initialization and the generation jumping, uniformly generated random points will be employed. In order to have a fair competition for this case, exactly like what we did for opposite points, the current interval (dynamic interval, $[\text{MIN}_j^p, \text{MAX}_j^p]$) of the variables are used to generate new random points in the generation jumping phase. So, line 4 in Algorithm 2 should be changed to:

$$RP_{0i,j} \longleftarrow l_j + (u_j - l_j) \times \text{RAND}(0,1),$$

where $\text{RAND}(0,1)$ generates a uniformly distributed random number on the interval $(0,1)$. In fact, instead of generating N_p, $2N_p$ random individuals are generated. In the same manner, line 30 in Algorithm 2 should be replaced with

$$RP_{i,j} \longleftarrow \text{MIN}_j^p + (\text{MAX}_j^p - \text{MIN}_j^p) \times \text{RAND}(0,1).$$

As mentioned before, the current boundaries of the variables ($[\text{MAX}_j^p, \text{MIN}_j^p]$) are used to generate random numbers for generation jumping. And finally, in order to have the same selection method, lines 7 and 33 in Algorithm 2 are substituted with

Select N_p fittest (best) individuals from P and RP as current population P;

After these modifications, the random version of ODE (called RDE) is established. Results for the current algorithm are presented in Table 8.3. As apparent, RDE can not outperform DE or ODE on any of benchmark function with respect to the success performance. This clearly demonstrates that the achieved improvements are due to usage of opposite points, and that the same level of improvement cannot be achieved via additional random sampling [14, 15].

8.6 Conclusions and Future Work

In this chapter, we briefly reviewed how opposition-based optimization can be employed to accelerate convergence speed of differential evolution by embedding opposition-based population initialization and opposition-based generation jumping. The experimental results confirmed that ODE provides a higher performance than the classical DE. However, opposition-based optimization is still in its infancy and future research is required to fully investigate its benefits and drawbacks.

By replacing opposite points with uniformly generated random points in the same variables' range, the resulted algorithm (RDE) performs slower than the parent algorithm (DE). Therefore, the contribution of opposite points to the acceleration process was confirmed and was not reproducible by additional random sampling.

The benefits of opposition-based optimization is most likely not the same for different problems. This is because of using fixed settings for the parameters and/or the different characteristics of each problem (e.g., modality, dimension, surface features, separability of the variables and so on). Similar to all optimization approaches, ODE does not present a consistent behavior over different problems. However, over the employed benchmark test suite, ODE performed better than classical DE.

The proposed opposition-based schemes are general enough to be applied to other population-based algorithms. The opposition-based schemes work at the population level and leave the evolutionary part of the algorithms untouched. This generality gives higher flexibility to these schemes to be embedded inside other population-based algorithms.

Opposition-based optimization opens new perspectives to accelerate optimization processes. For most practical applications, we are faced with constrained functions and also with multi-objective problems. So far, there are many approaches for handling constraints in DE and also for multi-objective optimization using DE. All of these proposals can be borrowed and investigated to generalize ODE to solve multi-objective constrained problems.

References

1. Feoktistov, V.: Differential Evolution. In: Search of Solutions. Springer, USA (2006)
2. Storn, R., Price, K.: Differential Evolution - a Simple and Efficient Adaptive Scheme for Global Optimization over Continuous Spaces, Technical Report in ICSI, TR-95-012 (1995)
3. Price, K., Storn, R.: Differential Evolution: Numerical Optimization Made Easy. Dr. Dobb's Journal 220, 18–24 (1997)
4. Storn, R., Price, K.: Differential evolution: A simple and efficient heuristic for global optimization over continuous spaces. Journal of Global Optimization 11(6), 341–359 (1997)
5. Price, K., Storn, R.M., Lampinen, J.A.: Differential Evolution: A Practical Approach to Global Optimization (Natural Computing Series), 1st edn. Springer, Heidelberg (2005)

6. Rahnamayan, S., Tizhoosh, H.R., Salama, M.M.A.: A Novel Population Initialization Method for Accelerating Evolutionary Algorithms. Elsevier Journal on Computers and Mathematics with Applications 53(10), 1605–1614 (2007)
7. Andre, J., Siarry, P., Dognon, T.: An Improvement of the Standard Genetic Algorithm Fighting Premature Convergence in Continuous Optimization. Advance in Engineering Software 32, 49–60 (2001)
8. Hrstka, O., Kučerová, A.: Improvement of Real Coded Genetic Algorithm Based on Differential Operators Preventing Premature Convergence. Advance in Engineering Software 35, 237–246 (2004)
9. Rahnamayan, S., Tizhoosh, H.R., Salama, M.M.A.: Opposition-Based Differential Evolution Algorithms. In: IEEE Congress on Evolutionary Computation (CEC 2006), IEEE World Congress on Computational Intelligence, Vancouver, Canada, pp. 7363–7370 (2006)
10. Suganthan, P.N., Hansen, N., Liang, J.J., Deb, K., Chen, Y.-P., Auger, A., Tiwari, S.: Problem Definitions and Evaluation Criteria for the CEC 2005 Special Session on Real-Parameter Optimization, Technical Report, Nanyang Technological University, Singapore And KanGAL Report Number 2005005 (Kanpur Genetic Algorithms Laboratory, IIT Kanpur) (2005)
11. Vesterstroem, J., Thomsen, R.: A Comparative Study of Differential Evolution, Particle Swarm Optimization, and Evolutionary Algorithms on Numerical Benchmark Problems. In: Proceedings of the Congress on Evolutionary Computation (CEC 2004), vol. 2, pp. 1980–1987. IEEE Publications, Los Alamitos (2004)
12. Tizhoosh, H.R.: Opposition-Based Learning: A New Scheme for Machine Intelligence. In: Proceedings of International Conference on Computational Intelligence for Modelling Control and Automation - CIMCA 2005, Vienna - Austria, vol. I, pp. 695–701 (2005)
13. Rahnamayan, S., Tizhoosh, H.R., Salama, M.M.A.: Opposition-Based Differential Evolution (ODE). Journal of IEEE Transactions on Evolutionary Computation 12(1), 64–79 (2008)
14. Rahnamayan, S., Tizhoosh, H.R., Salama, M.M.A.: Opposition versus Randomness in Soft Computing Techniques. Elsevier Journal on Applied Soft Computing 8, 906–918 (2008)
15. Rahnamayan, S.: Opposition-Based Differential Evolution, PhD Thesis, Department of Systems Design Engineering, University of Waterloo, Waterloo, Canada (2007)
16. Eiben, A.E., Hinterding, R.: Paramater Control in Evolutionary Algorithms. IEEE Transactions on Evolutionary Computation 3(2), 124–141 (1999)
17. Das, S., Konar, A., Chakraborty, U.K.: Two Improved Differential Evolution Schemes for Faster Global Search. In: Proceedings of the 2005 conference on Genetic and evolutionary computation, Washington, USA, pp. 991–998 (2005)
18. Rahnamayan, S., Tizhoosh, H.R., Salama, M.M.A.: Opposition-Based Differential Evolution (ODE) With Variable Jumping Rate. In: Proc. of IEEE Symposium on Foundations of Computational Intelligence, Honolulu, Hawaii, USA, pp. 81–88 (2007)
19. Rahnamayan, S., Tizhoosh, H.R., Salama, M.M.A.: Opposition-Based Differential Evolution for Optimization of Noisy Problems. In: IEEE Congress on Evolutionary Computation (CEC 2006), IEEE World Congress on Computational Intelligence, Vancouver, Canada, pp. 6756–6763 (2006)
20. Brest, J., Greiner, S., Bošković, B., Mernik, M., Žumer, V.: Self-Adapting Control Parameters in Differential Evolution: A Comparative Study on Numerical Benchmark Problems. Journal of IEEE Transactions on Evolutionary Computation 10(6), 646–657 (2006)
21. Lee, C.Y., Yao, X.: Evolutionary programming using mutations based on the Lvy probability distribution. IEEE Transactions on Evolutionary Computation 8(1), 1–13 (2004)
22. Yao, X., Liu, Y., Lin, G.: Evolutionary programming made faster. IEEE Transactions on Evolutionary Computation 3(2), 82–102 (1999)
23. Storn, R., Price, K.: Differential Evolution- A Simple and Efficient Heuristic for Global Optimization over Continuous Spaces. Journal of Global Optimization 11, 341–359 (1997)

24. Liu, J., Lampinen, J.: A fuzzy adaptive differential evolution algorithm. Soft Computing-A Fusion of Foundations, Methodologies and Applications 9(6), 448–462 (2005)
25. Ali, M.M., Trn, A.: Population set-based global optimization algorithms: Some modifications and numerical studies. Comput. Oper. Res. 31(10), 1703–1725 (2004)
26. Sun, J., Zhang, Q., Tsang, E.P.K.: DE/EDA: A new evolutionary algorithm for global optimization. Information Sciences 169, 249–262 (2005)
27. Onwubolu, G.C., Babu, B.V.: New Optimization Techniques in Engineering. Springer, Berlin (2004)
28. Brest, J., Greiner, S., Bošković, B., Mernik, M., Žumer, V.: Self-Adapting Control Parameters in Differential Evolution: A Comparative Study on Numerical Benchmark Problems. Journal of IEEE Transactions on Evolutionary Computation 10(6), 646–657 (2006)
29. Rahnamayan, S., Tizhoosh, H.R., Salama, M.M.A.: Quasi-Oppositional Differential Evolution. In: IEEE Congress on Evolutionary Computation (CEC 2007), Singapore, pp. 2229–2236 (September 2007)
30. Teo, J.: Exploring dynamic self-adaptive populations in differential evolution. Soft Computing - A Fusion of Foundations, Methodologies and Applications 10(8) (2006)
31. Tasoulis, D.K., Pavlidis, N.G., Plagianakos, V.P., Vrahatis, M.N.: Parallel Differential Evolution. In: Proceedings of the Congress on Evolutionary Computation (CEC 2004), vol. 2, pp. 2023–2029. IEEE Publications, Los Alamitos (2004)
32. Shi, Y.-J., Teng, H.-F., Li, Z.-Q.: Cooperative Co-evolutionary Differential Evolution for Function Optimization. In: Proceedings of First International Conference in Advances in Natural Computation (ICNC 2005), Changsha, China, pp. 1080–1088 (2005)
33. Fan, H.-Y., Lampinen, J.: A Trigonometric Mutation Operation to Differential Evolution. Global Optimization 27(1), 105–129 (2003)
34. Kaelo, P., Ali, M.M.: Probabilistic adaptation of point generation schemes in some global optimization algorithms. Optimization Methods and Software 27(3), 343–357 (2006)
35. Noman, N., Iba, H.: Enhancing differential evolution performance with local search for high dimensional function optimization. In: Proceedings of the 2005 conference on Genetic and evolutionary computation (GECCO 2005), Washington DC, USA, pp. 967–974 (2005)
36. Goldberg, D.E.: Genetic Algorithms in Search, Optimization, and Machine Learning. Addison-Wesley Longman Publishing Co., USA (2005)
37. Kennedy, J., Eberhart, R.: Particle swarm optimization. In: Proceedings of the IEEE International Conference on Neural Networks, Piscataway, NJ, pp. 1942–1948 (1995)

9

Evolving Opposition-Based Pareto Solutions: Multiobjective Optimization Using Competitive Coevolution

Tse Guan Tan and Jason Teo

School of Engineering and Information Technology, Universiti Malaysia Sabah, Malaysia
tseguantan@gmail.com, jtwteo@ums.edu.my

Summary. Recently a number of researchers have begun exploring the idea of combining Opposition-Based Learning (OBL) with evolutionary algorithms, reinforcement learning, neural networks, swarm intelligence and simulated annealing. However, an area of research that is still in infancy is the application of the OBL concept to coevolution. Hence, in this chapter, two new opposition-based competitive coevolution algorithms for multiobjective optimization called SPEA2-CE-HOF and SPEA2-CE-KR are discussed. These hybrid algorithms are the combination of Strength Pareto Evolutionary Algorithm 2 (SPEA2) with two types of the competitive fitness strategies, which are the Hall of Fame (HOF) and K-Random Opponents (KR), respectively. The selection of individuals as the opponents in the coevolutionary process strongly implements this opposition-based concept. Scalability tests have been conducted to evaluate and compare both algorithms against the original SPEA2 for seven Deb, Thiele, Laumanns, and Zitzler (DTLZ) test problems with 3 to 5 objectives. The experimental results show clearly that the performance scalability of the opposition-based SPEA2-CE-HOF and SPEA2-CE-KR were significantly better compared to the original non-opposition-based SPEA2 as the number of the objectives becomes higher in terms of the closeness to the true Pareto front, diversity maintenance and the coverage level.

9.1 Introduction

Coevolution is defined as the simultaneous evolution of two or more species with a coupled fitness [3]. One of the main advantages of the coevolutionary approach as a learning method is that coevolution does not necessarily need to specify a global fitness function to rank individuals in the population explicitly; rather only a relative fitness is needed [25]. At the population level, coevolution can be applied to the artificial evolution of individuals that compete with each other for superiority within their performing environment. The fitness of individuals thus depends on their superiority or otherwise when compared against opposing individuals that compete for survival within the population. Generally, coevolution can be divided into Cooperative Coevolution (CC) and Competitive Coevolution (CE). The cooperative coevolution [18] involves a number of individuals working together to solve the problem while the competitive coevolution involves individuals that compete against each other for dominance in the population. These are some well known examples using CE architecture and concept to solve

H.R. Tizhoosh, M. Ventresca (Eds.): Oppos. Concepts in Comp. Intel., SCI 155, pp. 161–201, 2008.

adversarial problems. Axelrod [13] used a Round Robin model or full competition strategy to obtain solutions to the Iterated Prisoner's Dilemma (IPD). Angeline and Pollack [12] presented a Single Elimination Tournament strategy to solve the Tic-Tac-Toe game. Hillis [31] used Bipartite tournament strategy to solve network sorting problems. Kim et al. [2] described an Entry Fee Change tournament topology to test some well-known problems such as sorting networks and the Nim game.

A variety of optimization solution techniques have been introduced for solving multiobjective problems or tasks [10]. Among these techniques, Evolutionary Algorithms (EAs) have been established as one of the most widely used methods. EAs are able to find a set of optimal solutions in a single run. Having this advantage, EAs are useful in the circumstance of multiobjective optimization, of which the task is to approximate the Pareto front of optimal solutions. Recently, numerous Multiobjective Evolutionary Algorithms (MOEAs) have been presented to solve real life problems [29]. However, a number of issues still remain with regards to MOEAs such as convergence to the true Pareto front as well as scalability to many objective problems rather than just bi-objective problems. The performance of these algorithms may be augmented by incorporating the coevolutionary concept. In this study, our objective is to conduct a comprehensive test for CE as an opposition-based method for evolving Pareto solutions to multiobjective problems. In our proposed algorithms, the Hall of Fame (HOF) and K-Random Opponents (KR) competitive fitness strategies will be integrated with the Strength Pareto Evolutionary Algorithm 2 (SPEA2). These modified SPEA2 algorithms will be referred to as SPEA2-CE-HOF and SPEA2-CE-KR. The performance of opposition-based coevolutionary SPEA2-CE-HOF and SPEA2-CE-KR algorithms will be compared against the original SPEA2 algorithm, which does not implement the opposition-based competitive coevolution, using the well-known multi-objective optimization DTLZ test problems. Generational distance, spacing and coverage metrics will be used to empirically validate the performance of the proposed opposition-based coevolutionary algorithms against its non-opposition-based counterpart.

The organization of this chapter is as follows. Section 9.2 provides a detailed description of Opposition-Based Learning. Then, the explanation on multiobjective optimization definitions and concepts are presented in Section 9.3. Section 9.4 presents the framework of SPEA2. Section 9.5 discusses the characteristics of each competitive fitness strategy. The integration between the CE and SPEA2 will be explained in Section 9.6. In Section 9.7, geometries of each of the test problems and the performance metrics used in the tests will be discussed, followed by the experimental settings in Section 9.8. The experimental results and discussions are given in Section 9.9. Finally, the conclusions and future work are presented in Section 9.10.

9.2   Proposed Use of Opposition

Opposition-Based Learning (OBL) was formally introduced by Tizhoosh [24]. OBL is a simple to understand yet powerful concept for machine learning. One can say that

whenever we are estimating, at the same time we ought to look at the opposite estimate. For instance, the quantity in genetic algorithms is chromosome, while the opposite quantity is anti-chromosome [33]. OBL has been used and shown to improve the
performances of evolutionary algorithms [4,5,19,20], reinforcement learning [6,22,23],
neural networks [27], swarm intelligence [15] and simulated annealing [28]. The opposite numbers [24] can be defined as follows:

Definition 1. *Let x be a real number in an interval $[a,b]$, ($x \in [a,b]$); the opposite
number $\breve{x}$ is defined by*

$$\breve{x} = a + b - x$$

In the same way, the definition of an opposite number in a multidimensional case can
be defined as follows [24]:

Definition 2. *Let $F(x_1, x_2, \ldots, x_n)$ be a point in n-dimensional space, where
$x_1, x_2, \ldots, x_n$
$\in \Re$ and $x_i \in [a_i, b_i]$, $\forall_i \in \{1, 2, \ldots, n\}$. The opposite number of F is defined by
$\breve{F}(\breve{x}_1, \breve{x}_2, \ldots, \breve{x}_n)$ where:*

$$\breve{x}_i = a_i + b_i - x_i$$

Now, by utilizing the above opposite number definition, the opposition-based competitive coevolution can be defined as follows:

Definition 3. *Let $P(I_1, I_2, \ldots, I_m)$ be a set of individuals in the population, where
$I_1, I_2, \ldots, I_m \in \mathbf{I}$ and m is the size of the population. Based on the opposite number
definition, the opposite set $\breve{P}$ can be defined by $\breve{P}(\breve{I}_1, \breve{I}_2, \ldots, \breve{I}_m)$ where*

$$\breve{I}_i = I_i - \mathbf{I}$$

The "−" denotes the operation "remove from". For example, if the population consist
of 4 individuals $P(I_1, I_2, I_3, I_4)$ and (I_1, I_2) are the individuals to be evaluated through
competitive coevolution, then

Non-opponent set $P(I_1, I_2)$

Opponent set $\breve{P}(\breve{I}_1, \breve{I}_2) = \breve{P}(I_3, I_4)$

These two sets of individuals will compete among each other for dominance. If the opponent individual is better than non-opponent individual, then the opponent individual
will replace the non-opponent individual, because competitive coevolution is based on
the law of survival of the fittest.

9.3 Multiobjective Optimization Problems

The common explanation of Multiobjective Optimization Problems (MOPs) [29] can be formally defined as:

minimize (or maximize) $z = f(v) = [f_1(v_1, \ldots, v_n), \ldots, f_m(v_1, \ldots, v_n)]$ subject to

$$\text{constraints } p \text{ are } \geq \text{ restriction } a_i(v) \geq 0 \quad (i = 1, 2, \ldots, p)$$

$$\text{constraints } q \text{ are } \leq \text{ restriction } b_i(v) \leq 0 \quad (i = 1, 2, \ldots, q)$$

$$\text{constraints } r \text{ are equality restriction } c_i(v) = 0 (i = 1, 2, \ldots, r)$$

where

$$v = (v_1, v_2, \ldots, v_n) \in V$$

$$z = (z_1, z_2, \ldots, z_m) \in Z$$

Vector v, $(v_1, v_2, \ldots, v_n)$ in the decision space V is called the decision vector and n is the number of decision variables. Additionally, the vector z, $(z_1, z_2, \ldots, z_m)$ in the objective space Z is called the objective vector and m is the number of the objective functions. The entire decision vector under consideration must satisfy a given set of constraints to produce an optimum solution values via the objective functions.

The concept of *Pareto dominance* [9] is defined as follows. An objective vector z^1 is said to dominate another objective vector z^2 or z^1 is said to be nondominated by z^2, $(z^1 \succ z^2)$ if z^1 is not worse than z^2 with respect to every objective and z^1 is strictly better than z^2 with respect to at least one objective. An optimal decision vector to a multiobjective problem is denoted as *Pareto optimal* if it is nondominated concerning the entire decision space, and at the same time its image in the objective space is denoted as a Pareto optimal objective vector. The whole of all Pareto optimal objective vectors is called the *Pareto front* and the set of all the Pareto optimal decision vectors is called the *Pareto optimal set*.

9.4 Multiobjective Evolutionary Algorithm: SPEA2

SPEA2 is a relatively new approach for finding or approximating the Pareto set in multiobjective optimization problems, which is the enhanced version of the original SPEA. SPEA2 proposed by Zitzler et al. in [11]. This algorithm is an elitist multiobjective evolutionary algorithm which incorporates a fine-grained fitness assignment strategy, a density estimation technique and an improved archive truncation method. SPEA2 is selected as the modified algorithm because it is one of the current state-of-the-art MOEAs. Furthermore, the performance of SPEA2 is better than other present day well-known algorithms, such as Pareto Envelope-based Selection Algorithm (PESA) and Nondominated Sorting Genetic Algorithm 2 (NSGA2) in higher dimensional objective spaces [11]. Frequently, the MOEAs must attempt to accomplish two core aims, that are to guide the search towards the global Pareto optimal region and maintain population diversity in Pareto optimal front [30]. The general framework of SPEA2 is introduced as below:

Algorithm 1. SPEA2 Algorithm

Input:
M (offspring population size)
N (archive size)
T (maximum number of generations)
Output:
A^* (nondominated set)

Step1: Initialization: Generate an initial population P_0 and create the empty archive (external set) $A_0 = \phi$. Set $t = 0$.

Step2: Fitness_assignment: Calculate fitness value of individuals in P_t and A_t.

Step3: Environmental_selection: Copy all nondominated individuals in P_t and A_t to A_{t+1}. If size of A_{t+1} exceeds N then reduce A_{t+1} by means of the truncation operator, otherwise if size A_{t+1} is less than N the fill A_{t+1} with dominated individuals in P_t and A_t.

Step4: Termination: If $t \geq T$ or another stopping criterion is satisfied then set A^* to the set of decision vectors represented by the nondominated individuals in A_{t+1}. Stop.

Step5: Mating_selection: Perform binary tournament selection with replacement on A_{t+1} in order to fill the mating pool.

Step6: Variation: Apply recombination and mutation operators to the mating pool and set P_{t+1} to the resulting population. Increment generation counter ($t = t + 1$) and go to Step 2.

9.5   Competitive Fitness Strategies

Thus far, there have been only a very limited number of studies that incorporate competitive coevolution into MOEAs. Parmee and Watson [17] introduced cooperative coevolution in evolutionary multiobjective optimization for bi-objective and tri-objective problems to design airframes. Lohn et al. [14] presented a competitive coevolution genetic algorithm to solve bi-objective problems. In his approach, the tournament is held between two populations, which are the trial population and target population.

In terms of canonical coevolution, a number of different competitive fitness strategies have been used to implement the competitive coevolution, such as Hall of Fame, K-Random Opponents, Round Robin and Single Elimination Tournament, as shown in Fig. 9.1 to Fig. 9.4. Rosin and Belew [21] introduced a Hall of Fame tournament concept, whereby each member from the current generation will be competing with every

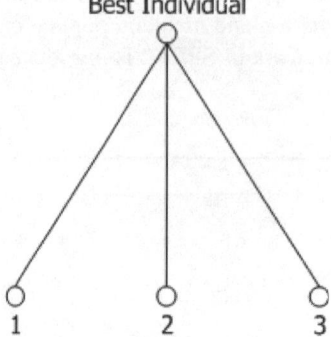

Fig. 9.1. Hall of fame

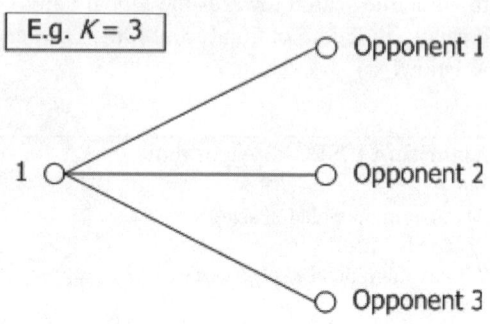

Fig. 9.2. *K*-random opponents

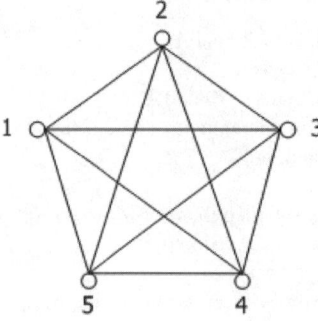

Fig. 9.3. Round robin

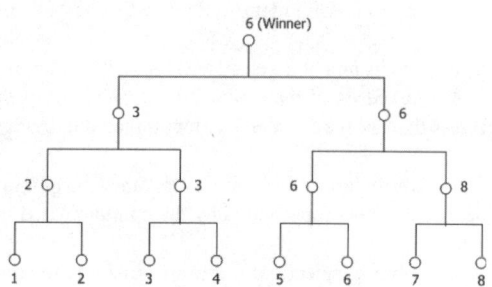

Fig. 9.4. Single elimination tournament

preserved best member from previous generations. Panait and Luke [16] explained a *K*-Random Opponents strategy, where each member will be competing against *K* other members of opponents. The opponents will be randomly selected from the same population without repeating the same opponents and rejecting self-play. Axelrod [13] described a Round Robin method as a full competition. Each member will be competing against every single member of the population and also disallowing self-play. Angeline and Pollack [12] presented a Single Elimination Tournament topology, whereby all members in the population will be randomly paired up and each pair will compete among themselves. For each competition, the loser will be eliminated, and the winner is preserved to the next level. After that, all the winners will also be randomly paired up and put to compete again, until a single winner is left.

9.6   Proposed Algorithms: SPEA2-CE-HOF and SPEA2-CE-KR

Two new opposition-based coevolutionary algorithms that integrate SPEA2 with HOF and KR respectively are presented. The resulting algorithms are referred to as

SPEA2-CE-HOF and SPEA2-CE-KR . Generally, the framework of the two hybrid algorithms are similar to the framework of SPEA2 with the exceptions of two additional methods, Opponents_selection and Reward_assignment as shown in below.

Algorithm 2. SPEA2-CE-HOF/SPEA2-CE-KR Algorithms

BEGIN
$gen = 0$
$Pop_s(gen)$ = randomly initialized population
Fitness_assignment $Pop_s(gen)$
Opponents_selection $Pop_s(gen)$ //choose the opponents based on HOF or KR
Reward_assignment $Pop_s(gen)$
Environmental_selection $Pop_s(gen)$

    WHILE Termination = False
      $gen = gen + 1$
      Mating_selection $Pop_s(gen)$
      Variation $Pop_s(gen)$
      Fitness_assignment $Pop_s(gen)$
      Opponents_selection $Pop_s(gen)$ //choose the opponents based on HOF or KR
      Reward_assignment $Pop_s(gen)$
      Environmental_selection $Pop_s(gen)$
    END

END SPEA2-CE-HOF / SPEA2-CE-KR

Fig. 9.5 shows the overall flow of SPEA2-CE-HOF and SPEA2-CE-KR. At the initialization state, the proposed algorithms randomly generate an initial population of real-valued vector (individuals). The individuals represent possible solutions to the problem. Then, the fitness values for each individual in the population are evaluated. Next, the Opponent_selection method will select individuals as the opponents based on the HOF or KR strategies. In the proposed algorithms, an archive is created to store all of the nondominated solutions. With this addition, SPEA2-CE-HOF's Opponents_selection method can easily select all the current best individuals as opponents from the archive. Meanwhile, SPEA2-CE-KR will randomly select opponents from the same population without repeating the identical opponents and prohibiting self-play. The K is tested with the values of 10, 20, 30, 40, 50, 60, 70, 80 and 90. After that, each individual will compete against the entire set of opponents. During the tournament, the reward value will be calculated for each competition by the reward function. Each reward value will be summed up as the fitness score for the individual using the Reward_assignment method. The number of competitions is based on the size of the archive or the K values. Subsequently, the archive update operation is executed. The archive is updated by copying nondominated individuals into the archive using the

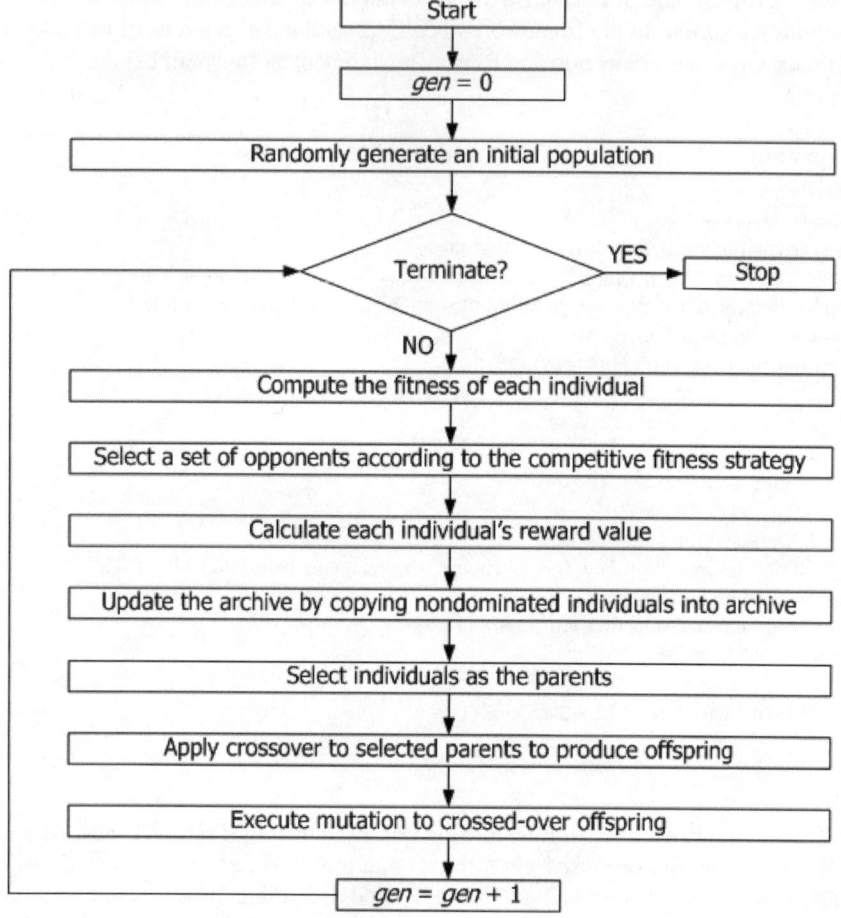

Fig. 9.5. Flowchart of SPEA2-CE-HOF and SPEA2-CE-KR

archive truncation method to maintain boundary solutions. Any dominated solutions will be removed from the archive during the update operation. Individuals which are then selected to form offspring according to their fitness by performing the genetic operations of simulated binary crossover and polynomial mutation. Each loop iteration is referred to as a generation. The run of SPEA2-CE-HOF and SPEA2-CE-KR terminates when the termination criterion is satisfied. The predefined maximum number of generations serves as the termination criterion of the loop.

Equation (9.1) illustrates the reward function . I represents the participating individual, while O represents the opponent. R is the raw fitness value, $\max(R)$ is the maximum raw fitness value and the $\min(R)$ is the minimum raw fitness value. The range for values in this function is within [-1, 1]. If *Reward*(I, O) = 0, it corresponds to the competition being a draw.

$$Reward(I, O) = \frac{R(O) - R(I)}{\max(R) - \min(R)} \tag{9.1}$$

9.7 Test Problems and Performance Metrics

The algorithms will be benchmarked using seven test problems, DTLZ1 to DTLZ7 for the following reasons [8]:

- It is one of the latest sets of test problems for multiobjective benchmarking problems but more importantly, these problems can be tested with varying numbers of decision variables and objectives.
- The structure of the test problems is easy created by using a bottom-up technique.
- The test problems have exact and known shapes as well as the location of the resulting true Pareto front. Also, the corresponding optimal decision variable values are known.

The true Pareto front for the entire seven test problems are displayed in Fig. 9.6 to Fig. 9.9. These problems are M-objective problems. Table 9.1 summarizes the geometrical properties of the DTLZ test problems.

Table 9.1. The geometries of DTLZ

Test Problem	Geometry
DTLZ1	Linear
DTLZ2, DTLZ3, DTLZ4	Concave
DTLZ5, DTLZ6	Curve
DTLZ7	Disconnected

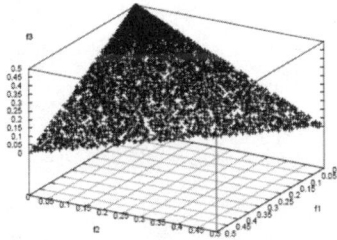

Fig. 9.6. True Pareto front of DTLZ1

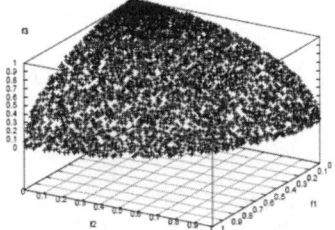

Fig. 9.7. True Pareto front of DTLZ2 to DTLZ4

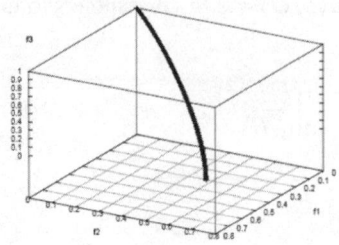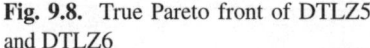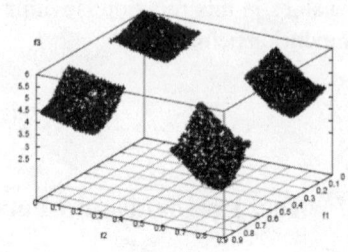

Fig. 9.8. True Pareto front of DTLZ5 and DTLZ6

Fig. 9.9. True Pareto front of DTLZ7

Three performance metrics were used to validate the proposed algorithms, namely generational distance, spacing and coverage metrics.

Generational distance (*GD*)**:** This metric was proposed by Van Veldhuizen and Lamont [26] and is used for estimating how far the elements in the Pareto front obtained are from the true Pareto front of the problem. This metric is defined as:

$$GD = \frac{\sqrt{\sum_{i=1}^{n} d_i^2}}{n} \tag{9.2}$$

where n is the number of nondominated vectors found by the algorithm being analyzed and d_i is the Euclidean distance (measured in objective space) between each of these and the nearest member of the true Pareto front. A value of $GD = 0$ indicates that all the elements generated are in the true Pareto front of the problem. Therefore, any other value will indicate how "far" the obtained solutions are from the global Pareto front of our problem.

Spacing (*SP*)**:** This metric was introduced by Schott [32] and measures how evenly the points in the approximation set are distributed in the objective space. This metric is defined as:

$$SP = \sqrt{\frac{1}{n-1} \sum_{i=1}^{n} (\bar{d} - d_i)^2} \tag{9.3}$$

where

$$d_i = \min_j (\sum_{k=1}^{m} |f_m^i - f_m^j|)$$

$i, j = 1, \ldots, n$ and m is the number of objectives, $\overline{d}$ is the mean of all d_i, and n refers to the number of elements of Pareto optimal set found so far. If $SP = 0$ it means that all the nondominated solutions found are equidistantly spaced.

Coverage metric (C): This metric was proposed by Zitzler et al. [7]. By using this metric, two sets of nondominated solutions can be compared to each other. Consider X', X'' as two sets of phenotype decision vectors. C is defined as the mapping of the ordered pair (X', X'') to the interval [0, 1]:

$$C(X', X'') \triangleq \frac{|\{a'' \in X''; \exists a' \in X' : a' \succeq a''\}|}{|X''|} \quad (9.4)$$

$a' \succeq a''$ if a' dominates a'' or a' equal to a''. $C(X', X'') = 1$ means that all the decision vectors in X'' are dominated by X'. Otherwise, $C(X', X'') = 0$ represents the situation when none of the points in X'' are dominated by X'. If $C(X', X'') > C(X'', X')$, then X' is better than X''. Fig. 9.10 and Fig. 9.11 show the graphical presentations for coverage metric.

Fig. 9.10. $C(X', X'') = 0$ **Fig. 9.11.** $C(X', X'') = 1$

9.8 Experimental Settings

The following common parameters and operators were based on the previous study of Deb [8]. In order to have a fair comparison across all algorithms, all runs considered are implemented with the same real-valued representation, simulated binary crossover (SBX) [1], polynomial mutation [1] and binary tournament selection as shown in Table 9.2. The number of evaluations in each run is fixed at 60,000. Table 9.3 lists all the parameter settings for each evolutionary multiobjective optimization algorithm.

Table 9.2. Evolutionary settings

Representation	: Real valued vectors
Crossover	: Simulated binary crossover
Mutation	: Polynomial mutation
Selection	: Binary tournament selection

Table 9.3. Parameter settings

Parameter	SPEA2, SPEA2-CE-HOF, SPEA2-CE-KR
Population size	100
Archive size	100
Number of decision variables per generation	12
Number of objectives	3 to 5
Number of generations	600
Mutation probability	0.08
Crossover probability	1
Polynomial mutation operator	20
SBX crossover operator	15
Number of repeated runs	30

9.9   Experimental Results and Discussions

The graphical presentations in box plot format of generational distance and spacing are shown in Fig. 9.12 to Fig. 9.14. The leftmost box plot relates to SPEA2, second left-most box plot relates to SPEA2-CE-HOF while from the third leftmost to the rightmost relate to SPEA2-CE-KR with a set of K values (10, 20, 30, 40, 50, 60, 70, 80 and 90). The dark dash is the median, the top of the box is the upper quartile, and the bottom of the box is the lower quartile. The graphical presentations for the test problems in terms of the coverage metric are shown in Fig. 9.15. Each rectangle contains seven box plots representing the distribution of the C value; the leftmost box plot relates to DTLZ1 while the rightmost box plot relates to DTLZ7. The box plots of SPEA2-CE-KR with 90 opponents are selected to be presented in Fig. 9.15, because the results show that the proposed algorithm with 90 opponents has a very good coverage level compared with other K values. In addition, Appendix presents the detailed description of the experimental results.

In the figures and tables, some symbols are utilized to represent the name of the the algorithms. The symbol S2 corresponds to the SPEA2 algorithm and the symbol CH represents the SPEA2-CE-HOF. The symbol CK indicates the SPEA2-CE-KR and the number refers to the size of the opponents. For example, CK10 corresponds to the SPEA2-CE-KR with 10 random opponents. From the results, the following can be observed.

Generational distance (GD): SPEA2-CE-KR is shown to be better than SPEA2 and SPEA2-CE-HOF in almost all the DTLZ test problems except DTLZ5 with 3 objectives. However for 4 and 5 objectives, the results reveal that SPEA2-CE-KR strongly outperformed SPEA2 and SPEA2-CE-HOF for the entire set of test problems. In addition, the mean GD values for most of the SPEA2-CE-KR are close to zero, which indicates that SPEA2-CE-KR is very near to the true Pareto front. Furthermore, a major enhancement in the search results produced by SPEA2-CE-KR can be obviously

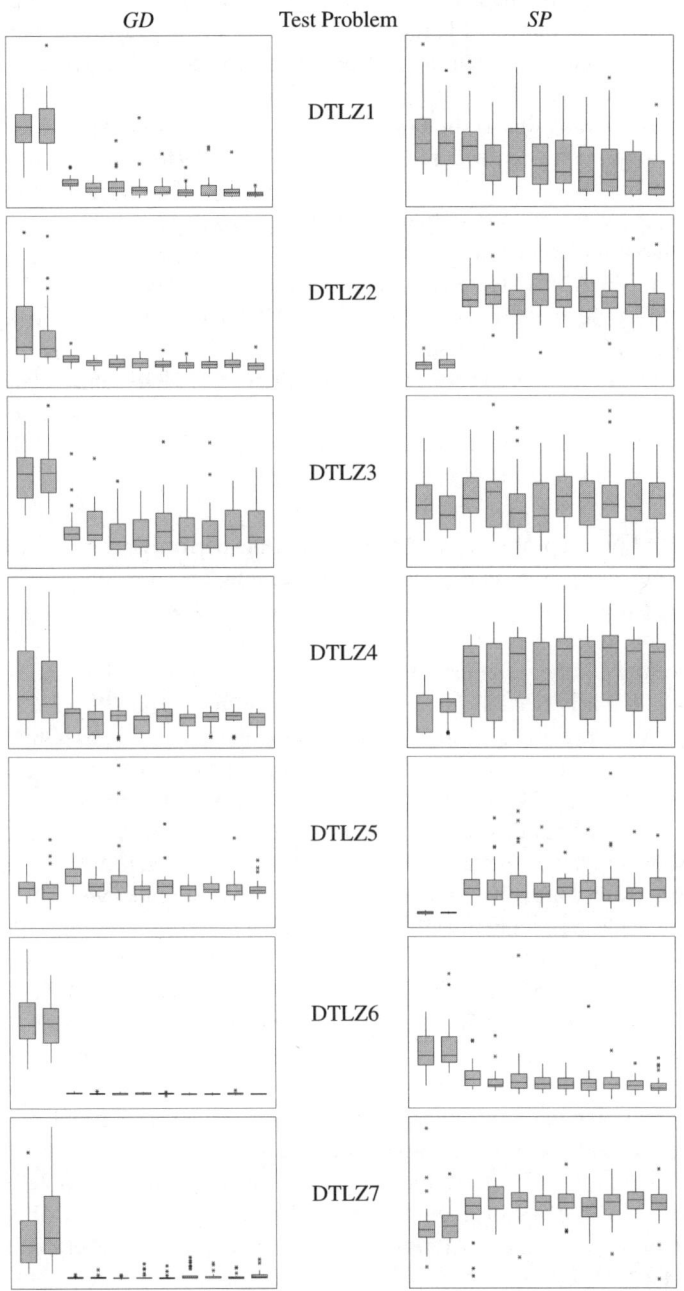

Fig. 9.12. Box plots of SPEA2, SPEA2-CE-HOF and SPEA2-CE-KR for generational distance (*GD*) and spacing (*SP*) with 3 objectives

noticed in the test problems DTLZ1, DTLZ3 and DTLZ7, which are problems with linear, concave or disconnected Pareto front. Overall, the performance of SPEA2-CE-KR significantly improved when the number of the objectives increased. It means that SPEA2-CE-KR has a far enhanced capability to escape from the sub-optimal local solutions in its exploration process for a more optimal solution. As observed in *GD* results, having larger opponent sizes will generally improve the performance of the augmented SPEA2 algorithm. Moreover, it scales very well in terms of convergence. It means that the KR competitive coevolution effect helps in creating a set of solutions that are very close the to the true Pareto front.

In addition, SPEA2-CE-HOF is better than SPEA2 in most of the DTLZ test problems with 3 to 5 objectives, in which SPEA2-CE-HOF performed better in 4 of 7 DTLZ problems. The SPEA2 with HOF strategy can achieve a good convergence, where a major improvement in the search performance of SPEA2-CE-HOF can be observed in the test problem DTLZ6. In the rest of the test problems, the proposed algorithm seems to only contribute a minor improvement, compared to the KR approach. Also, SPEA2-CE-HOF scales well in terms of convergence metric.

Spacing (*SP*): For test problems with 3 objectives, the performances of SPEA2-CE-KR against SPEA2 and SPEA2-CE-HOF are comparable, because SPEA2-CE-KR combined all the different objectives into a single raw fitness values. The tournament had focused only on some objectives caused by using raw fitness values as the competitive criteria. Once the number of objectives were increased to four and five, the performance of SPEA2-CE-KR is noticeably better than both algorithms. The SPEA2-CE-KR shows a slight improvement for the solution distribution in the test problems with three objectives. In contrast, a major improvement in the optimization results produced by the proposed algorithm can be noticed in the test problems with four and five objectives. Also, SPEA2-CE-KR scales very well in terms of the spacing metric. Here, the SPEA2-CE-KR is shown to have a better spread of solutions compared to SPEA2 and SPEA2-CE-HOF, especially for problems with a higher number of objectives. This means that SPEA2-CE-KR has an excellent ability to maintain a diverse set of final solutions.

SPEA2-CE-HOF performed moderately in most of the test problems for the diversity index. In the 3-objective problems, only two test problems, DTLZ1 and DTLZ3 performed better compared with the SPEA2. The reason may be that SPEA2-CE-HOF always preserves the existence of extreme solutions with higher reward scores, so this can cause worsening in the spacing values. But the performance of SPEA2-CE-HOF increased in the 4-objective problems, four of the DTLZ problems performed well, which are DTLZ1, DTLZ2, DTLZ5 and DTLZ6. Similarly, in the 5-objective problems, SPEA2-CE-HOF performed better for three DTLZ test problems, DTLZ1, DTLZ3 and DTLZ7. A major improvement in the search performance of SPEA2-CE-HOF can be observed in the test problem DTLZ1. However, in test problems DTLZ2, DTLZ3, DTLZ5, DTLZ6 and DTLZ7, there was only a minor enhancement. Moreover, SPEA2-CE-HOF scales less favorably in terms of the spacing metric.

Coverage (*C*): SPEA2-CE-KR shows regular coverage of nondominated solutions for DTLZ test problems with 3 objectives. But for 4 and 5 objectives, the obtained non-

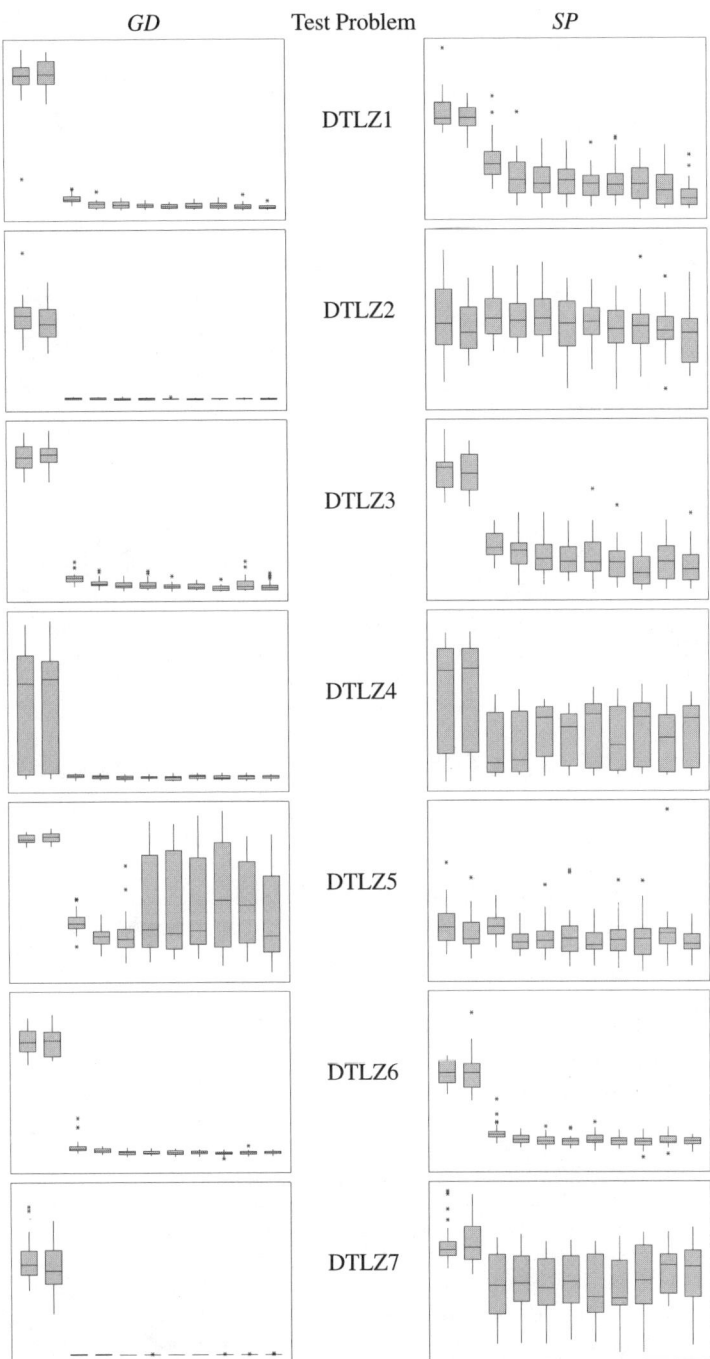

Fig. 9.13. Box plots of SPEA2, SPEA2-CE-HOF and SPEA2-CE-KR for generational distance (*GD*) and spacing (*SP*) with 4 objectives

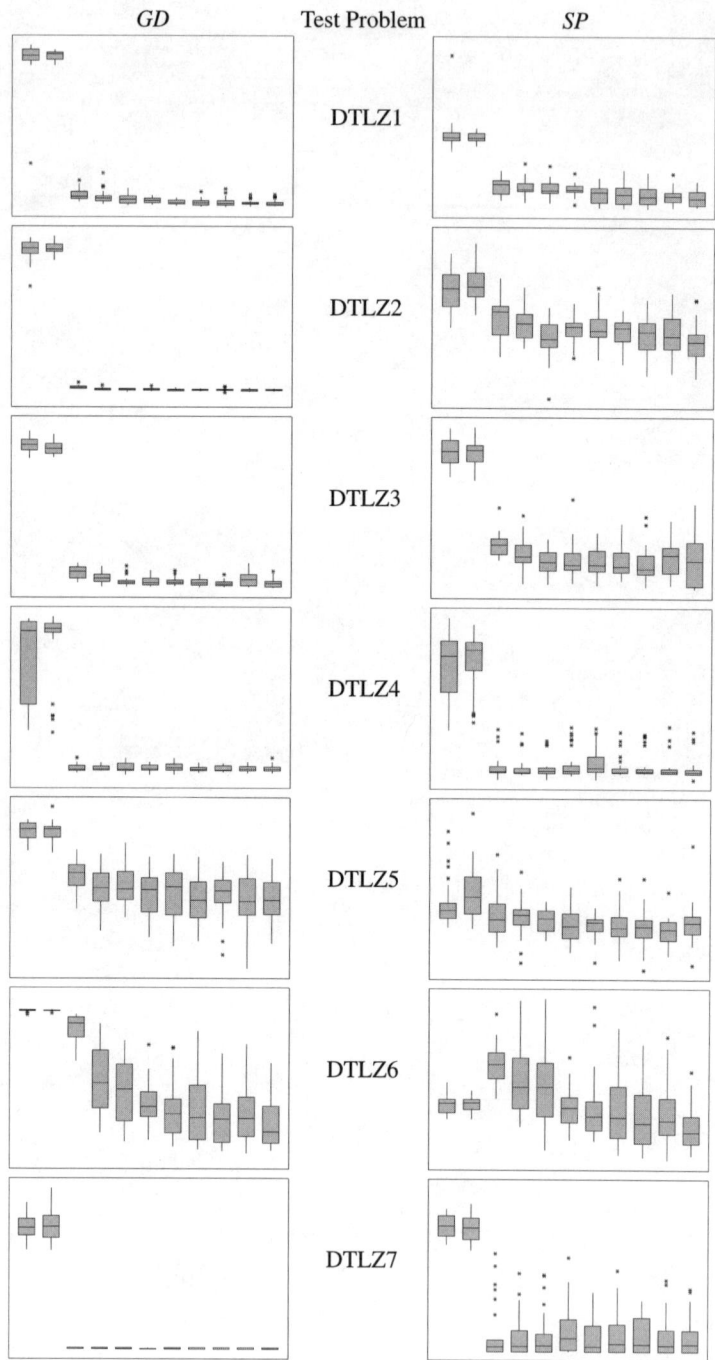

Fig. 9.14. Box plots of SPEA2, SPEA2-CE-HOF and SPEA2-CE-KR for generational distance (*GD*) and spacing (*SP*) with 5 objectives

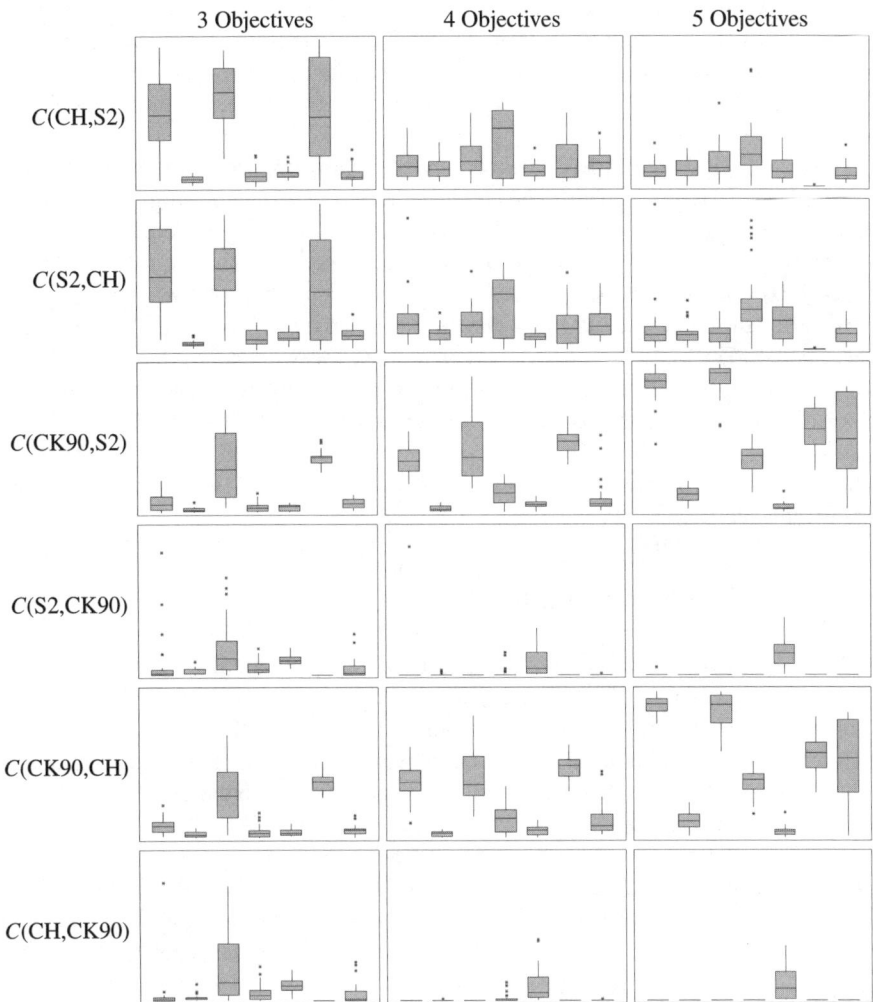

Fig. 9.15. Box plots of SPEA2, SPEA2-CE-HOF and SPEA2-CE-KR for coverage metric with 3 to 5 objectives

dominated solutions found by SPEA2-CE-KR clearly dominated the obtained nondominated solutions found by SPEA2 and SPEA2-CE-HOF in almost all of the DTLZ test problems, since C(SPEA2-CE-KR, SPEA2) $>$ C(SPEA2, SPEA2-CE-KR) and C(SPEA2-CE-KR, SPEA2-CE-HOF) $>$ C(SPEA2-CE-HOF, SPEA2-CE-KR). The SPEA2-CE-KR's weakest results are in DTLZ5 against SPEA2 and SPEA2-CE-HOF. As can be seen from the box plots of 4 and 5 objectives, some box plots in the column C(SPEA2, SPEA2-CE-KR) and C(SPEA2-CE-HOF, SPEA2-CE-KR) are at the bottom of the rectangle, which indicates that the proposed algorithm is completely free from the domination of SPEA2 and SPEA2-CE-HOF. Basically, it can be observed that with increasing values of K and the number of objectives, the dominance of SPEA2-CE-KR becomes greater over both algorithms. Additionally, SPEA2-CE-KR scales very well in terms of the coverage metric. A major improvement in the search performance of SPEA2-CE-KR can be observed in almost all the test problems, except for the DTLZ5 problem.

SPEA2-CE-HOF and SPEA2 have comparable performances for the coverage metric. SPEA2-CE-HOF is shown to be better in the 3- and 4-objective problems. But, the performance of this proposed algorithm decreased in the 5-objective problems. In the 3-objective problems, SPEA2-CE-HOF had a good performance for DTLZ2, DTLZ3 and DTLZ6, which have multiple local Pareto fronts, in order to attain good C values. The performance increased in the 4-objective problems for DTLZ2 to DTLZ6. On the other hand, SPEA2-CE-HOF performed better only for two test problems, DTLZ2 and DTLZ3 in the 5-objective problems. The results show that the use of HOF does have a minor improvement for the coverage level in test problems DTLZ4 to DTLZ6. In contrast, a major enhancement in the search results produced by SPEA2-CE-HOF can be noticed in the test problems DTLZ2 and DTLZ3. SPEA2-CE-HOF scales moderately well for this metric.

9.10   Conclusions and Future Work

In this chapter, two newly proposed opposition-based competitive coevolution algorithms, named SPEA2-CE-HOF and SPEA2-CE-KR, have been studied. A thorough comparison performance between these algorithms against the original SPEA2 were demonstrated using a set of DTLZ test problems. According to the experimental results, the analysis reveals that the overall performances of both the SPEA2-CE-KR and SPEA2-CE-HOF algorithms are superior to that of SPEA2. SPEA2-CE-KR was found to produce better results in terms of generational distance, spacing and coverage in almost all of the test problems while SPEA2-CE-HOF improved the performance of SPEA2 in some of the problems. This is a very promising indication that the inclusion of opposition-based competitive coevolution as an augmentation to the original SPEA2 algorithm is highly beneficial for improving its performance on evolutionary multi-objective optimization.

Several issues still remain for future research. It would be interesting to investigate whether a hybrid structure of competitive coevolution together with cooperative coevolution would be able to further improve the performance of MOEAs, since cooperative coevolution can focus on the individual objectives using speciation and competitive coevolution can focus on the opposition-based learning between individual solutions. The enhanced algorithm has been focusing on 3 to 5 objectives only, therefore it would also be highly informative to conduct further tests of scalability in higher dimensions for the proposed algorithms. Also, it would be worthwhile to test SPEA2-CE using different competitive fitness strategies, such as single elimination tournament. Based on Panait and Luke [16], the single elimination tournament is superior to K-Random opponents in noise free problems.

Acknowledgments

The first author is supported by a National Science Fellowship (NSF) from the Ministry of Science, Technology and Innovation (MOSTI), Malaysia.

References

1. Deb, K., Goyal, M.: A Combined Genetic Adaptive Search (GeneAS) for Engineering Design. Computer Science and Informatics 26(4), 30–45 (1996)
2. Kim, Y.K., Kim, J.Y., Kim, Y.H.: A Tournament-Based Competitive Coevolutionary Algorithm. Applied Intelligence 20(3), 267–281 (2004)
3. Paredis, J.: Coevolutionary computation. Artificial life 2(4), 355–375 (1995)
4. Rahnamayan, S., Tizhoosh, H.R., Salama, M.M.A.: A Novel Population Initialization Method for Accelerating Evolutionary Algorithms. Computers and Mathematics with Applications 53(10), 1605–1614 (2007)
5. Rahnamayan, S., Tizhoosh, H.R., Salama, M.M.A.: Opposition-Based Differential Evolution. IEEE Transactions on Evolutionary Computation 12(1), 64–79 (2008)
6. Tizhoosh, H.R.: Opposition-Based Reinforcement Learning. Journal of Advanced Computational Intelligence and Intelligent Informatics 10(4), 578–585 (2006)
7. Zitzler, E., Deb, K., Thiele, L.: Comparison of Multiobjective Evolutionary Algorithms: Empirical Results. Evolutionary Computation 8(2), 173–195 (2000)
8. Deb, K., Thiele, L., Laumanns, M., Zitzler, E.: Scalable Test Problems for Evolutionary Multi-Objective Optimization. KanGAL Report 2001001, Kanpur Genetic Algorithms Laboratory (KanGAL), Department of Mechanical Engineering, Indian Institute of Technology Kanpur, India (2001)
9. Knowles, J., Thiele, L., Zitzler, E.: A Tutorial on the Performance Assessment of Stochastic Multiobjective Optimizers. TIK Report 214, Computer Engineering and Networks Laboratory (TIK), Swiss Federal Institute of Technology (ETH) Zurich, Switzerland (2006)
10. Zheng, C., Wang, P.: Application of Flow and Transport Optimization Codes to Groundwater Pump-and-Treat Systems: Umatilla Army Depot, Oregon. Technical Report Revised 2/2003, University of Alabama, USA (2001)
11. Zitzler, E., Laumanns, M., Thiele, L.: SPEA2: Improving the Strength Pareto Evolutionary Algorithm. Technical Report 103, Computer Engineering and Network Laboratory (TIK), Swiss Federal Institute of Technology (ETH) Zurich, Switzerland (2001)
12. Angeline, P.J., Pollack, J.B.: Competitive Environments Evolve Better Solutions for Complex Tasks. In: Forrest, S. (ed.) Proc. 5th International Conference on Genetic Algorithm, pp. 264–270. Morgan Kaufmann, San Francisco (1993)
13. Axelrod, R.: The Evolution of Strategies in the Iterated Prisoner's Dilemma. In: Proc. Genetic Algorithms and Simulated Annealing, pp. 32–41. Morgan Kaufmann, San Francisco (1987)
14. Lohn, J.D., Kraus, W.F., Haith, G.L.: Comparing a Coevolutionary Genetic Algorithm for Multiobjective Optimization. In: Proc. 2002 Congress on Evolutionary Computation (CEC 2002), pp. 1157–1162. IEEE Press, Los Alamitos (2002)
15. Malisia, A.R., Tizhoosh, H.R.: Applying Opposition-Based Ideas to the Ant Colony System. In: Proc. IEEE Swarm Intelligence Symposium, pp. 182–189 (2007)
16. Panait, L., Luke, S.: A Comparative Study of Two Competitive Fitness Functions. In: Langdon, W.B., et al. (eds.) Proc. Genetic and Evolutionary Computation Conference (GECCO 2002), pp. 503–511. Morgan Kaufmann, San Francisco (2002)

17. Parmee, I.C., Watson, A.H.: Preliminary Airframe Design Using Co-evolutionary Multiob-jective Genetic Algorithms. In: Banzhaf, W., Daida, J., Eiben, A.E., Garzon, M.H., Honavar, V., Jakiela, M., Smith, R.E. (eds.) Proc. Genetic and Evolutionary Computation Conference (GECCO 1999), vol. 2, pp. 1657–1665. Morgan Kaufmann, San Francisco (1999)
18. Potter, M.A., De Jong, K.A.: A Cooperative Coevolutionary Approach to Function Optimiza-tion. In: Proc. 3rd Parallel Problem Solving From Nature, pp. 249–257. Springer, Heidelberg (1994)
19. Rahnamayan, S., Tizhoosh, H.R., Salama, M.M.A.: Opposition-Based Differential Evolution Algorithms. In: Proc. IEEE Congress on Evolutionary Computation, pp. 7363–7370 (2006)
20. Rahnamayan, S., Tizhoosh, H.R., Salama, M.M.A.: Opposition-Based Differential Evolution Algorithms for Optimization of Noisy Problems. In: Proc. IEEE Congress on Evolutionary Computation, pp. 6756–6763 (2006)
21. Rosin, C.D., Belew, R.K.: New Methods for Competitive Co-evolution. Evolutionary Com-putation 5(1), 1–29 (1997)
22. Shokri, M., Tizhoosh, H.R., Kamel, M.: Opposition-Based Q(λ) Algorithm. In: Proc. IEEE International Joint Conference on Neural Networks, pp. 646–653 (2006)
23. Tizhoosh, H.R.: Reinforcement Learning Based on Actions and Opposite Actions. In: Proc. International Conference on Artificial Intelligence and Machine Learning (2005)
24. Tizhoosh, H.R.: Opposition-Based Learning: A New Scheme for Machine Intelligence. In: Proc. International Conference on Computational Intelligence for Modelling Control and Automation - CIMCA 2005, vol. 1, pp. 695–701 (2005)
25. Tomassini, M.: Evolutionary Algorithms. In: Sanchez, E., Tomassini, M. (eds.) Proc. Interna-tional Workshop on Towards Evolvable Hardware, the Evolutionary Engineering Approach, pp. 19–47. Springer, Heidelberg (1996)
26. Van Veldhuizen, D.A., Lamont, G.B.: On Measuring Multiobjective Evolutionary Algorithm Performance. In: Proc. 2000 Congress on Evolutionary Computation, vol. 1, pp. 204–211 (2000)
27. Ventresca, M., Tizhoosh, H.R.: Improving the Convergence of Backpropagation by Opposite Transfer Functions. In: Proc. IEEE International Joint Conference on Neural Networks, pp. 9527–9534 (2006)
28. Ventresca, M., Tizhoosh, H.R.: Simulated Annealing with Opposite Neighbors. In: Proc. IEEE Symposium on Foundations of Computational Intelligence (FOCI 2007), pp. 186–192 (2007)
29. Coello Coello, C.A.: Evolutionary Multi-Objective Optimization: A Critical Review. In: Sarker, R., Mohammadian, M., Yao, X. (eds.) Evolutionary Optimization, pp. 117–146. Kluwer Academic, Massachusetts (2002)
30. Deb, K., Goel, T.: Multi-Objective Evolutionary Algorithms for Engineering Shape Design. In: Sarker, R., Mohammadian, M., Yao, X. (eds.) Evolutionary Optimization, pp. 147–175. Kluwer Academic, Massachusetts (2002)
31. Hillis, W.D.: Co-evolving Parasites Improve Simulated Evolution as an Optimization Proce-dure. Emergent Computation, pp. 228–234. MIT Press, Cambridge (1991)
32. Schott, J.R.: Fault Tolerant Design Using Single and Multicriteria Genetic Algorithm Opti-mization. Master's thesis, Department of Aeronautics and Astronautics, Massachusetts Insti-tute of Technology, Cambridge, Massachusetts (1995)
33. Opposition-Based Learning Research Group, Table of Opposites (2008),
    http://gauss.uwaterloo.ca/table_of_opposites.htm

Appendix

Table 9.4. Summarization of mean and standard deviation of generational distance (*GD*) between SPEA2, SPEA2-CE-HOF and SPEA2-CE-KR for DTLZ1 test problem with 3 to 5 objectives. The boldface indicates the best mean values.

Algorithm	3 Objectives		4 Objectives		5 Objectives	
	Mean	St Dev	Mean	St Dev	Mean	St Dev
S2	9.692000	2.987000	33.740000	5.610000	46.520000	6.510000
CH	10.082000	3.714000	34.836000	3.645000	47.342000	1.576000
CK10	2.306000	0.782000	3.247000	1.034000	3.910000	1.711000
CK20	1.619000	0.818000	1.817000	0.974000	3.194000	2.004000
CK30	1.948000	1.529000	1.666000	0.947000	2.566000	1.446000
CK40	1.619000	2.002000	1.439000	0.563000	2.190000	0.900000
CK50	1.309000	1.124000	1.212600	0.533600	1.614000	0.842000
CK60	1.114000	0.868000	1.301000	0.722000	1.575000	1.053000
CK70	1.394000	1.642000	1.399000	0.854000	1.659000	1.305000
CK80	1.046000	1.114000	1.113000	0.847000	1.402000	0.828000
CK90	**0.725600**	0.428700	**0.851700**	0.533500	**1.401000**	1.000000

Table 9.5. Summarization of mean and standard deviation of generational distance (*GD*) between SPEA2, SPEA2-CE-HOF and SPEA2-CE-KR for DTLZ2 test problem with 3 to 5 objectives. The boldface indicates the best mean values.

Algorithm	3 Objectives		4 Objectives		5 Objectives	
	Mean	St Dev	Mean	St Dev	Mean	St Dev
S2	0.003689	0.001755	0.062900	0.012560	0.141180	0.009030
CH	0.003291	0.001458	0.059010	0.012270	0.141950	0.005810
CK10	0.002249	0.000264	0.006647	0.000544	0.015267	0.001333
CK20	0.002051	0.000193	0.006527	0.000626	0.013589	0.001257
CK30	0.002007	0.000212	0.006280	0.000628	0.013094	0.000968
CK40	0.002036	0.000265	0.006388	0.000612	0.013241	0.001240
CK50	0.001973	0.000219	0.006352	0.000515	0.012942	0.000957
CK60	0.001942	0.000224	0.006105	0.000427	0.012947	0.000829
CK70	0.001948	0.000218	0.006107	0.000381	0.013064	0.001247
CK80	0.001964	0.000227	0.006164	0.000443	**0.012837**	0.000919
CK90	**0.001886**	0.000256	**0.006066**	0.000429	0.012938	0.000924

Table 9.6. Summarization of mean and standard deviation of generational distance (*GD*) between SPEA2, SPEA2-CE-HOF and SPEA2-CE-KR for DTLZ3 test problem with 3 to 5 objectives. The boldface indicates the best mean values.

Algorithm	3 Objectives		4 Objectives		5 Objectives	
	Mean	St Dev	Mean	St Dev	Mean	St Dev
S2	19.110000	5.381000	73.000000	7.870000	118.610000	5.700000
CH	19.330000	5.840000	74.210000	7.170000	116.330000	5.570000
CK10	6.941000	4.228000	7.842000	2.849000	13.031000	5.146000
CK20	7.545000	4.646000	5.356000	2.584000	8.401000	4.094000
CK30	**5.625000**	4.256000	4.178000	2.129000	5.783000	3.705000
CK40	6.056000	4.071000	4.459000	2.777000	5.627000	3.746000
CK50	7.120000	5.740000	3.533000	1.701000	5.793000	3.216000
CK60	6.262000	4.028000	3.222000	1.696000	5.331000	2.950000
CK70	6.673000	5.469000	**2.374000**	1.451000	**4.089000**	2.540000
CK80	7.371000	4.611000	4.371000	3.957000	8.040000	6.340000
CK90	7.569000	5.444000	3.154000	2.405000	4.422000	3.528000

Table 9.7. Summarization of mean and standard deviation of generational distance (*GD*) between SPEA2, SPEA2-CE-HOF and SPEA2-CE-KR for DTLZ4 test problem with 3 to 5 objectives. The boldface indicates the best mean values.

Algorithm	3 Objectives		4 Objectives		5 Objectives	
	Mean	St Dev	Mean	St Dev	Mean	St Dev
S2	0.003235	0.001623	0.047410	0.029130	0.125830	0.038940
CH	0.002993	0.001583	0.046880	0.028060	0.134240	0.033110
CK10	0.001949	0.000512	0.006746	0.001131	0.015870	0.003744
CK20	0.001778	0.000435	0.006551	0.001080	0.016043	0.002673
CK30	0.001937	0.000424	0.006018	0.001097	0.016413	0.004181
CK40	**0.001758**	0.000395	0.006022	0.001062	0.015638	0.003222
CK50	0.001917	0.000387	**0.005852**	0.001155	0.016380	0.003657
CK60	0.001799	0.000389	0.006244	0.001189	**0.015299**	0.002971
CK70	0.001870	0.000327	0.005869	0.001108	0.015377	0.003034
CK80	0.001903	0.000336	0.006196	0.001285	0.015659	0.003024
CK90	0.001823	0.000350	0.006131	0.001036	0.015561	0.003229

Table 9.8. Summarization of mean and standard deviation of generational distance (*GD*) between SPEA2, SPEA2-CE-HOF and SPEA2-CE-KR for DTLZ5 test problem with 3 to 5 objectives. The boldface indicates the best mean values.

Algorithm	3 Objectives		4 Objectives		5 Objectives	
	Mean	St Dev	Mean	St Dev	Mean	St Dev
S2	0.000378	0.000105	0.137610	0.003870	0.168450	0.005550
CH	0.000363	0.000146	0.138730	0.003970	0.168760	0.005290
CK10	0.000503	0.000102	0.067310	0.009430	0.140890	0.009030
CK20	0.000407	0.000081	**0.055040**	0.008940	0.131820	0.012600
CK30	0.000492	0.000290	0.056210	0.016760	0.133220	0.012570
CK40	0.000356	0.000067	0.080160	0.039710	0.128740	0.013820
CK50	0.000413	0.000142	0.078390	0.040110	0.128630	0.016290
CK60	**0.000349**	0.000065	0.078030	0.038570	**0.123830**	0.014280
CK70	0.000364	0.000062	0.088880	0.044920	0.127430	0.013990
CK80	0.000385	0.000151	0.083890	0.034450	0.124230	0.016860
CK90	0.000371	0.000085	0.070140	0.035950	0.125510	0.013660

Table 9.9. Summarization of mean and standard deviation of generational distance (*GD*) between SPEA2, SPEA2-CE-HOF and SPEA2-CE-KR for DTLZ6 test problem with 3 to 5 objectives. The boldface indicates the best mean values.

Algorithm	3 Objectives		4 Objectives		5 Objectives	
	Mean	St Dev	Mean	St Dev	Mean	St Dev
S2	0.105450	0.034230	0.516590	0.049630	0.806390	0.005660
CH	0.099750	0.027580	0.513270	0.054590	0.806890	0.003910
CK10	0.015030	0.001044	0.092410	0.028960	0.728800	0.060900
CK20	0.014478	0.001036	0.078690	0.009460	0.485100	0.150900
CK30	0.014149	0.001120	0.070310	0.010540	0.437000	0.140900
CK40	0.014515	0.001231	0.070240	0.009010	0.374600	0.095900
CK50	0.014206	0.000972	0.069210	0.010440	0.352300	0.123500
CK60	0.013800	0.001118	0.068540	0.007930	0.340600	0.139300
CK70	0.013869	0.001069	**0.065610**	0.008830	0.325100	0.124800
CK80	0.014164	0.001340	0.067870	0.009310	0.327900	0.124300
CK90	**0.013555**	0.000732	0.069020	0.006420	**0.291700**	0.108200

Table 9.10. Summarization of mean and standard deviation of generational distance (*GD*) between SPEA2, SPEA2-CE-HOF and SPEA2-CE-KR for DTLZ7 test problem with 3 to 5 objectives. The boldface indicates the best mean values.

Algorithm	3 Objectives		4 Objectives		5 Objectives	
	Mean	St Dev	Mean	St Dev	Mean	St Dev
S2	0.021630	0.015890	0.670800	0.138700	1.571800	0.148600
CH	0.029550	0.023300	0.619400	0.170400	1.583500	0.182700
CK10	0.001613	0.000451	0.006187	0.000827	0.024740	0.010030
CK20	0.001646	0.000846	0.005505	0.000789	0.021870	0.009090
CK30	**0.001419**	0.000488	0.005266	0.000930	0.023990	0.010150
CK40	0.002002	0.001753	0.005358	0.001250	**0.018890**	0.009220
CK50	0.001692	0.001306	0.005376	0.000776	0.023730	0.010280
CK60	0.002886	0.002978	0.005342	0.000761	0.021530	0.009880
CK70	0.002132	0.001894	0.005464	0.001020	0.021100	0.010060
CK80	0.001952	0.001602	**0.005124**	0.000923	0.021520	0.009880
CK90	0.002731	0.002518	0.005265	0.001200	0.023100	0.010140

Table 9.11. Summarization of mean and standard deviation of spacing (*SP*) between SPEA2, SPEA2-CE-HOF and SPEA2-CE-KR for DTLZ1 test problem with 3 to 5 objectives. The boldface indicates the best mean values.

Algorithm	3 Objectives		4 Objectives		5 Objectives	
	Mean	St Dev	Mean	St Dev	Mean	St Dev
S2	9.964000	5.157000	25.372000	4.639000	39.550000	8.690000
CH	9.177000	4.561000	24.241000	3.175000	37.895000	2.385000
CK10	9.519000	4.736000	13.180000	5.550000	12.148000	4.196000
CK20	6.225000	3.949000	8.590000	5.520000	11.346000	4.244000
CK30	7.748000	5.287000	7.904000	4.832000	10.914000	4.268000
CK40	5.677000	4.375000	8.019000	4.281000	10.389000	3.195000
CK50	5.791000	4.235000	6.780000	3.958000	8.201000	4.702000
CK60	4.890000	4.263000	7.552000	5.101000	8.030000	5.760000
CK70	5.255000	5.456000	7.025000	4.762000	7.173000	4.842000
CK80	3.684000	3.363000	5.832000	4.625000	6.866000	4.012000
CK90	**3.486000**	4.046000	**3.990000**	3.556000	**6.195000**	4.347000

Table 9.12. Summarization of mean and standard deviation of spacing (*SP*) between SPEA2, SPEA2-CE-HOF and SPEA2-CE-KR for DTLZ2 test problem with 3 to 5 objectives. The bold-face indicates the best mean values.

Algorithm	3 Objectives		4 Objectives		5 Objectives	
	Mean	St Dev	Mean	St Dev	Mean	St Dev
S2	**0.025130**	0.002828	0.105250	0.016970	0.188760	0.018550
CH	0.025798	0.002481	0.101090	0.013820	0.193700	0.017120
CK10	0.056490	0.006580	0.107560	0.011310	0.159290	0.021840
CK20	0.057880	0.009250	0.106140	0.011250	0.154670	0.016860
CK30	0.053980	0.007510	0.105410	0.012680	0.141060	0.019560
CK40	0.058720	0.010500	0.102490	0.013380	0.149950	0.013820
CK50	0.055440	0.007620	0.103920	0.010090	0.151160	0.016610
CK60	0.056230	0.007960	0.101030	0.012330	0.147530	0.013810
CK70	0.054930	0.007000	0.100560	0.012550	0.142850	0.017940
CK80	0.055560	0.009810	0.100240	0.010830	0.144540	0.021450
CK90	0.053190	0.007890	**0.096240**	0.013010	**0.136720**	0.018290

Table 9.13. Summarization of mean and standard deviation of spacing (*SP*) between SPEA2, SPEA2-CE-HOF and SPEA2-CE-KR for DTLZ3 test problem with 3 to 5 objectives. The bold-face indicates the best mean values.

Algorithm	3 Objectives		4 Objectives		5 Objectives	
	Mean	St Dev	Mean	St Dev	Mean	St Dev
S2	19.650000	8.920000	64.150000	10.360000	103.780000	9.920000
CH	**15.660000**	6.590000	63.760000	10.670000	102.400000	8.830000
CK10	21.280000	7.800000	23.910000	7.340000	33.260000	8.470000
CK20	20.240000	10.850000	21.290000	8.960000	27.340000	11.530000
CK30	16.460000	10.330000	18.430000	9.130000	20.450000	8.580000
CK40	16.830000	10.420000	17.910000	9.260000	22.400000	13.090000
CK50	20.680000	9.640000	19.010000	11.840000	20.780000	9.710000
CK60	18.440000	9.180000	15.940000	10.480000	21.660000	11.800000
CK70	19.740000	10.910000	**11.150000**	8.300000	**19.580000**	12.840000
CK80	19.120000	10.130000	16.240000	11.350000	24.010000	11.930000
CK90	18.630000	9.130000	13.550000	10.160000	20.650000	18.050000

Table 9.14. Summarization of mean and standard deviation of spacing (*SP*) between SPEA2, SPEA2-CE-HOF and SPEA2-CE-KR for DTLZ4 test problem with 3 to 5 objectives. The bold-face indicates the best mean values.

Algorithm	3 Objectives		4 Objectives		5 Objectives	
	Mean	St Dev	Mean	St Dev	Mean	St Dev
S2	0.020280	0.011180	0.077580	0.042130	0.170310	0.046350
CH	**0.020150**	0.009290	0.080380	0.039380	0.177090	0.041840
CK10	0.039890	0.022820	**0.035320**	0.025380	0.019710	0.018860
CK20	0.036810	0.026750	0.035830	0.025080	**0.016780**	0.014820
CK30	0.048220	0.022330	0.046900	0.021310	0.017550	0.015340
CK40	0.038040	0.026630	0.040350	0.020760	0.022970	0.021730
CK50	0.050300	0.024360	0.047040	0.024660	0.026470	0.022980
CK60	0.042800	0.025890	0.038070	0.026050	0.018260	0.015630
CK70	0.051110	0.024040	0.047170	0.025620	0.020430	0.018760
CK80	0.044430	0.024800	0.036720	0.024380	0.017460	0.016090
CK90	0.043890	0.024900	0.044530	0.024310	0.016910	0.016340

Table 9.15. Summarization of mean and standard deviation of spacing (*SP*) between SPEA2, SPEA2-CE-HOF and SPEA2-CE-KR for DTLZ5 test problem with 3 to 5 objectives. The bold-face indicates the best mean values.

Algorithm	3 Objectives		4 Objectives		5 Objectives	
	Mean	St Dev	Mean	St Dev	Mean	St Dev
S2	**0.005045**	0.000556	0.087290	0.023240	0.153300	0.034530
CH	0.005081	0.000351	0.076880	0.021160	0.167570	0.037590
CK10	0.015992	0.005078	0.086360	0.012440	0.135960	0.032300
CK20	0.016060	0.008690	0.070680	0.012950	0.135090	0.026560
CK30	0.018090	0.010840	0.072470	0.016850	0.130680	0.016560
CK40	0.015450	0.007410	0.075970	0.024600	0.122080	0.021270
CK50	0.016158	0.004952	0.068730	0.016130	0.123490	0.016690
CK60	0.015590	0.006710	0.072170	0.019250	0.122920	0.024530
CK70	0.016270	0.011220	0.072260	0.023260	0.119010	0.024650
CK80	0.014040	0.005860	0.078800	0.028770	**0.115720**	0.017900
CK90	0.016820	0.007700	**0.068610**	0.014970	0.125250	0.028930

Table 9.16. Summarization of mean and standard deviation of spacing (*SP*) between SPEA2, SPEA2-CE-HOF and SPEA2-CE-KR for DTLZ6 test problem with 3 to 5 objectives. The boldface indicates the best mean values.

Algorithm	3 Objectives		4 Objectives		5 Objectives	
	Mean	St Dev	Mean	St Dev	Mean	St Dev
S2	0.047210	0.013340	0.329610	0.040750	0.533700	0.056900
CH	0.048660	0.015810	0.330500	0.064200	0.532490	0.048650
CK10	0.029610	0.008310	0.125610	0.030270	0.788100	0.126100
CK20	0.026600	0.007370	0.103040	0.017620	0.683500	0.212400
CK30	0.029700	0.016570	0.097870	0.018060	0.646300	0.241600
CK40	0.025280	0.004798	0.094650	0.016940	0.514200	0.125600
CK50	0.024665	0.004728	0.100610	0.021540	0.501600	0.198200
CK60	0.025970	0.010340	0.094400	0.016750	0.498100	0.212400
CK70	0.025350	0.006090	**0.090630**	0.018780	0.454800	0.200300
CK80	**0.024084**	0.004731	0.097290	0.018670	0.470100	0.188200
CK90	0.024180	0.005930	0.094290	0.014670	**0.396500**	0.144000

Table 9.17. Summarization of mean and standard deviation of spacing (*SP*) between SPEA2, SPEA2-CE-HOF and SPEA2-CE-KR for DTLZ7 test problem with 3 to 5 objectives. The boldface indicates the best mean values.

Algorithm	3 Objectives		4 Objectives		5 Objectives	
	Mean	St Dev	Mean	St Dev	Mean	St Dev
S2	**0.046800**	0.022270	0.164590	0.029300	0.294940	0.026320
CH	0.047580	0.013370	0.169140	0.036380	0.291100	0.030800
CK10	0.060870	0.018470	0.101470	0.054240	0.038600	0.065400
CK20	0.068610	0.013110	0.109700	0.044780	0.032790	0.047390
CK30	0.068020	0.013500	0.099190	0.043070	0.035700	0.054800
CK40	0.066900	0.011590	0.108500	0.042540	0.054100	0.059500
CK50	0.067150	0.012790	**0.095710**	0.048720	0.033130	0.043700
CK60	0.063200	0.013920	0.100220	0.045890	0.038870	0.050960
CK70	0.065690	0.014890	0.112270	0.053140	0.038200	0.041560
CK80	0.070240	0.011120	0.125730	0.033250	0.036680	0.048520
CK90	0.064500	0.017210	0.118580	0.044200	**0.032510**	0.042730

Table 9.18. Summarization of coverage (C) between SPEA2 and SPEA2-CE-HOF for DTLZ test problems with 3 to 5 objectives

If $C(X',X'') > C(X'',X')$, then X' is better than X''				
DTLZ	C Metric	3 Objectives	4 Objectives	5 Objectives
1	C(CH,S2) C(S2,CH)	S2	S2	S2
2	C(CH,S2) C(S2,CH)	CH	CH	CH
3	C(CH,S2) C(S2,CH)	CH	CH	CH
4	C(CH,S2) C(S2,CH)	S2	CH	S2
5	C(CH,S2) C(S2,CH)	CH	CH	S2
6	C(CH,S2) C(S2,CH)	CH	CH	S2
7	C(CH,S2) C(S2,CH)	S2	S2	S2

Table 9.19. Summarization of coverage (C) between SPEA2 and SPEA2-CE-KR (10, 20, 30 opponents) for DTLZ test problems with 3 objectives

If $C(X',X'') > C(X'',X')$, then X' is better than X''				
DTLZ	C Metric	10 Opponents	20 Opponents	30 Opponents
1	C(CK,S2) C(S2,CK)	S2	S2	S2
2	C(CK,S2) C(S2,CK)	S2	S2	S2
3	C(CK,S2) C(S2,CK)	S2	S2	S2
4	C(CK,S2) C(S2,CK)	S2	S2	S2
5	C(CK,S2) C(S2,CK)	S2	S2	S2
6	C(CK,S2) C(S2,CK)	CK	CK	CK
7	C(CK,S2) C(S2,CK)	CK	CK	CK

Table 9.20. Summarization of coverage (C) between SPEA2 and SPEA2-CE-KR (40, 50, 60 opponents) for DTLZ test problems with 3 objectives

		If $C(X',X'') > C(X'',X')$, then X' is better than X''		
DTLZ	C Metric	40 Opponents	50 Opponents	60 Opponents
1	C(CK,S2) C(S2,CK)	S2	S2	S2
2	C(CK,S2) C(S2,CK)	S2	S2	S2
3	C(CK,S2) C(S2,CK)	S2	CK	CK
4	C(CK,S2) C(S2,CK)	S2	S2	S2
5	C(CK,S2) C(S2,CK)	S2	S2	S2
6	C(CK,S2) C(S2,CK)	CK	CK	CK
7	C(CK,S2) C(S2,CK)	CK	CK	CK

Table 9.21. Summarization of coverage (C) between SPEA2 and SPEA2-CE-KR (70, 80, 90 opponents) for DTLZ test problems with 3 objectives

		If $C(X',X'') > C(X'',X')$, then X' is better than X''		
DTLZ	C Metric	70 Opponents	80 Opponents	90 Opponents
1	C(CK,S2) C(S2,CK)	S2	S2	CK
2	C(CK,S2) C(S2,CK)	S2	CK	S2
3	C(CK,S2) C(S2,CK)	CK	CK	CK
4	C(CK,S2) C(S2,CK)	S2	S2	S2
5	C(CK,S2) C(S2,CK)	S2	S2	S2
6	C(CK,S2) C(S2,CK)	CK	CK	CK
7	C(CK,S2) C(S2,CK)	CK	CK	CK

Table 9.22. Summarization of coverage (C) between SPEA2 and SPEA2-CE-KR (10, 20, 30 opponents) for DTLZ test problems with 4 objectives

	If $C(X',X'') > C(X'',X')$, then X' is better than X''		
DTLZ C Metric	10 Opponents	20 Opponents	30 Opponents
1 C(CK,S2) C(S2,CK)	CK	CK	CK
2 C(CK,S2) C(S2,CK)	CK	CK	CK
3 C(CK,S2) C(S2,CK)	CK	CK	CK
4 C(CK,S2) C(S2,CK)	CK	CK	CK
5 C(CK,S2) C(S2,CK)	S2	CK	CK
6 C(CK,S2) C(S2,CK)	CK	CK	CK
7 C(CK,S2) C(S2,CK)	CK	CK	CK

Table 9.23. Summarization of coverage (C) between SPEA2 and SPEA2-CE-KR (40, 50, 60 opponents) for DTLZ test problems with 4 objectives

	If $C(X',X'') > C(X'',X')$, then X' is better than X''		
DTLZ C Metric	40 Opponents	50 Opponents	60 Opponents
1 C(CK,S2) C(S2,CK)	CK	CK	CK
2 C(CK,S2) C(S2,CK)	CK	CK	CK
3 C(CK,S2) C(S2,CK)	CK	CK	CK
4 C(CK,S2) C(S2,CK)	CK	CK	CK
5 C(CK,S2) C(S2,CK)	S2	S2	S2
6 C(CK,S2) C(S2,CK)	CK	CK	CK
7 C(CK,S2) C(S2,CK)	CK	CK	CK

Table 9.24. Summarization of coverage (C) between SPEA2 and SPEA2-CE-KR (70, 80, 90 opponents) for DTLZ test problems with 4 objectives

	If $C(X',X'') > C(X'',X')$, then X' is better than X''			
DTLZ	C Metric	70 Opponents	80 Opponents	90 Opponents
1	C(CK,S2) C(S2,CK)	CK	CK	CK
2	C(CK,S2) C(S2,CK)	CK	CK	CK
3	C(CK,S2) C(S2,CK)	CK	CK	CK
4	C(CK,S2) C(S2,CK)	CK	CK	CK
5	C(CK,S2) C(S2,CK)	S2	S2	S2
6	C(CK,S2) C(S2,CK)	CK	CK	CK
7	C(CK,S2) C(S2,CK)	CK	CK	CK

Table 9.25. Summarization of coverage (C) between SPEA2 and SPEA2-CE-KR (10, 20, 30 opponents) for DTLZ test problems with 5 objectives

	If $C(X',X'') > C(X'',X')$, then X' is better than X''			
DTLZ	C Metric	10 Opponents	20 Opponents	30 Opponents
1	C(CK,S2) C(S2,CK)	CK	CK	CK
2	C(CK,S2) C(S2,CK)	CK	CK	CK
3	C(CK,S2) C(S2,CK)	CK	CK	CK
4	C(CK,S2) C(S2,CK)	CK	CK	CK
5	C(CK,S2) C(S2,CK)	S2	S2	S2
6	C(CK,S2) C(S2,CK)	CK	CK	CK
7	C(CK,S2) C(S2,CK)	CK	CK	CK

Table 9.26. Summarization of coverage (*C*) between SPEA2 and SPEA2-CE-KR (40, 50, 60 opponents) for DTLZ test problems with 5 objectives

		If $C(X',X'') > C(X'',X')$, then X' is better than X''		
DTLZ	*C* Metric	40 Opponents	50 Opponents	60 Opponents
1	C(CK,S2) C(S2,CK)	CK	CK	CK
2	C(CK,S2) C(S2,CK)	CK	CK	CK
3	C(CK,S2) C(S2,CK)	CK	CK	CK
4	C(CK,S2) C(S2,CK)	CK	CK	CK
5	C(CK,S2) C(S2,CK)	S2	S2	S2
6	C(CK,S2) C(S2,CK)	CK	CK	CK
7	C(CK,S2) C(S2,CK)	CK	CK	CK

Table 9.27. Summarization of coverage (*C*) between SPEA2 and SPEA2-CE-KR (70, 80, 90 opponents) for DTLZ test problems with 5 objectives

		If $C(X',X'') > C(X'',X')$, then X' is better than X''		
DTLZ	*C* Metric	70 Opponents	80 Opponents	90 Opponents
1	C(CK,S2) C(S2,CK)	CK	CK	CK
2	C(CK,S2) C(S2,CK)	CK	CK	CK
3	C(CK,S2) C(S2,CK)	CK	CK	CK
4	C(CK,S2) C(S2,CK)	CK	CK	CK
5	C(CK,S2) C(S2,CK)	S2	S2	S2
6	C(CK,S2) C(S2,CK)	CK	CK	CK
7	C(CK,S2) C(S2,CK)	CK	CK	CK

Table 9.28. Summarization of coverage (C) between SPEA2-CE-HOF and SPEA2-CE-KR (10, 20, 30 opponents) for DTLZ test problems with 3 objectives

	If $C(X',X'') > C(X'',X')$, then X' is better than X''		
DTLZ	C Metric	10 Opponents	20 Opponents	30 Opponents
1	C(CK,CH) C(CH,CK)	CH	CH	CH
2	C(CK,CH) C(CH,CK)	CH	CH	CH
3	C(CK,CH) C(CH,CK)	CH	CH	CH
4	C(CK,CH) C(CH,CK)	CH	CH	CH
5	C(CK,CH) C(CH,CK)	CH	CH	CH
6	C(CK,CH) C(CH,CK)	CK	CK	CK
7	C(CK,CH) C(CH,CK)	CK	CK	CK

Table 9.29. Summarization of coverage (C) between SPEA2-CE-HOF and SPEA2-CE-KR (40, 50, 60 opponents) for DTLZ test problems with 3 objectives

	If $C(X',X'') > C(X'',X')$, then X' is better than X''		
DTLZ	C Metric	40 Opponents	50 Opponents	60 Opponents
1	C(CK,CH) C(CH,CK)	CH	CH	CH
2	C(CK,CH) C(CH,CK)	CH	CH	CH
3	C(CK,CH) C(CH,CK)	CH	CK	CK
4	C(CK,CH) C(CH,CK)	CH	CH	CH
5	C(CK,CH) C(CH,CK)	CK	CK	CH
6	C(CK,CH) C(CH,CK)	CK	CK	CK
7	C(CK,CH) C(CH,CK)	CK	CK	CK

Table 9.30. Summarization of coverage (C) between SPEA2-CE-HOF and SPEA2-CE-KR (70, 80, 90 opponents) for DTLZ test problems with 3 objectives

		If $C(X',X'') > C(X'',X')$, then X' is better than X''		
DTLZ	C Metric	70 Opponents	80 Opponents	90 Opponents
1	C(CK,CH) C(CH,CK)	CH	CH	CK
2	C(CK,CH) C(CH,CK)	CH	CH	CK
3	C(CK,CH) C(CH,CK)	CK	CK	CK
4	C(CK,CH) C(CH,CK)	CH	CH	CH
5	C(CK,CH) C(CH,CK)	CH	CH	CH
6	C(CK,CH) C(CH,CK)	CK	CK	CK
7	C(CK,CH) C(CH,CK)	CK	CK	CK

Table 9.31. Summarization of coverage (C) between SPEA2-CE-HOF and SPEA2-CE-KR (10, 20, 30 opponents) for DTLZ test problems with 4 objectives

		If $C(X',X'') > C(X'',X')$, then X' is better than X''		
DTLZ	C Metric	10 Opponents	20 Opponents	30 Opponents
1	C(CK,CH) C(CH,CK)	CK	CK	CK
2	C(CK,CH) C(CH,CK)	CK	CK	CK
3	C(CK,CH) C(CH,CK)	CK	CK	CK
4	C(CK,CH) C(CH,CK)	CK	CK	CK
5	C(CK,CH) C(CH,CK)	CH	CH	CH
6	C(CK,CH) C(CH,CK)	CK	CK	CK
7	C(CK,CH) C(CH,CK)	CK	CK	CK

Table 9.32. Summarization of coverage (C) between SPEA2-CE-HOF and SPEA2-CE-KR (40, 50, 60 opponents) for DTLZ test problems with 4 objectives

If $C(X',X'') > C(X'',X')$, then X' is better than X''				
DTLZ	C Metric	40 Opponents	50 Opponents	60 Opponents
1	C(CK,CH) C(CH,CK)	CK	CK	CK
2	C(CK,CH) C(CH,CK)	CK	CK	CK
3	C(CK,CH) C(CH,CK)	CK	CK	CK
4	C(CK,CH) C(CH,CK)	CK	CK	CK
5	C(CK,CH) C(CH,CK)	CH	CH	CH
6	C(CK,CH) C(CH,CK)	CK	CK	CK
7	C(CK,CH) C(CH,CK)	CK	CK	CK

Table 9.33. Summarization of coverage (C) between SPEA2-CE-HOF and SPEA2-CE-KR (70, 80, 90 opponents) for DTLZ test problems with 4 objectives

If $C(X',X'') > C(X'',X')$, then X' is better than X''				
DTLZ	C Metric	70 Opponents	80 Opponents	90 Opponents
1	C(CK,CH) C(CH,CK)	CK	CK	CK
2	C(CK,CH) C(CH,CK)	CK	CK	CK
3	C(CK,CH) C(CH,CK)	CK	CK	CK
4	C(CK,CH) C(CH,CK)	CK	CK	CK
5	C(CK,CH) C(CH,CK)	CH	CH	CH
6	C(CK,CH) C(CH,CK)	CK	CK	CK
7	C(CK,CH) C(CH,CK)	CK	CK	CK

Table 9.34. Summarization of coverage (*C*) between SPEA2-CE-HOF and SPEA2-CE-KR (10, 20, 30 opponents) for DTLZ test problems with 5 objectives

		If $C(X',X'') > C(X'',X')$, then X' is better than X''		
DTLZ	*C* Metric	10 Opponents	20 Opponents	30 Opponents
1	C(CK,CH) C(CH,CK)	CK	CK	CK
2	C(CK,CH) C(CH,CK)	CK	CK	CK
3	C(CK,CH) C(CH,CK)	CK	CK	CK
4	C(CK,CH) C(CH,CK)	CK	CK	CK
5	C(CK,CH) C(CH,CK)	CH	CH	CH
6	C(CK,CH) C(CH,CK)	CK	CK	CK
7	C(CK,CH) C(CH,CK)	CK	CK	CK

Table 9.35. Summarization of coverage (*C*) between SPEA2-CE-HOF and SPEA2-CE-KR (40, 50, 60 opponents) for DTLZ test problems with 5 objectives

		If $C(X',X'') > C(X'',X')$, then X' is better than X''		
DTLZ	*C* Metric	40 Opponents	50 Opponents	60 Opponents
1	C(CK,CH) C(CH,CK)	CK	CK	CK
2	C(CK,CH) C(CH,CK)	CK	CK	CK
3	C(CK,CH) C(CH,CK)	CK	CK	CK
4	C(CK,CH) C(CH,CK)	CK	CK	CK
5	C(CK,CH) C(CH,CK)	CH	CH	CH
6	C(CK,CH) C(CH,CK)	CK	CK	CK
7	C(CK,CH) C(CH,CK)	CK	CK	CK

Table 9.36. Summarization of coverage (C) between SPEA2-CE-HOF and SPEA2-CE-KR (70, 80, 90 opponents) for DTLZ test problems with 5 objectives

DTLZ	C Metric	70 Opponents	80 Opponents	90 Opponents
	If $C(X',X'') > C(X'',X')$, then X' is better than X''			
1	$C(\text{CK,CH})$ $C(\text{CH,CK})$	CK	CK	CK
2	$C(\text{CK,CH})$ $C(\text{CH,CK})$	CK	CK	CK
3	$C(\text{CK,CH})$ $C(\text{CH,CK})$	CK	CK	CK
4	$C(\text{CK,CH})$ $C(\text{CH,CK})$	CK	CK	CK
5	$C(\text{CK,CH})$ $C(\text{CH,CK})$	CH	CH	CH
6	$C(\text{CK,CH})$ $C(\text{CH,CK})$	CK	CK	CK
7	$C(\text{CK,CH})$ $C(\text{CH,CK})$	CK	CK	CK

Table 9.37. Summarization of mean and standard deviation of coverage (C) between SPEA2 and SPEA2-CE-HOF for DTLZ test problems with 3 to 5 objectives. $C(X',X'') > C(X'',X')$, then X' is better than X''.

DTLZ	C Metric	3 Objectives		4 Objectives		5 Objectives	
		Mean	St Dev	Mean	St Dev	Mean	St Dev
1	$C(\text{CH,S2})$	0.497800	0.254100	0.156300	0.099100	0.105900	0.061200
	$C(\text{S2,CH})$	0.533300	0.265400	0.198000	0.159400	0.133000	0.173600
2	$C(\text{CH,S2})$	0.052330	0.025010	0.126000	0.068700	0.120700	0.069000
	$C(\text{S2,CH})$	0.043330	0.022180	0.109000	0.051820	0.110300	0.076100
3	$C(\text{CH,S2})$	0.621000	0.211700	0.196300	0.111300	0.162000	0.111500
	$C(\text{S2,CH})$	0.540300	0.225000	0.176700	0.109600	0.105700	0.067400
4	$C(\text{CH,S2})$	0.075700	0.059400	0.318300	0.213500	0.248300	0.191500
	$C(\text{S2,CH})$	0.081000	0.056800	0.303000	0.207200	0.325300	0.226500
5	$C(\text{CH,S2})$	0.095670	0.040660	0.110670	0.052450	0.122000	0.090900
	$C(\text{S2,CH})$	0.090330	0.038100	0.087670	0.034510	0.187000	0.123400
6	$C(\text{CH,S2})$	0.502300	0.364000	0.175000	0.133500	0.001667	0.003790
	$C(\text{S2,CH})$	0.419700	0.352400	0.162700	0.135500	0.002333	0.004302
7	$C(\text{CH,S2})$	0.083300	0.058400	0.170300	0.071300	0.090700	0.061300
	$C(\text{S2,CH})$	0.103670	0.052090	0.172300	0.089200	0.100700	0.061700

Table 9.38. Summarization of mean and standard deviation of coverage (C) between SPEA2 and SPEA2-CE-KR (10, 20, 30 opponents) for DTLZ test problems with 3 objectives. $C(X',X'') > C(X'',X')$, then X' is better than X''.

3 Objectives		10 Opponents		20 Opponents		30 Opponents	
DTLZ	C Metric	Mean	St Dev	Mean	St Dev	Mean	St Dev
1	C(CK,S2)	0.079100	0.055500	0.065500	0.063700	0.086100	0.078700
	C(S2,CK)	0.255300	0.261000	0.257700	0.323300	0.242000	0.295100
2	C(CK,S2)	0.020000	0.016610	0.023000	0.016220	0.021000	0.015610
	C(S2,CK)	0.046330	0.028220	0.029000	0.022950	0.033670	0.024700
3	C(CK,S2)	0.090000	0.074700	0.156000	0.152200	0.155000	0.173900
	C(S2,CK)	0.326000	0.325700	0.328000	0.312300	0.233300	0.279700
4	C(CK,S2)	0.032330	0.040740	0.050670	0.048630	0.039330	0.044020
	C(S2,CK)	0.105300	0.115000	0.065700	0.076300	0.055000	0.051780
5	C(CK,S2)	0.024670	0.018140	0.031000	0.020230	0.033330	0.019880
	C(S2,CK)	0.169300	0.079800	0.126700	0.056000	0.122700	0.056400
6	C(CK,S2)	0.404300	0.062000	0.397300	0.064800	0.388700	0.075100
	C(S2,CK)	0.000000	0.000000	0.000000	0.000000	0.000000	0.000000
7	C(CK,S2)	0.080000	0.042260	0.067330	0.033110	0.074330	0.037200
	C(S2,CK)	0.012670	0.012300	0.015670	0.029090	0.010000	0.017020

Table 9.39. Summarization of mean and standard deviation of coverage (C) between SPEA2 and SPEA2-CE-KR (40, 50, 60 opponents) for DTLZ test problems with 3 objectives. $C(X',X'') > C(X'',X')$, then X' is better than X''.

3 Objectives		40 Opponents		50 Opponents		60 Opponents	
DTLZ	C Metric	Mean	St Dev	Mean	St Dev	Mean	St Dev
1	C(CK,S2)	0.070060	0.049160	0.085300	0.061100	0.088100	0.103000
	C(S2,CK)	0.190700	0.289500	0.150700	0.277800	0.154000	0.295600
2	C(CK,S2)	0.020000	0.013390	0.020670	0.015070	0.015670	0.012230
	C(S2,CK)	0.029670	0.026190	0.033000	0.018030	0.025670	0.018880
3	C(CK,S2)	0.201300	0.179300	0.286000	0.207600	0.292300	0.210700
	C(S2,CK)	0.248300	0.319600	0.183000	0.229900	0.213000	0.258500
4	C(CK,S2)	0.059000	0.049010	0.033330	0.037260	0.041000	0.038000
	C(S2,CK)	0.060700	0.062300	0.060300	0.062300	0.053700	0.057400
5	C(CK,S2)	0.034670	0.025430	0.036670	0.020730	0.037330	0.024200
	C(S2,CK)	0.105670	0.037750	0.113700	0.059500	0.089330	0.047700
6	C(CK,S2)	0.387700	0.062900	0.368300	0.058400	0.387700	0.064000
	C(S2,CK)	0.000000	0.000000	0.000000	0.000000	0.000000	0.000000
7	C(CK,S2)	0.062000	0.039860	0.057000	0.039050	0.057000	0.034560
	C(S2,CK)	0.031300	0.064500	0.011670	0.028050	0.045300	0.071300

Table 9.40. Summarization of mean and standard deviation of coverage (C) between SPEA2 and SPEA2-CE-KR (70, 80, 90 opponents) for DTLZ test problems with 3 objectives. $C(X',X'') > C(X'',X')$, then X' is better than X''.

3 Objectives		70 Opponents		80 Opponents		90 Opponents	
DTLZ	C Metric	Mean	St Dev	Mean	St Dev	Mean	St Dev
1	C(CK,S2)	0.071300	0.075400	0.083600	0.121100	0.072100	0.060500
	C(S2,CK)	0.192000	0.309000	0.171300	0.316900	0.069700	0.174600
2	C(CK,S2)	0.023330	0.017290	0.024000	0.017730	0.018670	0.015480
	C(S2,CK)	0.028670	0.020800	0.023000	0.016850	0.021330	0.022700
3	C(CK,S2)	0.320300	0.228500	0.314300	0.211900	0.335000	0.227800
	C(S2,CK)	0.173300	0.187600	0.204300	0.245500	0.179300	0.189700
4	C(CK,S2)	0.032000	0.030330	0.037330	0.043230	0.038670	0.036360
	C(S2,CK)	0.061700	0.063100	0.070300	0.080600	0.050670	0.052650
5	C(CK,S2)	0.035330	0.020800	0.032670	0.024490	0.032330	0.022080
	C(S2,CK)	0.126670	0.046490	0.098700	0.057800	0.100000	0.041850
6	C(CK,S2)	0.376300	0.063500	0.359300	0.068900	0.368700	0.056700
	C(S2,CK)	0.000000	0.000000	0.000000	0.000000	0.000000	0.000000
7	C(CK,S2)	0.060330	0.038730	0.054670	0.035790	0.063330	0.033150
	C(S2,CK)	0.034700	0.069400	0.032300	0.065100	0.046700	0.078800

Table 9.41. Summarization of mean and standard deviation of coverage (C) between SPEA2 and SPEA2-CE-KR (10, 20, 30 opponents) for DTLZ test problems with 4 objectives. $C(X',X'') > C(X'',X')$, then X' is better than X''.

4 Objectives		10 Opponents		20 Opponents		30 Opponents	
DTLZ	C Metric	Mean	St Dev	Mean	St Dev	Mean	St Dev
1	C(CK,S2)	0.471500	0.119000	0.418600	0.126700	0.423300	0.113200
	C(S2,CK)	0.017000	0.089400	0.028000	0.153400	0.018300	0.098500
2	C(CK,S2)	0.042330	0.020120	0.034670	0.027000	0.037670	0.023290
	C(S2,CK)	0.005670	0.008980	0.005670	0.007740	0.004000	0.008940
3	C(CK,S2)	0.532700	0.146300	0.508700	0.158500	0.506700	0.176300
	C(S2,CK)	0.002000	0.004068	0.000330	0.001826	0.000333	0.001826
4	C(CK,S2)	0.158700	0.096700	0.158000	0.090800	0.145000	0.111700
	C(S2,CK)	0.014670	0.048900	0.019330	0.048350	0.022000	0.045140
5	C(CK,S2)	0.046330	0.028100	0.045670	0.029440	0.049330	0.029350
	C(S2,CK)	0.069670	0.049650	0.044330	0.032130	0.042330	0.044460
6	C(CK,S2)	0.455000	0.094600	0.472300	0.068600	0.460700	0.080000
	C(S2,CK)	0.000000	0.000000	0.000000	0.000000	0.000000	0.000000
7	C(CK,S2)	0.153000	0.137900	0.123700	0.114700	0.137700	0.111900
	C(S2,CK)	0.001333	0.003457	0.000000	0.000000	0.001000	0.004026

Table 9.42. Summarization of mean and standard deviation of coverage (*C*) between SPEA2 and SPEA2-CE-KR (40, 50, 60 opponents) for DTLZ test problems with 4 objectives. $C(X',X'') > C(X'',X')$, then X' is better than X''.

4 Objectives	40 Opponents		50 Opponents		60 Opponents	
DTLZ *C* Metric	Mean	St Dev	Mean	St Dev	Mean	St Dev
1 *C*(CK,S2)	0.395200	0.090700	0.407700	0.099800	0.364000	0.107100
1 *C*(S2,CK)	0.031000	0.169800	0.025300	0.138800	0.018300	0.100400
2 *C*(CK,S2)	0.032000	0.022030	0.040330	0.026060	0.030000	0.019650
2 *C*(S2,CK)	0.004330	0.008580	0.005000	0.010420	0.005330	0.009000
3 *C*(CK,S2)	0.489300	0.178100	0.445300	0.200200	0.404000	0.176900
3 *C*(S2,CK)	0.000000	0.000000	0.000330	0.001826	0.000000	0.000000
4 *C*(CK,S2)	0.138700	0.090700	0.141700	0.097200	0.143300	0.075800
4 *C*(S2,CK)	0.012000	0.028090	0.013670	0.041310	0.019670	0.052880
5 *C*(CK,S2)	0.054330	0.028000	0.056670	0.034070	0.055330	0.025290
5 *C*(S2,CK)	0.121700	0.151200	0.118700	0.143500	0.099700	0.112000
6 *C*(CK,S2)	0.481700	0.090900	0.463300	0.074200	0.448300	0.074200
6 *C*(S2,CK)	0.000000	0.000000	0.000000	0.000000	0.000000	0.000000
7 *C*(CK,S2)	0.094700	0.090400	0.152000	0.127400	0.119700	0.136000
7 *C*(S2,CK)	0.000330	0.001826	0.000000	0.000000	0.001000	0.003051

Table 9.43. Summarization of mean and standard deviation of coverage (*C*) between SPEA2 and SPEA2-CE-KR (70, 80, 90 opponents) for DTLZ test problems with 4 objectives. $C(X',X'') > C(X'',X')$, then X' is better than X''.

4 Objectives	70 Opponents		80 Opponents		90 Opponents	
DTLZ *C* Metric	Mean	St Dev	Mean	St Dev	Mean	St Dev
1 *C*(CK,S2)	0.390700	0.114300	0.368500	0.110300	0.355200	0.098800
1 *C*(S2,CK)	0.027000	0.147900	0.026700	0.146100	0.029000	0.158800
2 *C*(CK,S2)	0.030330	0.018840	0.035000	0.034910	0.029670	0.021730
2 *C*(S2,CK)	0.003670	0.011590	0.002000	0.004842	0.002670	0.006910
3 *C*(CK,S2)	0.427300	0.190900	0.456300	0.193700	0.441300	0.229600
3 *C*(S2,CK)	0.000000	0.000000	0.000333	0.001826	0.000000	0.000000
4 *C*(CK,S2)	0.132000	0.085400	0.147700	0.085900	0.132300	0.075900
4 *C*(S2,CK)	0.021300	0.056700	0.013330	0.038710	0.013000	0.036020
5 *C*(CK,S2)	0.059670	0.036530	0.057330	0.026120	0.052670	0.025720
5 *C*(S2,CK)	0.162300	0.162100	0.128700	0.117200	0.085300	0.100600
6 *C*(CK,S2)	0.468300	0.086300	0.464700	0.075500	0.474000	0.084100
6 *C*(S2,CK)	0.000000	0.000000	0.000000	0.000000	0.000000	0.000000
7 *C*(CK,S2)	0.132000	0.156100	0.078700	0.055900	0.094700	0.114400
7 *C*(S2,CK)	0.000000	0.000000	0.000333	0.001826	0.000667	0.002537

Table 9.44. Summarization of mean and standard deviation of coverage (C) between SPEA2 and SPEA2-CE-KR (10, 20, 30 opponents) for DTLZ test problems with 5 objectives. $C(X',X'') > C(X'',X')$, then X' is better than X''.

5 Objectives	10 Opponents		20 Opponents		30 Opponents	
DTLZ C Metric	Mean	St Dev	Mean	St Dev	Mean	St Dev
1 C(CK,S2)	0.911000	0.119000	0.911400	0.122400	0.906400	0.112300
1 C(S2,CK)	0.018300	0.100400	0.009330	0.051120	0.010000	0.054800
2 C(CK,S2)	0.148300	0.068100	0.120000	0.053820	0.122700	0.063400
2 C(S2,CK)	0.000330	0.001826	0.000000	0.000000	0.000000	0.000000
3 C(CK,S2)	0.909700	0.063500	0.890700	0.092400	0.912700	0.072500
3 C(S2,CK)	0.000000	0.000000	0.000000	0.000000	0.000000	0.000000
4 C(CK,S2)	0.363000	0.085000	0.329300	0.112300	0.342700	0.103500
4 C(S2,CK)	0.000000	0.000000	0.000000	0.000000	0.000000	0.000000
5 C(CK,S2)	0.061000	0.039770	0.049000	0.026310	0.034000	0.024300
5 C(S2,CK)	0.205700	0.112400	0.209000	0.100300	0.187000	0.101200
6 C(CK,S2)	0.067700	0.116500	0.384300	0.215800	0.433000	0.163900
6 C(S2,CK)	0.000000	0.000000	0.000000	0.000000	0.000000	0.000000
7 C(CK,S2)	0.591300	0.267300	0.522300	0.252900	0.562700	0.296100
7 C(S2,CK)	0.000000	0.000000	0.000000	0.000000	0.000000	0.000000

Table 9.45. Summarization of mean and standard deviation of coverage (C) between SPEA2 and SPEA2-CE-KR (40, 50, 60 opponents) for DTLZ test problems with 5 objectives. $C(X',X'') > C(X'',X')$, then X' is better than X''.

5 Objectives	40 Opponents		50 Opponents		60 Opponents	
DTLZ C Metric	Mean	St Dev	Mean	St Dev	Mean	St Dev
1 C(CK,S2)	0.888600	0.112800	0.889800	0.124600	0.886300	0.106600
1 C(S2,CK)	0.007670	0.041990	0.021700	0.118700	0.010000	0.054800
2 C(CK,S2)	0.110670	0.052780	0.105000	0.043850	0.121000	0.054900
2 C(S2,CK)	0.000000	0.000000	0.000000	0.000000	0.000000	0.000000
3 C(CK,S2)	0.897300	0.099400	0.883300	0.080500	0.875300	0.135100
3 C(S2,CK)	0.000000	0.000000	0.000000	0.000000	0.000000	0.000000
4 C(CK,S2)	0.338000	0.106400	0.346700	0.106100	0.353000	0.093500
4 C(S2,CK)	0.000330	0.001826	0.000000	0.000000	0.000000	0.000000
5 C(CK,S2)	0.039670	0.037180	0.040000	0.025330	0.037670	0.026480
5 C(S2,CK)	0.163000	0.103000	0.163000	0.090800	0.160700	0.111900
6 C(CK,S2)	0.472000	0.113700	0.513700	0.158900	0.529300	0.157800
6 C(S2,CK)	0.000000	0.000000	0.000000	0.000000	0.000000	0.000000
7 C(CK,S2)	0.408700	0.280400	0.580300	0.256000	0.504700	0.282400
7 C(S2,CK)	0.000000	0.000000	0.000000	0.000000	0.000000	0.000000

Table 9.46. Summarization of mean and standard deviation of coverage (C) between SPEA2 and SPEA2-CE-KR (70, 80, 90 opponents) for DTLZ test problems with 5 objectives. $C(X',X'') > C(X'',X')$, then X' is better than X''.

5 Objectives		70 Opponents		80 Opponents		90 Opponents	
DTLZ	C Metric	Mean	St Dev	Mean	St Dev	Mean	St Dev
1	C(CK,S2)	0.861200	0.122200	0.863800	0.134900	0.867600	0.105400
	C(S2,CK)	0.000000	0.000000	0.030300	0.166100	0.001670	0.009130
2	C(CK,S2)	0.107330	0.053880	0.126000	0.054430	0.113330	0.050330
	C(S2,CK)	0.000000	0.000000	0.000000	0.000000	0.000000	0.000000
3	C(CK,S2)	0.897300	0.077400	0.901000	0.091700	0.899300	0.107300
	C(S2,CK)	0.000000	0.000000	0.000000	0.000000	0.000000	0.000000
4	C(CK,S2)	0.354300	0.092500	0.349000	0.106000	0.354700	0.096400
	C(S2,CK)	0.000000	0.000000	0.000000	0.000000	0.000000	0.000000
5	C(CK,S2)	0.043330	0.027210	0.033670	0.021730	0.034670	0.029560
	C(S2,CK)	0.159300	0.092800	0.163000	0.104100	0.151000	0.105900
6	C(CK,S2)	0.524300	0.147300	0.513300	0.172300	0.568300	0.133500
	C(S2,CK)	0.000000	0.000000	0.000000	0.000000	0.000000	0.000000
7	C(CK,S2)	0.479300	0.273700	0.502700	0.273500	0.506000	0.277000
	C(S2,CK)	0.000000	0.000000	0.000000	0.000000	0.000000	0.000000

Table 9.47. Summarization of mean and standard deviation of coverage (C) between SPEA2-CE-HOF and SPEA2-CE-KR (10, 20, 30 opponents) for DTLZ test problems with 3 objectives. $C(X',X'') > C(X'',X')$, then X' is better than X''.

3 Objectives		10 Opponents		20 Opponents		30 Opponents	
DTLZ	C Metric	Mean	St Dev	Mean	St Dev	Mean	St Dev
1	C(CK,CH)	0.076000	0.068700	0.072300	0.074700	0.087700	0.068900
	C(CH,CK)	0.223700	0.240600	0.258300	0.343900	0.168300	0.258900
2	C(CK,CH)	0.016670	0.011240	0.023330	0.018630	0.022330	0.015240
	C(CH,CK)	0.054000	0.028840	0.036000	0.024150	0.033000	0.022770
3	C(CK,CH)	0.083300	0.069600	0.129000	0.123600	0.142000	0.139100
	C(CH,CK)	0.363700	0.360600	0.355300	0.319000	0.291300	0.336700
4	C(CK,CH)	0.042000	0.047730	0.054000	0.051230	0.028670	0.041250
	C(CH,CK)	0.092000	0.106300	0.057300	0.078800	0.062300	0.062100
5	C(CK,CH)	0.027330	0.023030	0.033670	0.022200	0.037670	0.023290
	C(CH,CK)	0.180300	0.063700	0.116000	0.052890	0.110700	0.056500
6	C(CK,CH)	0.416000	0.064500	0.405670	0.053670	0.394000	0.057300
	C(CH,CK)	0.000000	0.000000	0.000000	0.000000	0.000000	0.000000
7	C(CK,CH)	0.085000	0.042810	0.077330	0.030280	0.067670	0.021760
	C(CH,CK)	0.009000	0.012420	0.014000	0.031360	0.013330	0.021710

Table 9.48. Summarization of mean and standard deviation of coverage (C) between SPEA2-CE-HOF and SPEA2-CE-KR (40, 50, 60 opponents) for DTLZ test problems with 3 objectives. $C(X',X'') > C(X'',X')$, then X' is better than X''.

3 Objectives		40 Opponents		50 Opponents		60 Opponents	
DTLZ	C Metric	Mean	St Dev	Mean	St Dev	Mean	St Dev
1	C(CK,CH)	0.073000	0.062000	0.064670	0.036740	0.092700	0.100500
	C(CH,CK)	0.173000	0.290800	0.114000	0.215800	0.168000	0.286300
2	C(CK,CH)	0.019000	0.015390	0.016000	0.013800	0.020670	0.017410
	C(CH,CK)	0.031670	0.019840	0.029670	0.021570	0.024330	0.019600
3	C(CK,CH)	0.197300	0.178500	0.239000	0.184100	0.270000	0.208600
	C(CH,CK)	0.266300	0.330100	0.170000	0.175200	0.241000	0.250300
4	C(CK,CH)	0.050670	0.048990	0.039330	0.049610	0.039330	0.043700
	C(CH,CK)	0.053700	0.067000	0.053300	0.056800	0.051300	0.057700
5	C(CK,CH)	0.041330	0.022850	0.040330	0.025390	0.044000	0.027240
	C(CH,CK)	0.106000	0.046360	0.127700	0.065800	0.093670	0.045820
6	C(CK,CH)	0.398300	0.065500	0.381000	0.069000	0.398300	0.065200
	C(CH,CK)	0.000000	0.000000	0.000000	0.000000	0.000000	0.000000
7	C(CK,CH)	0.061330	0.034410	0.065330	0.033500	0.057000	0.030870
	C(CH,CK)	0.031000	0.061200	0.014000	0.024010	0.046700	0.073000

Table 9.49. Summarization of mean and standard deviation of coverage (C) between SPEA2-CE-HOF and SPEA2-CE-KR (70, 80, 90 opponents) for DTLZ test problems with 3 objectives. $C(X',X'') > C(X'',X')$, then X' is better than X''.

3 Objectives		70 Opponents		80 Opponents		90 Opponents	
DTLZ	C Metric	Mean	St Dev	Mean	St Dev	Mean	St Dev
1	C(CK,CH)	0.083000	0.104100	0.090300	0.120400	0.080330	0.050620
	C(CH,CK)	0.167000	0.315600	0.125000	0.265700	0.038700	0.144600
2	C(CK,CH)	0.015330	0.011960	0.018670	0.014320	0.023330	0.019000
	C(CH,CK)	0.032330	0.022850	0.029330	0.020160	0.021000	0.021550
3	C(CK,CH)	0.275300	0.185800	0.292000	0.205900	0.304700	0.194900
	C(CH,CK)	0.225000	0.230100	0.259000	0.279400	0.225700	0.234700
4	C(CK,CH)	0.033670	0.037920	0.043330	0.050610	0.041670	0.042680
	C(CH,CK)	0.055000	0.054310	0.061700	0.076400	0.050700	0.057200
5	C(CK,CH)	0.034330	0.023440	0.041330	0.029210	0.037000	0.025350
	C(CH,CK)	0.115000	0.042160	0.098330	0.047860	0.103330	0.048020
6	C(CK,CH)	0.386700	0.057700	0.370000	0.061200	0.377300	0.065200
	C(CH,CK)	0.000000	0.000000	0.000000	0.000000	0.000000	0.000000
7	C(CK,CH)	0.055670	0.031810	0.062000	0.030220	0.051330	0.034610
	C(CH,CK)	0.032700	0.060000	0.031700	0.071200	0.046300	0.072300

Table 9.50. Summarization of mean and standard deviation of coverage (C) between SPEA2-CE-HOF and SPEA2-CE-KR (10, 20, 30 opponents) for DTLZ test problems with 4 objectives. $C(X',X'') > C(X'',X')$, then X' is better than X''.

4 Objectives	10 Opponents		20 Opponents		30 Opponents	
DTLZ C Metric	Mean	St Dev	Mean	St Dev	Mean	St Dev
1 C(CK,CH)	0.499300	0.144800	0.446300	0.121600	0.444700	0.146700
C(CH,CK)	0.002670	0.007850	0.000000	0.000000	0.001000	0.005480
2 C(CK,CH)	0.032670	0.019990	0.038670	0.027380	0.032670	0.017410
C(CH,CK)	0.005330	0.008190	0.005670	0.012780	0.004000	0.008940
3 C(CK,CH)	0.512700	0.125100	0.496300	0.157100	0.491700	0.161600
C(CH,CK)	0.001000	0.003051	0.001000	0.004026	0.000667	0.002537
4 C(CK,CH)	0.149000	0.095900	0.143300	0.085800	0.128000	0.093800
C(CH,CK)	0.015000	0.052960	0.014330	0.044460	0.013330	0.028450
5 C(CK,CH)	0.050330	0.024980	0.039330	0.022730	0.046670	0.020230
C(CH,CK)	0.079000	0.052220	0.056670	0.039420	0.051000	0.046490
6 C(CK,CH)	0.451000	0.062700	0.482700	0.067200	0.466300	0.080300
C(CH,CK)	0.000000	0.000000	0.000000	0.000000	0.000000	0.000000
7 C(CK,CH)	0.164300	0.141700	0.128000	0.107300	0.131700	0.106800
C(CH,CK)	0.002000	0.004842	0.002000	0.005510	0.001667	0.004611

Table 9.51. Summarization of mean and standard deviation of coverage (C) between SPEA2-CE-HOF and SPEA2-CE-KR (40, 50, 60 opponents) for DTLZ test problems with 4 objectives. $C(X',X'') > C(X'',X')$, then X' is better than X''.

4 Objectives	40 Opponents		50 Opponents		60 Opponents	
DTLZ C Metric	Mean	St Dev	Mean	St Dev	Mean	St Dev
1 C(CK,CH)	0.417700	0.095700	0.425000	0.123400	0.389700	0.117800
C(CH,CK)	0.000000	0.000000	0.000000	0.000000	0.000333	0.001826
2 C(CK,CH)	0.034670	0.021610	0.035670	0.019770	0.025330	0.015480
C(CH,CK)	0.004330	0.008170	0.002000	0.004842	0.001333	0.003457
3 C(CK,CH)	0.455300	0.170600	0.423700	0.162100	0.387300	0.185200
C(CH,CK)	0.000333	0.001826	0.000000	0.000000	0.000000	0.000000
4 C(CK,CH)	0.129000	0.086600	0.142700	0.085300	0.161300	0.090000
C(CH,CK)	0.005330	0.013830	0.010670	0.025450	0.011330	0.031810
5 C(CK,CH)	0.048330	0.022450	0.050670	0.025590	0.043330	0.027960
C(CH,CK)	0.122300	0.146100	0.135000	0.150000	0.115000	0.127300
6 C(CK,CH)	0.480300	0.068200	0.461300	0.060800	0.450300	0.088400
C(CH,CK)	0.000000	0.000000	0.000000	0.000000	0.000000	0.000000
7 C(CK,CH)	0.105000	0.093200	0.146000	0.120800	0.140000	0.134500
C(CH,CK)	0.002000	0.004842	0.000333	0.001826	0.000667	0.002537

Table 9.52. Summarization of mean and standard deviation of coverage (C) between SPEA2-CE-HOF and SPEA2-CE-KR (70, 80, 90 opponents) for DTLZ test problems with 4 objectives. $C(X',X'') > C(X'',X')$, then X' is better than X''.

4 Objectives		70 Opponents		80 Opponents		90 Opponents	
DTLZ	C Metric	Mean	St Dev	Mean	St Dev	Mean	St Dev
1	C(CK,CH)	0.385700	0.121700	0.385700	0.110000	0.377000	0.115900
	C(CH,CK)	0.000000	0.000000	0.000333	0.001826	0.000000	0.000000
2	C(CK,CH)	0.030330	0.016710	0.029000	0.020570	0.028330	0.017040
	C(CH,CK)	0.002333	0.005040	0.002333	0.005040	0.002000	0.004068
3	C(CK,CH)	0.416000	0.178600	0.436700	0.156000	0.425700	0.199600
	C(CH,CK)	0.000333	0.001826	0.000667	0.002537	0.000000	0.000000
4	C(CK,CH)	0.124700	0.085100	0.149300	0.085100	0.131700	0.098900
	C(CH,CK)	0.015330	0.034710	0.007670	0.020960	0.012330	0.029560
5	C(CK,CH)	0.053000	0.030070	0.049330	0.026770	0.049000	0.030210
	C(CH,CK)	0.171700	0.161100	0.143700	0.137700	0.099300	0.114700
6	C(CK,CH)	0.464300	0.086800	0.454700	0.081000	0.471300	0.079600
	C(CH,CK)	0.000000	0.000000	0.000000	0.000000	0.000000	0.000000
7	C(CK,CH)	0.128000	0.163800	0.083000	0.068700	0.119700	0.113600
	C(CH,CK)	0.000667	0.002537	0.001000	0.004026	0.001000	0.003051

Table 9.53. Summarization of mean and standard deviation of coverage (C) between SPEA2-CE-HOF and SPEA2-CE-KR (10, 20, 30 opponents) for DTLZ test problems with 5 objectives. $C(X',X'') > C(X'',X')$, then X' is better than X''.

5 Objectives		10 Opponents		20 Opponents		30 Opponents	
DTLZ	C Metric	Mean	St Dev	Mean	St Dev	Mean	St Dev
1	C(CK,CH)	0.937700	0.059100	0.927000	0.072100	0.928000	0.057100
	C(CH,CK)	0.000000	0.000000	0.000000	0.000000	0.000000	0.000000
2	C(CK,CH)	0.152000	0.076200	0.121000	0.056400	0.123300	0.054900
	C(CH,CK)	0.000000	0.000000	0.000000	0.000000	0.000000	0.000000
3	C(CK,CH)	0.873300	0.107900	0.863000	0.133000	0.895000	0.089500
	C(CH,CK)	0.000000	0.000000	0.000000	0.000000	0.000000	0.000000
4	C(CK,CH)	0.372300	0.100100	0.362700	0.062000	0.350300	0.076300
	C(CH,CK)	0.000000	0.000000	0.000000	0.000000	0.000000	0.000000
5	C(CK,CH)	0.076670	0.045740	0.058330	0.035440	0.043330	0.022940
	C(CH,CK)	0.152300	0.091600	0.160000	0.111400	0.146000	0.112500
6	C(CK,CH)	0.068700	0.108500	0.376300	0.208800	0.427000	0.155500
	C(CH,CK)	0.000000	0.000000	0.000000	0.000000	0.000000	0.000000
7	C(CK,CH)	0.594300	0.266100	0.512300	0.265000	0.566700	0.274400
	C(CH,CK)	0.000000	0.000000	0.000000	0.000000	0.000000	0.000000

Table 9.54. Summarization of mean and standard deviation of coverage (C) between SPEA2-CE-HOF and SPEA2-CE-KR (40, 50, 60 opponents) for DTLZ test problems with 5 objectives. $C(X',X'') > C(X'',X')$, then X' is better than X''.

5 Objectives		40 Opponents		50 Opponents		60 Opponents	
DTLZ	C Metric	Mean	St Dev	Mean	St Dev	Mean	St Dev
1	C(CK,CH)	0.901000	0.087600	0.895700	0.077200	0.919000	0.058200
	C(CH,CK)	0.000000	0.000000	0.000000	0.000000	0.000000	0.000000
2	C(CK,CH)	0.117330	0.047560	0.114300	0.063900	0.121300	0.059300
	C(CH,CK)	0.000000	0.000000	0.000000	0.000000	0.000000	0.000000
3	C(CK,CH)	0.878300	0.140400	0.883300	0.090800	0.882300	0.122100
	C(CH,CK)	0.000000	0.000000	0.000000	0.000000	0.000000	0.000000
4	C(CK,CH)	0.354000	0.111600	0.359300	0.102400	0.380700	0.102600
	C(CH,CK)	0.000000	0.000000	0.000000	0.000000	0.000000	0.000000
5	C(CK,CH)	0.045000	0.027640	0.056000	0.023870	0.045330	0.024880
	C(CH,CK)	0.127700	0.089800	0.126000	0.108100	0.112300	0.099100
6	C(CK,CH)	0.475000	0.126600	0.508700	0.142000	0.522000	0.163300
	C(CH,CK)	0.000000	0.000000	0.000000	0.000000	0.000000	0.000000
7	C(CK,CH)	0.425000	0.272000	0.570300	0.258000	0.484700	0.283500
	C(CH,CK)	0.000000	0.000000	0.000000	0.000000	0.000000	0.000000

Table 9.55. Summarization of mean and standard deviation of coverage (C) between SPEA2-CE-HOF and SPEA2-CE-KR (70, 80, 90 opponents) for DTLZ test problems with 5 objectives. $C(X',X'') > C(X'',X')$, then X' is better than X''.

5 Objectives		70 Opponents		80 Opponents		90 Opponents	
DTLZ	C Metric	Mean	St Dev	Mean	St Dev	Mean	St Dev
1	C(CK,CH)	0.886300	0.085400	0.901000	0.076700	0.897000	0.056700
	C(CH,CK)	0.000000	0.000000	0.000000	0.000000	0.000000	0.000000
2	C(CK,CH)	0.109000	0.073500	0.121670	0.049690	0.119000	0.062900
	C(CH,CK)	0.000000	0.000000	0.000000	0.000000	0.000000	0.000000
3	C(CK,CH)	0.878300	0.135500	0.887300	0.088800	0.867700	0.116000
	C(CH,CK)	0.000000	0.000000	0.000000	0.000000	0.000000	0.000000
4	C(CK,CH)	0.363700	0.088200	0.355300	0.082400	0.372700	0.088100
	C(CH,CK)	0.000000	0.000000	0.000000	0.000000	0.000000	0.000000
5	C(CK,CH)	0.051330	0.028370	0.047000	0.028910	0.042330	0.032450
	C(CH,CK)	0.120000	0.092600	0.130000	0.108700	0.107300	0.100900
6	C(CK,CH)	0.534300	0.142800	0.510700	0.162200	0.570700	0.132900
	C(CH,CK)	0.000000	0.000000	0.000000	0.000000	0.000000	0.000000
7	C(CK,CH)	0.488000	0.277800	0.519700	0.271200	0.527300	0.263000
	C(CH,CK)	0.000000	0.000000	0.000000	0.000000	0.000000	0.000000

Part IV

Learning

10

Bayesian Ying-Yang Harmony Learning for Local Factor Analysis: A Comparative Investigation

Lei Shi

Department of Computer Science and Engineering,
The Chinese University of Hong Kong, China
shil@cse.cuhk.edu.hk

Summary. For unsupervised learning by Local Factor Analysis (LFA), it is important to determine both the component number and the local hidden dimensions appropriately, which is a typical example of model selection. One conventional approach for model selection is to implement a two-phase procedure with the help of model selection criteria, such as AIC, CAIC, BIC(MDL), SRM, CV, etc.. Although all working well given large enough samples, they still suffer from two problems. First, their performances will deteriorate greatly on a small sample size. Second, two-phase procedure requires intensive computation. To tackle the second problem, one type of efforts has been made in the literature, featured by an incremental implementation, e.g. IMoFA and VBMFA. Bayesian Ying-Yang (BYY) harmony learning provides not only a BYY harmony data smoothing criterion (BYY-C) in a two-phase implementation for the first problem, but also an algorithm called automatic BYY harmony learning (BYY-A) that have automatic model selection ability during parameter learning and thus can reduce the computational expense significantly. The lack of systematic comparisons in the literature motivates this work. Comparative experiments are first conducted on synthetic data considering not only different settings including noise, dimension and sample size, but also different evaluations including model selection accuracy and three other applied performances. Thereafter, comparisons are also made on several real world classification datasets. In two-phase implementation, the observations show that BIC and CAIC generally outperform AIC, SRM and CV, while BYY-C is the best for small sample sizes. Moreover, in the cases of a sufficiently large sample size, IMoFA, VBMFA, and BYY-A produce similar performances but with much reduced computational costs, where, still, BYY-A provides better or at least comparably good performances.

Keywords: Local Factor Analysis, Model selection, Two-phase implementation, Incremental methods, Automatic model selection, Bayesian Ying-Yang harmony learning, Data smoothing, Small sample size.

10.1 Model Selection for Local Factor Analysis

As a useful and widely-used multivariate analysis model, Local Factor Analysis (LFA), also called Mixture of Factor Analyzers (MFA), has received wide applications including pattern recognition, bioinformatics, and financial engineering [10, 11, 33, 34]. One important task for LFA model is to determine both the component number and the local hidden dimensions appropriately, which is a typical example of model selection. Under this topic, many efforts have been made in the literature [9, 10, 11, 16, 17, 25, 33, 37].

H.R. Tizhoosh, M. Ventresca (Eds.): Oppos. Concepts in Comp. Intel., SCI 155, pp. 209–232, 2008.
springerlink.com   © Springer-Verlag Berlin Heidelberg 2008

However, there still lacks enough comparative investigations on them. This chapter is such a comparative study organized as follows: In Section 1, after introducing LFA and its model selection problem, we briefly describe two kinds of efforts in the literature, i.e., the two-phase implementation with the help of model selection criteria, and incremental methods, respectively. In Section 2, a sketch of Bayesian Ying-Yang (BYY) harmony learning will be given on LFA, including both an automatic BYY harmony learning algorithm (BYY-A) and a BYY harmony learning criterion (BYY-C) in a two-phase implementation. Thereafter, comparative experiments are conducted on simulated datasets in Section 3 and on real world datasets in Section 4. Finally, concluding remarks are given in Section 5.

10.1.1   Local Factor Analysis and Its Model Selection Problem

Local Factor Analysis (LFA) combines Gaussian Mixture Model (GMM) with Factor Analysis (FA), performing clustering and dimension reduction in each component simultaneously. As a whole, LFA implements nonlinear dimension reduction by modeling the density as a mixture of local linear subspaces. Provided a d-dimensional observable variable vector $\mathbf{x}$, LFA assumes that $\mathbf{x}$ is distributed according to a mixture of k underlying probability distributions $p(\mathbf{x}) = \sum_{l=1}^{k} \alpha_l p(\mathbf{x}|l)$, where α_l is the prior of the l-th component with $\alpha_l \geq 0, l = 1, \ldots, k, \sum_{l=1}^{k} \alpha_l = 1$, and $p(\mathbf{x}|l)$ is the conditional probability of $\mathbf{x}$ on the l-th component modeled by a single FA [10, 11, 33]. That is,

$$p(\mathbf{x}|\mathbf{y}, l) = G(\mathbf{x}|\mathbf{A}_l\mathbf{y} + \boldsymbol{\mu}_l, \boldsymbol{\Psi}_l), \; p(\mathbf{y}|l) = G(\mathbf{y}|\mathbf{0}, \mathbf{I}_{m_l}),$$

$$p(\mathbf{x}|l) = \int p(\mathbf{x}|\mathbf{y}, l)p(\mathbf{y}|l)d\mathbf{y} = G(\mathbf{x}|\boldsymbol{\mu}_l, \mathbf{A}_l\mathbf{A}_l^T + \boldsymbol{\Psi}_l), \quad (10.1)$$

where $G(\mathbf{x}|\boldsymbol{\mu}, \boldsymbol{\Sigma})$ is the multivariate Gaussian distribution of $\mathbf{x}$ with the mean $\boldsymbol{\mu}$ and the covariance matrix $\boldsymbol{\Sigma}$. For the l-th component, $\mathbf{y}$ is a m_l-dimensional unobservable latent vector, $\mathbf{A}_l$ is a $d \times m_l$ loading matrix, $\boldsymbol{\mu}_l$ is a d-dimensional mean vector, $\boldsymbol{\Psi}_l$ is a diagonal covariance matrix, $l = 1, 2, \ldots, k$.

For a set of observations $\{\mathbf{x}_t\}_{t=1}^{N}$, supposing that the component number k and the local hidden dimensions $\{m_l\}$ are given, one widely used method to estimate parameters $\boldsymbol{\Theta} = \{\alpha_l, \mathbf{A}_l, \boldsymbol{\mu}_l, \boldsymbol{\Psi}_l\}_{l=1}^{k}$ is the maximum-likelihood (ML) learning, which aims for maximizing the log-likelihood function $L(\mathbf{X}|\hat{\boldsymbol{\Theta}})$ and can be effectively implemented by the well-known Expectation-Maximization (EM) algorithm [11, 14]. Throughout this chapter, $\hat{\boldsymbol{\Theta}}$ is the estimate of $\boldsymbol{\Theta}$.

However, two important issues are still left, i.e., how to determine the Gaussian component number k and how to determine the local hidden dimensions $\{m_l\}_{l=1}^{k}$, or say in general, how to determine an appropriate scale tuple $\mathbf{k} = \{k, \{m_l\}_{l=1}^{k}\}$. This is a typical *model selection* problem to avoid both under-fitting and over-fitting, for which many efforts have been made in the literature, including three major types of methods, i.e., two-phase procedure [1, 2, 3, 5, 6, 7, 9, 18, 19, 21, 22, 24, 36, 38], incremental [10, 11, 17] and automatic [25, 36, 38]. The former two types will be discussed in the current section, while a representative of the last one will be introduced in Section 2.

10.1.2 Two-Phase Procedure and Several Typical Criteria

In the literature, one conventional approach for model selection is featured by a two-phase procedure [1, 2, 3, 5, 11, 18, 19, 21, 22]. In the first phase, a range of model complexity for $\mathbf{k} = \{k, \{m_l\}\}$ is pre-specified and assumed to include the optimal complexity $\mathbf{k}^*$. For each specific $\mathbf{k}$ in this range, parameters $\Theta_{\mathbf{k}}$ are estimated under the ML principle, usually implemented by EM algorithm [11, 12, 14]. In the second phase, among all trained candidate models $\{\hat{\Theta}_{\mathbf{k}}\}$ within the pre-determined model range, the optimal model scale $\hat{\mathbf{k}}^*$ is selected as the one with the minimum criterion[1] value $J(\mathbf{k}, \hat{\Theta}_{\mathbf{k}})$, i.e.:

$$\hat{\mathbf{k}}^* = \arg\min_{\mathbf{k}} J(\mathbf{k}, \hat{\Theta}_{\mathbf{k}}). \tag{10.2}$$

Several typical model selection criteria $J(\mathbf{k}, \hat{\Theta}_{\mathbf{k}})$ have been investigated for the two-phase implementation in the literature, including Akaike's Information Criterion (AIC) [1], Bozdogan's Consistent Akaike's Information Criterion (CAIC) [5], Schwarz's Bayesian Inference Criterion (BIC) [18] which coincides with Rissanen's Minimum Description Length (MDL) criterion [2], Structural Risk Minimization (SRM) [21, 22] principle based on Vapnik's VC dimension theory, and cross-validation (CV) [3, 19]. The mathematical expressions for these criteria are summarized in Table 10.1 and will be briefly explained in the following. Here $L(\mathbf{X}_N|\hat{\Theta}_{\mathbf{k}})$ is the log-likelihood of training samples $\mathbf{X}_N$ based on the estimated parameters $\hat{\Theta}_{\mathbf{k}}$ under a given model scale $\mathbf{k} = \{k, \{m_l\}\}$ for LFA.

For criteria AIC, CAIC, and BIC, they share a similar form, where $D(\mathbf{k})$ is the number of free parameters in the referred model. For a LFA model with parameters $\Theta_{\mathbf{k}} = \Theta_{k, \{m_l\}_{l=1}^{k}} = \{\alpha_l, \boldsymbol{\mu}_l, \boldsymbol{\Psi}_l, \mathbf{A}_l\}_{l=1}^{k}$, the free parameter number for $\{\alpha_l\}$ is $k-1$, for $\{\boldsymbol{\mu}_l\}$ and $\{\boldsymbol{\Psi}_l\}$ are both kd, while for $\{\mathbf{A}_l\}$ is $\sum_{l=1}^{k}(dm_l - m_l(m_l-1)/2)$. Finally, the total number of free parameters is:

$$D(\mathbf{k}) = D(k, \{m_l\}) = k - 1 + kd + kd + \sum_{l=1}^{k}[dm_l - \frac{m_l(m_l - 1)}{2}]. \tag{10.3}$$

Based on the well-known VC (Vapnik-Chervonenkis) dimension theory, Structure Risk Minimization (SRM) principle [21, 22] trades off between the empirical error and hypothesis space complexity, which leads to finding the model with the minimum structural risk $R(\Theta)$. Due to the impossibility of directly minimizing the risk, Vapnik [21, 22] suggested to minimize its upper bound. However, finding a computable expression of this upper bound for a given specific learning model is usually quite a difficult task. For our task on LFA, we have tried several academic search engines, but found only one result on GMM in [24], as briefly adopted in Table 10.1. In order to simply apply this GMM result onto LFA, we revisit the original result on the VC dimension type

---

[1] Actually, given an infinite or a sufficiently large size of samples, the log-likelihood value $L(\mathbf{X}|\hat{\Theta}_{\mathbf{k}})$ can work asymptotically to select the component number, i.e., $\hat{\mathbf{k}}^* = \arg\min_{\mathbf{k}} L(\mathbf{X}|\hat{\Theta}_{\mathbf{k}})$. However, given a finite or small size of samples, it does not work. Instead, a better model selection criterion $J(\mathbf{k}, \hat{\Theta}_{\mathbf{k}})$ is needed.

Table 10.1. Description of typical model selection criteria

Criteria	Mathematical Description
AIC	$J_{AIC}(\hat{\Theta}_{\mathbf{k}}, \mathbf{k}) = -2L(\mathbf{X}|\hat{\Theta}_{\mathbf{k}}) + 2D(\mathbf{k})$
CAIC	$J_{CAIC}(\hat{\Theta}_{\mathbf{k}}, \mathbf{k}) = -2L(\mathbf{X}|\hat{\Theta}_{\mathbf{k}}) + [\ln(N) + 1]D(\mathbf{k})$
BIC	$J_{BIC}(\hat{\Theta}_{\mathbf{k}}, \mathbf{k}) = -2L(\mathbf{X}|\hat{\Theta}_{\mathbf{k}}) + \ln(N)D(\mathbf{k})$
SRM	$J_{SRM}(\hat{\Theta}_{\mathbf{k}}, \mathbf{k}) = \dfrac{-L(\mathbf{Z}|\hat{\Theta}_{\mathbf{k}})}{1 - \{\frac{w(d+m,k)[\ln\frac{N}{w(d+m,k)}+1]+\frac{1}{2}\ln N}{N}\}^{\frac{1}{2}}}$ $w(d+m,k) = k \cdot \max\{\frac{(d+m)[\ln(d+m)-1]}{\ln 2}, 2(d+m) + 3\}$
CV-m	$J_{CV(m)}(\hat{\Theta}_{\mathbf{k}}, \mathbf{k}) = -\sum_{i=1}^{m} L(\mathbf{D}_i|\hat{\Theta}_{\mathbf{D}_{-i}, \mathbf{k}})$

integer $w(k)$, i.e. just with respect to k, and extract the dimension d as another variable to rename it $w(k, d)$, i.e. with respect to both k and d. Thus after defining a new variable $\mathbf{z} = (\mathbf{x}^T, \mathbf{y}^T)^T$, we just need to think about GMM for $(d + m_l)$ dimensional vector $\mathbf{z}$, i.e.,

$$p(\mathbf{z}|\Theta) = \sum_{l=1}^{k} \alpha_l G(\mathbf{z}|\varphi_l, \Gamma_l), \quad \varphi_l = \begin{pmatrix} \mu_l \\ \mathbf{0}_{m_l} \end{pmatrix},$$
$$\Gamma_l = \begin{pmatrix} \mathbf{A}_l \mathbf{A}_l^T + \Psi_l & \mathbf{A}_l \\ \mathbf{A}_l^T & \mathbf{I}_{m_l} \end{pmatrix}. \tag{10.4}$$

In order to apply this on LFA, given one sample $\mathbf{x}_t$, for each component l we can get $\mathbf{y}_{l,t} = \mathbf{A}_l^T(\mathbf{A}_l \mathbf{A}_l^T + \Psi_l)^{-1}(\mathbf{x}_t - \mu_l)$, which easily results from ML estimation. Consequently, $\mathbf{Z}_l$ is obtained from training samples $\{\mathbf{x}_t\}_{t=1}^{N}$ for each l-th component via $\mathbf{Z}_l = \{(\mathbf{x}_t^T, \mathbf{y}_{l,t}^T)^T\}_{t=1}^{N}$. Then used to calculate the criterion value. Due to the above arbitrary simplification, the local hidden dimensions $\{m_l\}_{l=1}^{k}$ have to be the same, i.e., $m_l = m$, for $l = 1, \ldots, k$.

Another well-known model selection technique is called cross-validation (CV) [3, 19]. Here we consider CV based on the maximum-likelihood estimation. First the data are randomly and evenly divided into m parts, namely $\mathbf{X}_N = \{\mathbf{D}_i\}_{i=1}^{m}$. For the i-th partition, let $\mathbf{D}_i$ be the subset used for testing and $\mathbf{D}_{-i}$ be the rest used for training, the CV criterion is given as follows:

$$J_{CV(m)}(\hat{\Theta}_{\mathbf{k}}, \mathbf{k}) = -\sum_{i=1}^{m} L(\mathbf{D}_i|\hat{\Theta}_{\mathbf{D}_{-i}, \mathbf{k}}) \tag{10.5}$$

where $\hat{\mathbf{\Theta}}_{\mathbf{D}_{-i},\mathbf{k}}$ is the ML estimate based on $\mathbf{D}_{-i}$ given the model scale $\mathbf{k}$, and $L(\mathbf{D}_i|\hat{\mathbf{\Theta}}_{\mathbf{D}_{-i},\mathbf{k}})$ is the log-likelihood on $\mathbf{D}_i$ given $\hat{\mathbf{\Theta}}_{\mathbf{D}_{-i},\mathbf{k}}$. Featured by m, it is usually referred as m-fold CV and in this chapter shortly denoted as CV-m.

As discussed in [25, 31, 32, 37, 38], the two-phase implementation still has two problems. First, in the cases that the sample size N is small or not large enough, each criterion actually provides a rough estimate that can not guarantee to give $\hat{\mathbf{k}}^* = \mathbf{k}^*$, and even results in a wrong selection especially when $\mathbf{k}$ consists of several integers to enumerate. Second, in addition to the performance problem, another difficulty is its feasibility in implementation, because the two-phase implementation is so extensive in computing that it may be infeasible in many real applications, letting alone the difficulty to appropriately pre-decide the candidate model range.

10.1.3 Approaches with an Incremental Implementation

To tackle the second problem of a two-phase implementation, i.e. the intensive computational cost problem, one type of efforts have been made in the literature, featured by an incremental implementation [10, 17]. Briefly speaking, instead of enumerating every candidate model, this type of efforts conduct parameter learning incrementally in a sense that it attempts to incorporate as much as possible what learned at $\mathbf{k}$ into the learning procedure at $\mathbf{k} + 1$. Also, the calculation of criterion $J(\mathbf{k}, \hat{\mathbf{\Theta}}_{\mathbf{k}})$ is made incrementally. As discussed in [37], although such an incremental implementation can indeed save computing costs to a certain extent, parameter learning has to be made still by enumerating the values of $\mathbf{k}$, and computing costs are still very high. Moreover, as $\mathbf{k}$ increases to $\mathbf{k} + 1$, an incremental implementation of parameter learning may also lead to suboptimal performance because not only those newly added parameters but also the old parameters have to be re-learned. Two approaches under such an incremental implementation are considered in this chapter, namely Incremental Mixture of Factor Analysers (IMoFA) [17] and Variational Bayesian Mixture of Factor Analyzers (VBMFA) [10].

Incremental Mixture of Factor Analysers

The first incremental algorithm is Incremental Mixture of Factor Analysers (IMoFA) [17], which adds in both components and factors iteratively. IMoFA aims at making model selection by a procedure component splitting or factor adding according to the validation likelihood, which is terminated when there is no improvement on the validation likelihood. It starts with 1-factor, 1-component mixture and proceeds by adding new factors or new components until some stopping condition is satisfied. There are two variants IMoFA-L and IMoFA-A for unsupervised and supervised approaches, respectively. In this chapter, we consider unsupervised learning by IMoFA-L, shortly denoted by IMoFA, and the validation set is randomly selected out from the original data with 10% size. The sketch of IMoFA is listed in Algorithm 1, while the detailed procedures are referred to [17].

Algorithm 1. Sketch for IMoFA algorithm

Initialization: Divide the original data $\mathbf{X}_N$ into training set $\mathbf{X}_T$ and validation set $\mathbf{X}_V$. Initialize a 1-component 1-factor model based on $\mathbf{X}_T$. Set the old validation likelihood $L_{old} = -\infty$.
repeat
    Splitting: select a component l_1 and split it. Train Θ_1 on $\mathbf{X}_T$ by EM(split l_1), and calculate the validation likelihood L_1 on $\mathbf{X}_V$, i.e, $L_1 = L(\mathbf{X}_V|\Theta_1)$.
    Factor Adding: select a component l_2 and add one hidden dimension to it, then train Θ_2 on $\mathbf{X}_T$ by EM(add a factor to l_2), calculate the validation likelihood L_2 on $\mathbf{X}_V$, , i.e, $L_2 = L(\mathbf{X}_V|\Theta_2)$.
    Select Action: select either splitting or adding factor according to $z = \arg\max_i(L_i)$. Let $\Theta = \Theta_z$, $L_{new} = L_z$, and $L_{old} = L_{new}$.
until $L_{new} < L_{old}$

Variational Bayesian Mixture of Factor Analyzers

The other approach with such an incremental model selection nature is the Variational Bayesian Mixture of Factor Analyzers (VBMFA) [10]. Under the Bayesian framework, the VBMFA method targets at maximizing the variational function F, an approximated lower bound of the marginal log-likelihood, which is motivated by avoiding the integration difficulty in help of the Jensen's inequality [10]. Similar to IMoFA, still starting from 1-component full factor model, VBMFA makes parameter learning in an incremental tendency for the component number via stochastically selecting out and splitting a component l with probability proportional to $e^{-\zeta F_l}$, where F_l is component l's contribution to F, reject this attempt if F does not recover. During this procedure, VBMFA also deletes component l if its responsibility approaches zero and deletes the i-th hidden dimension of component l if its corresponding hyper-parameter $\nu_l^i \to \infty$. The iteration stops if the variational function F has no improvement. The sketched algorithm of VBMFA is listed in Algorithm 2, while details about VBMFA are referred to [10].

Algorithm 2. Sketch for VBMFA algorithm

Initialization: Initialize a 1-component full factor model with hyper-parameters that indicate the priors. Set the old variational function $F_{old} = -\infty$.
repeat
    E Step: Partially fix the current model and change the model scales by:
        (1) Stochastically pick a component l with probability proportional to $e^{-\zeta F_l}$.
            Split it into two components, and reject this attempt if F does not recover;
        (2) Regard component l dead and delete it if its responsibility is too small;
        (3) Delete the i-th hidden dimension of the l-th component if $\nu_l^i \to \infty$.
    M Step: Fixing current model scales k, adjust both the model parameters and hyper-parameters so as to maximize the variational function F. Calculate the new variational function F_{new}.
until $F_{new} < F_{old}$

10.2   Bayesian Ying-Yang Harmony Learning for LFA

Bayesian Ying-Yang (BYY) harmony learning was firstly proposed in 1995 and then systematically developed in the past decade [25, 26, 33, 34, 35, 37, 38]. The BYY harmony learning with typical structures leads us to a set of new model selection criteria, new techniques for implementing regularization and a class of algorithms with automatic model selection ability during parameter learning. Considering the LFA model, in order to remove the rotational indeterminacy and conduct automatic model selection, BYY harmony learning adopts the alternative but probabilistically equivalent model compared with (10.1), where the l-th component's distribution is assumed as follows:

$$p(l) = \alpha_l, \quad p(\mathbf{y}|l) = G(\mathbf{y}|\mathbf{0}, \mathbf{\Lambda}_l), \quad p(\mathbf{x}|\mathbf{y}, l) = G(\mathbf{x}|\mathbf{U}_l\mathbf{y} + \boldsymbol{\mu}_l, \mathbf{\Psi}_l),$$
$$p(\mathbf{x}|l) = G(\mathbf{x}|\boldsymbol{\mu}_l, \mathbf{U}_l\mathbf{\Lambda}_l\mathbf{U}_l^T + \mathbf{\Psi}_l), \quad \mathbf{U}_l^T\mathbf{U}_l = \mathbf{I}_{m_l}, \tag{10.6}$$

where both $\mathbf{\Lambda}_l$ and $\mathbf{\Psi}_l$ are diagonal, while loading matrix $\mathbf{U}_l$ is constrained on the Stiefel manifold, i.e., $\mathbf{U}_l^T\mathbf{U}_l = \mathbf{I}_{m_l}$. Furthermore, data smoothing based regularization combining parametric model with Gaussian-kernel Parzen window is adopted, i.e., $p_h(\mathbf{x}) = \frac{1}{N}\sum_{t=1}^{N} G(\mathbf{x}|\mathbf{x}_t, h^2\mathbf{I}_d)$.

Applying BYY harmony learning to LFA, the harmony function $H(\mathbf{\Theta}, h)$ targeted to be maximized is adopted as follows:

$$H(\mathbf{\Theta}, h) = L_h(\mathbf{\Theta}) + Z(h)$$
$$Z(h) = -\ln \sum_{t=1}^{N} p_h(\mathbf{x}_t) = -\ln[\frac{1}{N}\sum_{t=1}^{N}\sum_{\tau=1}^{N} G(\mathbf{x}_t|\mathbf{x}_\tau, h^2\mathbf{I}_d)]$$
$$L_h(\mathbf{\Theta}) = \sum_{l=1}^{k} \int p(\mathbf{x}, \mathbf{y}, l) \ln q(\mathbf{x}, \mathbf{y}, l)d\mathbf{x}d\mathbf{y}$$
$$= \sum_{l=1}^{k} \int p_h(\mathbf{x})p(l|\mathbf{x})p(\mathbf{y}|\mathbf{x}, l) \ln[\alpha_l p(\mathbf{y}|l)p(\mathbf{x}|\mathbf{y}, l)]d\mathbf{x}d\mathbf{y} \tag{10.7}$$

where $L_h(\mathbf{\Theta})$ is the harmony measure between the Yang machine $p(\mathbf{x}, \mathbf{y}, l)$ and the Ying machine $q(\mathbf{x}, \mathbf{y}, l)$. And $Z(h)$ can be understood as a regularization prior. Maximizing the harmony function will lead us the best matching between Yang and Ying machines, the simplest structure, and an appropriate smoothing window width h [36, 38].

Finally, we can obtain an algorithm with automatic model selection ability during parameter learning. Shortly named as BYY-A, this algorithm can greatly save computational cost. Moreover, a model selection criterion in a two-phase implementation is also provided and shortly named as BYY-C, which aims to tackle the first problem discussed in Section 10.1.2. Moreover, [37] discussed about the relationship not only between BYY-A and incremental methods, but also between BYY-C and other typical model selection criteria. The details of BYY harmony system and best harmony learning are referred to [25, 37, 38] for recent systematical summaries.

10.2.1   Automatic BYY Harmony Learning

Briefly, the automatic BYY harmony learning (BYY-A) works under the trend of efforts on seeking fast model selection. Stating in a sense that parameter learning on a model begins with a large enough scale to include the correct one, BYY-A implements model selection automatically during parameter learning, by shrinking the model scale appropriately and discarding those extra substructures. In detail, after initialization at large enough $k = k_{init}$ and $m_l = m_{init}$ for $l = 1, \ldots, k$ with $m_{init} < d$ and during maximizing the harmony function, BYY-A will push α_l towards zero if component l is extra. Thus we can delete component l if α_l approaches zero. Also, if the latent dimension $\mathbf{y}^{(j)}$ is extra, maximizing $\ln p(\mathbf{y}|l)$ will push the variance $\Lambda_l^{(j)}$ towards zero, thus factor j can be removed. As long as k and $\{m_l\}$ are initialized at values large enough, they will be determined appropriately and automatically during parameter learning [25, 31, 33, 37, 38].

Algorithm 3. Sketch for BYY-A algorithm

Initialization: Randomly initialize the component number k and local hidden dimensions $\{m_l\}_{l=1}^{k}$ with large enough integer values. Set $\tau = 0$ and the initial harmony function $H(\tau) = -\infty$.

repeat

Yang-Step: Randomly selecting a sample $\mathbf{x}_t$

for $l = 1, \ldots, k$ **do**

$$\mathbf{e}_{t,l} = \mathbf{x}_t - \boldsymbol{\mu}_l, \hat{\mathbf{y}}_l(\mathbf{x}_t) = \arg\max_y \ln[p(\mathbf{x}_t|\mathbf{y}, l)p(\mathbf{y}|l)] = \mathbf{W}_l(\mathbf{x}_t - \boldsymbol{\mu}_l),$$

$$p(l|\mathbf{x}_t) = \begin{cases} 1, & \text{if } l = l_t, \\ 0, & \text{otherwise.} \end{cases}, \quad l_t = \arg\max_l \{F(\mathbf{x}_t, l) + \frac{h^2}{2} Tr[\frac{\partial^2 F(\mathbf{x},l)}{\partial \mathbf{x}\mathbf{x}^T}|_{\mathbf{x}=\mathbf{x}_t}]\},$$

where $\mathbf{W}_l = \Lambda_l \mathbf{U}_l^T \mathbf{M}_l$, $\mathbf{M}_l = (\mathbf{U}_l \Lambda_l \mathbf{U}_l^T + \Psi_l)^{-1}$,
$F(\mathbf{x}, l) = \ln[\alpha_l p(\hat{\mathbf{y}}_l(\mathbf{x})|l)p(\mathbf{x}|\hat{\mathbf{y}}_l(\mathbf{x}), l)]$.

end for

Ying-Step: Update the parameters and conduct model selection as follows

for $l = 1, \ldots, k$ **do**

$$\alpha_l^{new} = \begin{cases} \frac{\alpha_l + \eta_0}{1+\eta_0}, & \text{if } l = l_t, \\ \frac{\alpha_l}{1+\eta_0}, & \text{otherwise.} \end{cases}$$

end for

$\boldsymbol{\mu}_{l_t}^{new} = \boldsymbol{\mu}_{l_t} + \eta_0 \mathbf{e}_{t,l_t}, \; \varepsilon_{t,l} = \mathbf{e}_{t,l} - \mathbf{U}_l \hat{\mathbf{y}}_l(\mathbf{x}_t)$
$\Lambda_{l_t}^{new} = (1 - \eta_0)\Lambda_{l_t} + \eta_0 \{h^2 diag[\mathbf{W}_{l_t} \mathbf{W}_{l_t}^T] + diag[\hat{\mathbf{y}}_{l_t}(\mathbf{x}_t)\hat{\mathbf{y}}_{l_t}(\mathbf{x}_t)^T]\}$
$\Psi_{l_t}^{new} = (1 - \eta_0)\Psi_{l_t} + \eta_0 \{h^2 diag[(\mathbf{I}_d - \mathbf{U}_{l_t}\mathbf{W}_{l_t})(\mathbf{I}_d - \mathbf{U}_{l_t}\mathbf{W}_{l_t})^T] + diag[\varepsilon_{t,l}\varepsilon_{t,l}^T]\}$
$\mathbf{G}_{\mathbf{U}_{l_t}} = \mathbf{M}_{l_t}\mathbf{e}_{t,l_t}\hat{\mathbf{y}}_{l_t}(\mathbf{x}_t)^T + h^2 \mathbf{M}_{l_t}\mathbf{W}_{l_t}^T, \; \mathbf{U}_{l_t}^{new} = \mathbf{U}_{l_t} + \eta_0(\mathbf{G}_{\mathbf{U}_{l_t}} - \mathbf{U}_{l_t}\mathbf{G}_{\mathbf{U}_{l_t}}^T\mathbf{U}_{l_t})$.
(#) Discard the l-th component if α_l approaches 0.
(#) Discard the j-th factor of component l if Λ_l's j-th element approaches 0.

Smoothing-Step:
$h^{2^{new}} = h^{new^2}, \; h^{new} = h + \eta_0 \Delta h, \; \Delta h = \frac{d}{h} - h\sum_{l=1}^{k} \alpha_l Tr[\mathbf{M}_l] - \frac{\beta}{h^3}.$

where, $\beta = \frac{\sum_{t=1}^{N}\sum_{\tau=1}^{N}(\mathbf{x}_\tau - \mathbf{x}_t)^T(\mathbf{x}_\tau - \mathbf{x}_t)\exp[-\frac{1}{2h^2}(\mathbf{x}_\tau - \mathbf{x}_t)^T(\mathbf{x}_\tau - \mathbf{x}_t)]}{\sum_{t=1}^{N}\sum_{\tau=1}^{N}\exp[-\frac{1}{2h^2}(\mathbf{x}_\tau - \mathbf{x}_t)^T(\mathbf{x}_\tau - \mathbf{x}_t)]}$

(*) If without data smoothing, h can be simply assigned and held as $h = 0$.

if Another $N/5$ iterations have passed **then**

$\tau = \tau + 1$
$H(\tau) = \frac{1}{2}\sum_{l=1}^{k}\alpha_l\{2\ln\alpha_l - m_l\ln(2\pi) - \ln|\Lambda_l| - \ln|\Psi_l| - h^2 Tr[\mathbf{M}_l^{-1}]\}.$

end if

until $H(\tau) - H(\tau - 1) < \epsilon H(\tau - 1)$, with $\epsilon = 10^{-5}$ in our implementation

---

The algorithm of BYY-A is sketched in Algorithm 3, which implements automatic model selection during parameter learning via iterating through three steps named Yang-Step, Ying-Step and Smoothing-Step, respectively [25, 36, 37, 38]. At the Yang-Step, l_t, $\hat{y}_l(\mathbf{x}_t)$, and $\mathbf{W}_l$ are estimated, while the parameters $\theta \in \{\alpha_l, \boldsymbol{\mu}_l, \boldsymbol{\Lambda}_l, \boldsymbol{\Psi}_l, \mathbf{U}_l\}$ are updated at the Ying-Step via computing the gradient $\nabla_\theta H$, subject to the constraints that $\alpha_l > 0$, $\sum_{l=1}^{k} \alpha_l = 1$, $\boldsymbol{\Lambda}_l$, $\boldsymbol{\Psi}_l$ are positive definite diagonal, and $\mathbf{U}_l^T \mathbf{U}_l = \mathbf{I}_{m_l}$. Moreover, the smoothing parameter h is adjusted in Smoothing-Step.

For details about BYY harmony learning please refer to [25, 31, 32, 33, 36, 37, 38]. Particularly, the derivation of BYY-A is referred to [25] and Section 3.3.2 in [38] for recent summaries. Three notes during the BYY-A implementation need to be mentioned here:

1. During computing $F(\mathbf{x}, l)$, both $\det(\mathbf{U}_l \boldsymbol{\Lambda}_l \mathbf{U}_l^T + \boldsymbol{\Psi}_l)$ and $\mathbf{M}_l = (\mathbf{U}_l \boldsymbol{\Lambda}_l \mathbf{U}_l^T + \boldsymbol{\Psi}_l)^{-1}$ can be calculated via the following cheaper ways instead of direct calculation, especially in case of $m_l << d$:
$$\det(\mathbf{U}_l \boldsymbol{\Lambda}_l \mathbf{U}_l^T + \boldsymbol{\Psi}_l) = \det(\boldsymbol{\Lambda}_l^{-1} + \mathbf{U}_l^T \boldsymbol{\Psi}_l^{-1} \mathbf{U}_l) \det(\boldsymbol{\Lambda}_l) \det(\boldsymbol{\Psi}_l),$$
$$\mathbf{M}_l = (\mathbf{U}_l \boldsymbol{\Lambda}_l \mathbf{U}_l^T + \boldsymbol{\Psi}_l)^{-1} = \boldsymbol{\Psi}_l^{-1} - \boldsymbol{\Psi}_l^{-1} \mathbf{U}_l (\boldsymbol{\Lambda}_l^{-1} + \mathbf{U}_l^T \boldsymbol{\Psi}_l^{-1} \mathbf{U}_l)^{-1} \mathbf{U}_l^T \boldsymbol{\Psi}_l^{-1}.$$

2. The learning rate η_0 in different parameters' updating equations may be different. Actually, in our experience, different learning rates usually perform better than a constant value.

3. The value $N/5$ is heuristically set in our experiments.

10.2.2 BYY Harmony Criterion

Besides BYY-A, a BYY harmony learning criterion (BYY-C) is also adopted for LFA model selection, implemented in a two-phase procedure. In the first phase, the learning algorithm is similar with BYY-A listed in Algorithm 3, except those steps marked with (#) are all absent, i.e., without any automatic model selection operations. In the second phase, the BYY harmony criterion from [25, 33, 38] is taken as follows:

$$J_{BYY}(\hat{\boldsymbol{\Theta}}_{\mathbf{k}}, \mathbf{k}) = -H(\hat{\boldsymbol{\Theta}}_{\mathbf{k}}, h) + 0.5D(\mathbf{k}).$$
$$H(\hat{\boldsymbol{\Theta}}_{\mathbf{k}}, h) = \frac{1}{2} \sum_{l=1}^{k} \alpha_l \{2 \ln \alpha_l - m_l[\ln(2\pi) + 1] - \ln|\boldsymbol{\Lambda}_l| - \ln|\boldsymbol{\Psi}_l| - h^2 Tr[\mathbf{M}_l^{-1}]\}$$
(10.8)

where $J_{BYY}(\hat{\boldsymbol{\Theta}}_{\mathbf{k}}, \mathbf{k})$ consists of the negative of harmony function and an additional term $0.5D(\mathbf{k})$, which comes from the consideration to regain the Hessian information of $p(\boldsymbol{\Theta}|\mathbf{X}_N)$. More details are referred to [25, 36, 38] and skipped here to save space.

The $D(\mathbf{k})$ in (10.8) is still the number of free parameters in the referred model with scale $\mathbf{k}$. For an LFA model with parameters $\boldsymbol{\Theta}_{\mathbf{k}} = \boldsymbol{\Theta}_{k, \{m_l\}_{l=1}^k} = \{\alpha_l, \boldsymbol{\mu}_l, \boldsymbol{\Psi}_l, \mathbf{U}_l, \boldsymbol{\Lambda}_l\}_{l=1}^k$, the free parameter number for $\{\alpha_l\}$ is $k - 1$, for $\{\boldsymbol{\mu}_l\}$ and $\{\boldsymbol{\Psi}_l\}$ are both kd, while for $\{\mathbf{U}_l\}$ is $\sum_{l=1}^{k}(dm_l - m_l(m_l + 1)/2)$, and for $\{\boldsymbol{\Lambda}_l\}$ is $\sum_{l=1}^{k} m_l$. Finally, the total number of free parameters is as follows, identical to that given in (10.3):

$$D(\mathbf{k}) = D(k, \{m_l\}) = k - 1 + kd + kd + \sum_{l=1}^{k}[dm_l - \frac{m_l(m_l + 1)}{2} + m_l]. \quad (10.9)$$

10.3   Experimental Results on Simulated Datasets

In the following two sections, we evaluate the performance of LFA by BYY harmony learning (BYY-A and BYY-C) on simulated and real world datasets respectively, compared with not only the two-phase implementation with the help of criteria AIC, CAIC, BIC, SRM and CV, but also the incremental methods IMoFA and VBMFA. For the cross-validation (CV) method, we consider the 5-fold CV, shortly named as CV-5 in the following. Trying to avoid local optima caused by inappropriate initialization, we implement the parameter training in the two-phase implementation for 10 times on each candidate model **k**, as well as the automatic or incremental methods for 10 times on each simulation. Among the 10 rounds' learned models, we choose the one with the best target function based on the training data as the resulted model.

In this section, after generating data sets from designed LFA models, we compare each method on both the *model selection accuracy*, and three other evaluations including the *clustering accuracy*, the *testing samples' log-likelihood*, and the *classification accuracy*[2]. The comparison is designed systematically, mainly considering and changing three key scales for the training data, i.e., *noise*, *data dimension*, and *sample size*. That is, the environment in our experiments will deteriorate as the noise increases, the data dimension increases, and the sample size decreases. Each simulation in a specific environment is repeated for 100 times.

10.3.1   Results of Model Selection Accuracy

In this part, we focus on comparing the model selection accuracy by each method based on simulated datasets. We generate 7 series simulated data sets from LFA models with the same component number $k = 3$ and the same local hidden dimension $m_l = 3$ for each $l = 1, \ldots, k$. The design of the 7 series are shown in Figure 10.1 and Table 10.2, where the former shows the three environmental settings, noise, data dimension d and sample size N, and the latter shows each series trends. The Gaussian noise is added with the covariance $\psi^2 \mathbf{I}_d$ determined based on ζ, where $\zeta = \min_l \zeta_l$, and ζ_l denotes the mean eigenvalue of component l's covariance $\mathbf{U}_l \mathbf{\Lambda}_l \mathbf{U}_l^T + \mathbf{\Psi}_l$. In an increasing trend, the noise is set as $\psi^2 = 0.2\zeta, 0.5\zeta, 0.8\zeta$. The dimension d in an increasing trend is set $d = 5, 7, 9$. The sample size N in a decreasing trend is set $N = 1000, 200, 40$. Since we know the true model complexity, i.e. $k^* = 3$ and each $m_l^* = m^* = 3$, the candidate model range for the two-phase implementation is pre-determined with the component number $k \in [1, 5]$ and the same local hidden dimension $m_l = m \in [1, 5]$. BYY-A is initialized with the component number $k_{init} = 5$ and each local hidden dimension equal to 5, while for IMoFA and VBMFA the upper bounds for both the component number and local hidden dimension is set as 5.

---

[2] Actually, the model selection accuracy is the major focus of this chapter. Whereas, the other three reflect different application of the LFA model, which are not directly based on but at outer levels of the model selection. Since often considered in the machine learning society, they are also evaluated here, trying to offer some valuable and related observations for future research.

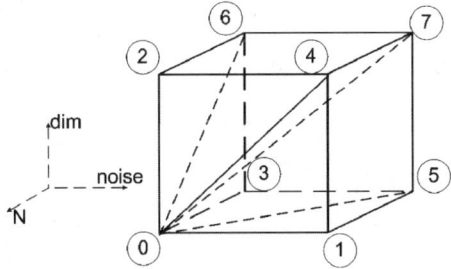

Fig. 10.1. Cubic expression of the environmental settings for 7 simulated experimental series. Three scales are considered, including noise, dimension and sample size, respectively.

Table 10.2. Description of 7 experimental series configuration. Each environmental setting sequence is listed in the same order as Figure 10.1.

series	noise ψ^2	dimension d	sample size N
1	$0.2\zeta, 0.5\zeta, 0.8\zeta$	5	1000
2	0.2ζ	$5, 7, 9$	1000
3	0.2ζ	5	$1000, 200, 40$
4	$0.2\zeta, 0.5\zeta, 0.8\zeta$	$5, 7, 9$	1000
5	$0.2\zeta, 0.5\zeta, 0.8\zeta$	5	$1000, 200, 40$
6	0.2ζ	$5, 7, 9$	$1000, 200, 40$
7	$0.2\zeta, 0.5\zeta, 0.8\zeta$	$5, 7, 9$	$1000, 200, 40$

After running 100 times for each situation, the selection rates for each method on each experimental environment are investigated. We show the rates of the total correct selection, i.e., correct on both component number $k = 3$ and the hidden dimensions $m = 3$, in Figure 10.2 also in the same cubic expression as Figure 10.1. The results of SRM are expressed in the form of $r_1(r_2)$, which means that SRM makes total correct selection r_1 times out of correct selection rate r_2 on only component number k. To save space, only the detailed selection rates along Series 6 are reported in Figure 10.3 and Figure 10.4 for example, where Figure 10.3 corresponds to the case with $\psi^2 = 0.2\zeta$, $d = 7$, and $N = 200$, and Figure 10.4 for the case with $\psi^2 = 0.2\zeta$, $d = 9$, and $N = 40$. Each pole represents a selection rate out of 100 at the corresponding model complexity, ranging through the scale space of $k \in [1, 5]$ and $m_l \in [1, 5]$. Each labeled value at the center pole corresponds to the correct selection rate.

The results show that, when the experimental case design is at the baseline, i.e., a comparably good enough environment, all these different methods perform well, without considerable difference. However, when the environment is changing worse along the three configuration axes, they all suffer from a problem in model selection, while different methods behave differently for the changing effects. Analytically, on one hand, for the two-phase methods as the environment deteriorates: AIC tends to over-select both the hidden dimension and components number; CAIC tends to have the risk of under-selection; BIC and CV perform generally better than AIC and CAIC, where 5-fold CV owns a relatively better robustness as the environment

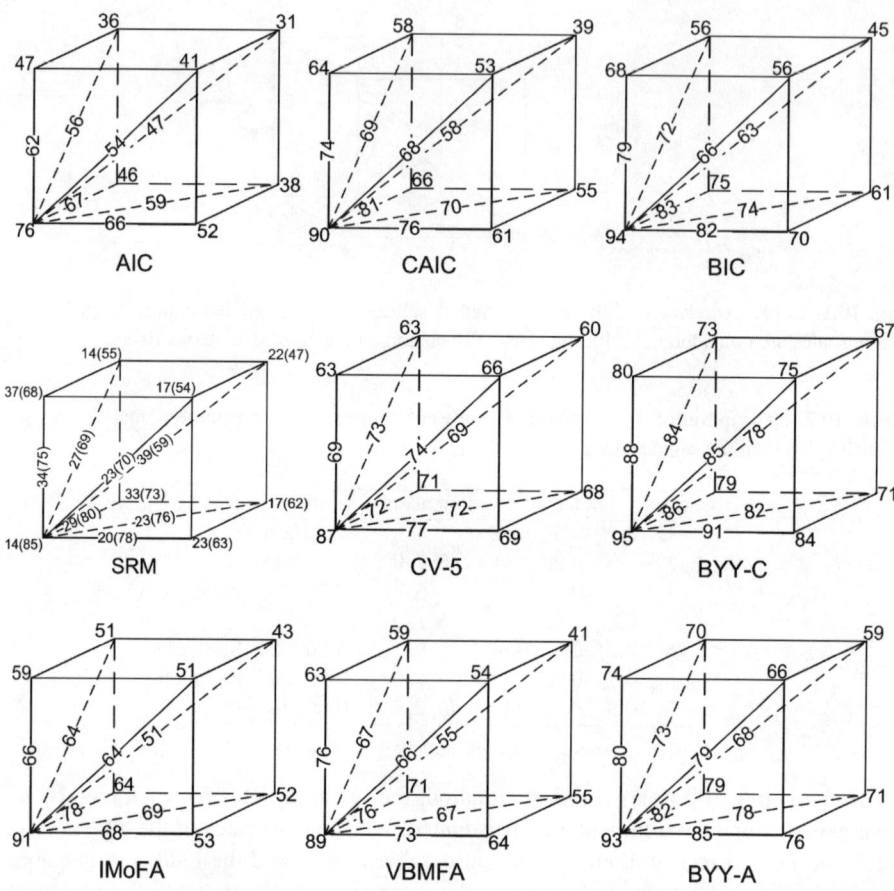

Fig. 10.2. Correct selection rates out of 100 implementations by different methods. The experiments are conducted in 7 series simulations. The cube is constructed in the same way as in Figure 10.1, where each labeled value is the correction selection rate in that particular corresponding environment. The component number is fixed as $k = 3$ and each local hidden dimension is the same as $m = 3$. For SRM, the numbers in the parentheses represent the rates for correct selection purely on component number k, while those outside represent the correct rates on both component number and local hidden dimensions.

shades worse; SRM also performs well in selecting the component number, while over-selecting the local hidden dimensions, which probably originates from the simply direct adoption of the results on GMM; BYY-C generally outperforms all other methods, especially in case of either small enough sample size or high dimension. On the other hand, among the automatic methods, as the environment deteriorates, IMoFA and VBMFA tend to make wrong selections, especially when noise increases; although owning a tend to over-select, BYY-A still obviously outperforms IMoFA and VBMFA.

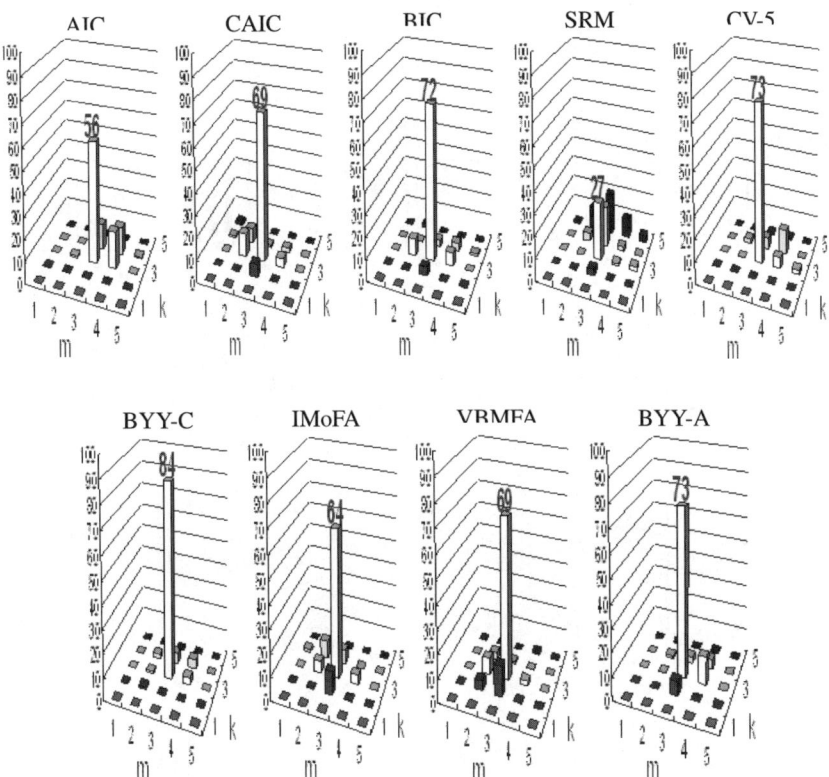

Fig. 10.3. Histogram representation of selection rates by different methods in Series 6 with $\psi^2 = 0.2\zeta$, $d = 7$, and $N = 200$. The two axes against the selection rates are the component number $k \in \{1, \dots, 5\}$ and the identical local hidden dimension $m \in \{1, \dots, 5\}$, respectively.

Computational Cost

We also list the computational cost of each method during the 7 simulated series experiments in Table 10.3 for comparison, in terms of the average cost and standard deviation along the simulated series. The whole procedure is computed by MATLAB 7.0.1 (R14) on a P4 3.2GHz 1GB DRAM PC. We can find that, on one hand, since sharing the EM implementation, AIC, CAIC, BIC and SRM cost almost the same time, and BYY-C costs a little more mainly due to the data smoothing computation, while CV needs the most computational cost due to the cross-folder repetitions. On the other, BYY-A, IMoFA and VBMFA cost similar amounts but far less than the criteria.

10.3.2 Results of Clustering Accuracy

One widely-used application of LFA is to do clustering analysis, or unsupervised learning. Given a set of data consisting of unlabeled samples from C classes, the task of clustering analysis is to label every sample in the same class by the same symbol such

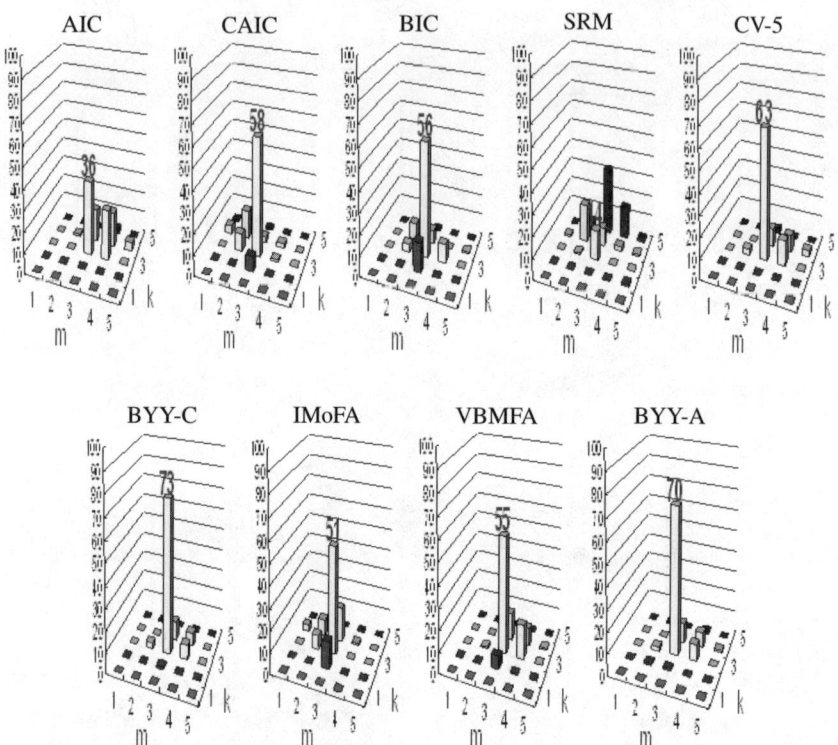

Fig. 10.4. Histogram representation of selection rates by different methods in Series 6 with $\psi^2 = 0.2\zeta$, $d = 9$, and $N = 40$. The two axes against the selection rates are the component number $k \in \{1, \ldots, 5\}$ and the identical local hidden dimension $m \in \{1, \ldots, 5\}$, respectively.

that the data set is divided into several clusters (classes), each associated with a different symbol. This task by LFA is carried out in two steps. First, the LFA model is obtained based on the samples. Second, each component l in the model is named as the center of a possible class, and each sample $\mathbf{x}_t$ is labeled to $\hat{l}(\mathbf{x}_t)$ as follows:

$$\hat{l}(\mathbf{x}_t) = \arg\max_l p(\mathbf{x}_t|l). \tag{10.10}$$

As a result, all the samples with the same label l constitute a class, and the component number k is the number of labeled classes. Since the samples are generated, we thus know the actual true clustering result, i.e., the true class number C and the true labels $\{c(\mathbf{x}_t)\}$, with each $c(\mathbf{x}_t) \in \{1, \ldots, C\}$. Once each sample $\mathbf{x}_t$ is labeled by a cluster $\hat{l}(\mathbf{x}_t)$, we can construct a connection from the true class label $c \in [1, \ldots, C]$ the cluster label $l \in [1, \ldots, k]$ via the following mapping:

$$l(c) = \arg\max_l |\mathbb{X}(l, c)|, \quad \mathbb{X}(l, c) = \{\mathbf{x}_t|c(\mathbf{x}_t) = c \text{ and } \hat{l}(\mathbf{x}_t) = l\} \tag{10.11}$$

Table 10.3. Comparison of computational cost by each method on simulated data (in minutes). Each item is reported in terms of averages and standard deviations along the 7 series. The item *Crit* refers to the average of AIC, CAIC, BIC, and SRM, due to their computational costs' similarity. The experiments were carried out by MATLAB 7.4.0 (R2007a) on a P4 3.2 GHz 1 GB RAM PC.

Methods	CPU Time (in minutes)
Crit	10.5 ± 5.7
CV-5	53.7 ± 28.0
BYY-C	12.8 ± 4.7
IMoFA	2.6 ± 2.6
VBMFA	2.9 ± 1.3
BYY-A	2.5 ± 1.2

That is, we first construct sample subsets $\mathbb{X}(l, c)$, for $l = 1, \ldots, k$ and $c = 1, \ldots, C$. Then cluster $l(c)$ with the maximum sized $\mathbb{X}(l, c)$ is selected as the optimal expression of the true class c. Then the clustering accuracy is evaluated by the optimal described samples' percentage as follows:

$$Cluster\,Accuracy = \frac{\sum_{c=1}^{C} |\mathbb{X}(l(c), c)|}{N} \qquad (10.12)$$

Actually, this kind of evaluation is widely used for clustering analysis, usually expressed in form of a confusion matrix. However, we conduct the simulations for 100 times and select the same cubic expression for experimental statistics instead. One difference in this part from the previous two is, since the mapping from $l = 1, \ldots, k$ to $c = 1, \ldots, C$ needs $k \geq C$, such that each true class has one corresponding cluster, we revise the candidate model scale for k from $k = 1, \ldots, 5$ to $k = 3, \ldots, 7$, with the true class number known as $C = 3$. The samples and the remaining settings are the same as Section 10.3.1.

After running 100 simulations for each case, the average clustering accuracies (in percentage) are shown in Figure 10.5, still in the same cubic expression as Figure 10.1. From these results, we can briefly have the following observations. First, as the data environment deteriorates, BIC and BYY-C provide similarly better values than all the other criteria. Again, BYY-A outperforms IMoFA and VBMFA. Second, due to the revised candidate model range $k \in [3, 7]$ bounded with the true class number $c = 3$, the criterion CAIC, which has an under-estimate tendency of k, usually obtains the best performance in good data environments. However, these values cannot reflects the modeling performance in total.

10.3.3 Results of Testing Samples' Log-Likelihood

Since the LFA model can also be used for density description, this part evaluates each method on the testing samples' log-likelihood, given the trained and selected LFA models. The settings of models and datasets are the same as in Section 10.3.1. The testing set is randomly generated from the same LFA model as that for constructing the corresponding training set. The testing sample size is fixed as 200, no matter what the training

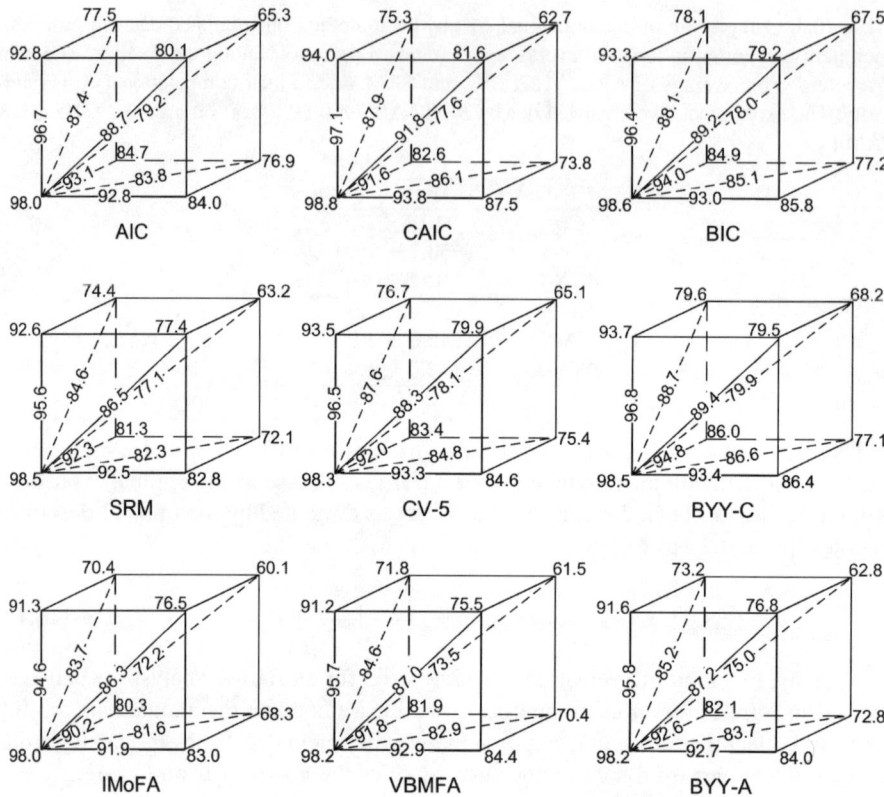

Fig. 10.5. Average clustering accuracies (in percentage) on the testing set out of 100 implementations by each method. The experiments are conducted in 7 series simulations, where each labeled value is the average clustering accuracy in that particular corresponding environment. The component number is fixed as $k = 3$ and each local hidden dimension is the same as $m = 3$.

sample size N is. Then, after running 100 times for each situation, the testing samples' log-likelihood values based on all learned and selected LFA models are investigated, where the probability $p(\mathbf{x}_t|\mathbf{\Theta})$ of a testing sample $\mathbf{x}_t$ is computed according to (10.1) or (10.6). We show the mean log-likelihood values on the testing set in Figure 10.6 still in the same cubic expression as Figure 10.1. It is different from Figure 10.2 since different LFA models may generate data with quite different likelihood value-levels, so that the values of a cube in Figure 10.6 do not have the vertical tendency comparison but tendency comparisons in the lateral plane and among different methods on the same corresponding position.

These results briefly show the following observations. First, on the same position, the values' difference by all methods are not that obvious especially for a good enough environment, compared with the correct selection rates as shown in Figure 10.2. Second, as the data environment deteriorates, we can observe that BYY-A and VBMFA outperforms IMoFA, with BYY-A a little better than VBMFA. Moreover, BIC, CAIC,

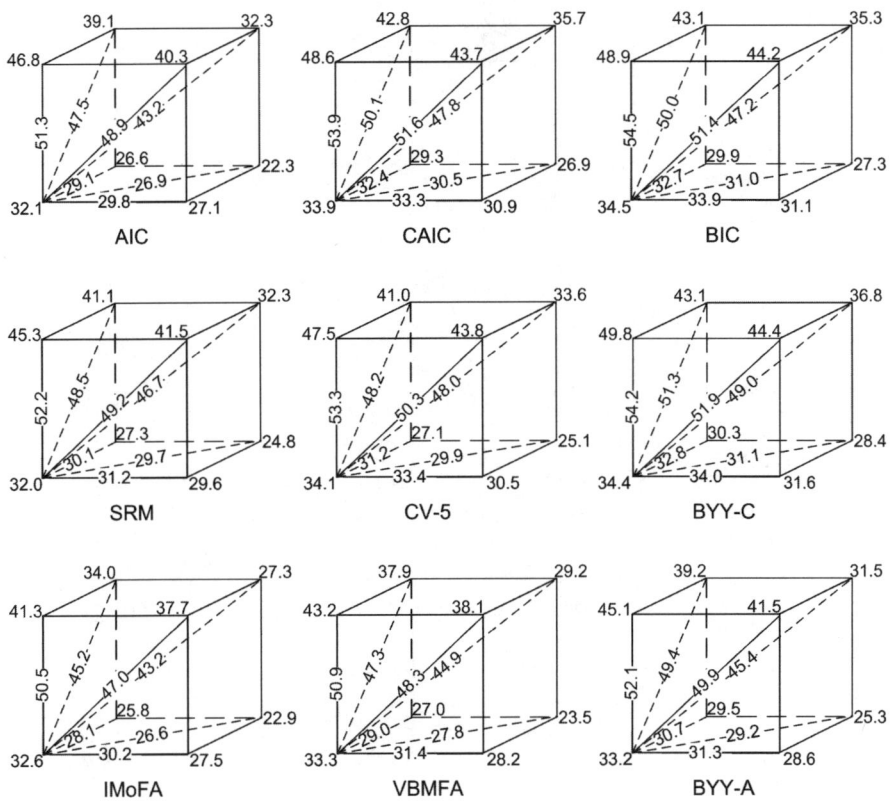

Fig. 10.6. Average log-likelihood values on the testing set out of 100 implementations by each method. The experiments are conducted in 7 series simulations, where each labeled value is the average log-likelihood value in that particular corresponding environment. The component number is fixed as $k = 3$ and each local hidden dimension is the same as $m = 3$.

and CV provides similarly better values than AIC and SRM, while interestingly BYY-C still outperforms them.

10.3.4 Results of Classification Accuracy

In this part, the comparison is conducted on the classification accuracy by each method's obtained LFA classifier. For classification on samples of C classes, one LFA model $\mathbb{M}_c$ is obtained on each class c, with $c = 1, \ldots, C$. As a test sample $\mathbf{x}_t$ comes, the classification is conducted via a Bayesian way, i.e., $\mathbf{x}_t$ is assigned to the class with the maximum posterior probability $\hat{c}(\mathbf{x}_t) = \arg\max_{c=1,\ldots,C} p(\mathbb{M}_c|\mathbf{x}_t)$, with $p(\mathbb{M}_c|\mathbf{x}_t) \propto p(\mathbb{M}_c)p(\mathbf{x}_t|\mathbb{M}_c)$. Here $p(\mathbb{M}_c)$ is the prior probability, which can be simply assigned as proportional to the training sample size of the c-th class. And $p(\mathbf{x}_t|\mathbb{M}_c)$ is the probability of $\mathbf{x}_t$ on the trained model $\mathbb{M}_c$, which is computed according to (10.1) or (10.6).

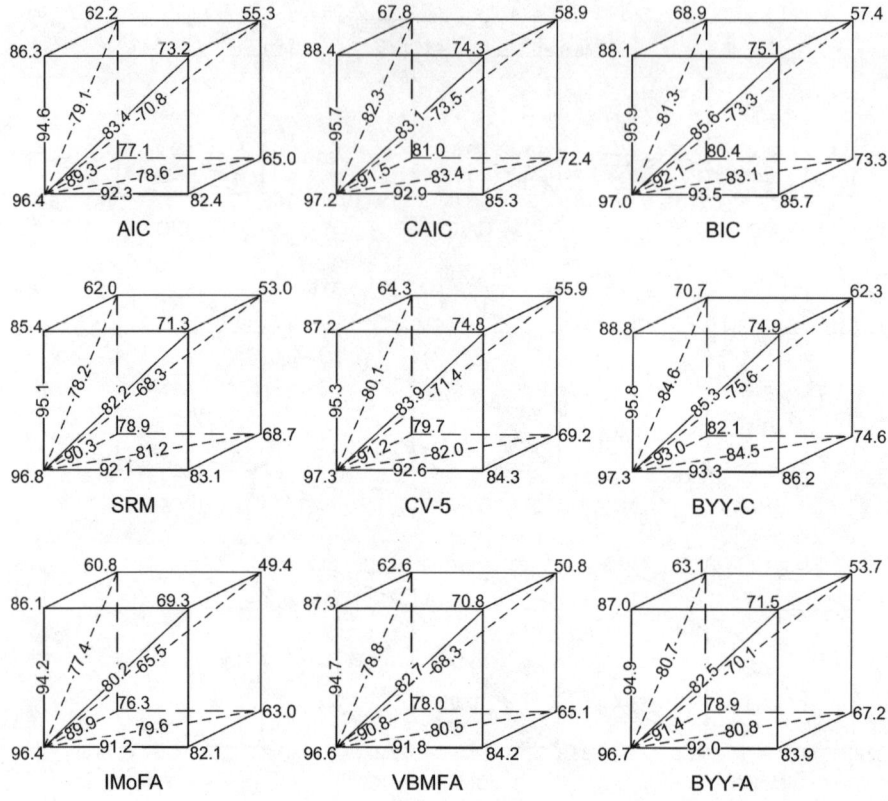

Fig. 10.7. Average correct classification rates (in percentage) with LFA classifiers on the testing set out of 100 implementations by each method. The experiments are conducted in 7 series simulations, where each labeled value is the correction classification rate in that particular corresponding environment. The component number is fixed as $k = 3$ and each local hidden dimension is the same as $m = 3$.

The simulated datasets used here for classification are different from those in previous three subsections. We generate a series of three-class datasets, where the data generating LFA model of each class has the same model complexity as $k = 3$ and each $m_l = m = 3$. The environmental settings are still taken in the same tendency as shown in Figure 10.1 and Table 10.2, considering again the noise, dimension and training sample size. The testing sets are randomly generated from the same LFA models, with a fixed testing sample size of 200, no matter what the training sample size N is. Then, after running 100 times for each situation, the average correct classification rates (in percentage) of testing samples based on LFA classifiers are reported in Figure 10.7.

From the results of average classification accuracies on testing samples, we can briefly have the following observations. When the training sample sizes are large, IMoFA, VBMFA and BYY-A produce performances similar to or even better than the two-phase methods. As the training sample sizes decrease, all methods' performances deteriorate considerably. The two-phase procedure generally produces better classification accuracies than the incremental and automatic methods. Interestingly, BYY-C generally outperforms all the other criteria, especially for the small-sample-size cases. Still, BYY-A produces better or at least similar results compared with IMoFA and VBMFA.

10.4 Experimental Results on Real World Datasets

In this section, experiments are conducted on several real world datasets with a decreasing ratio between training and testing sample sizes. Since every true model complexity k^* is unknown, the performance of each method is indirectly evaluated based on the classification accuracies on testing set out of 100 independent implementations. The organization and classification procedure by a LFA classifier is the same as that described in Section 10.3.4.

The candidate model range for a two-phase procedure, the model complexity upper bound for incremental methods, and the initial model scale for BYY-A are chosen as follows:

- Since the true component number k^* is unknown in the real world data, k_{init} for BYY-A and k_{max} for incremental methods are assigned a same and large enough integer. Similarly, we assign a same large enough integer to m_{init} for BYY-A and m_{max} for IMoFA and VBMFA.
- For a two-phase procedure it is here not easy to determine the candidate model range. Heuristically, we first implement the automatic and incremental methods, and obtain the mostly selected component number $\tilde{k}$. Then, for the two-phase procedure, we let $k_{max} = 2\tilde{k} + 1$ and $k_{min} = 2$. Moreover, $m_{min} = 1$ and $m_{max} = m_{init}$ are assigned, where m_{init} is the initial hidden dimension for BYY-A.
- For the method using SRM as discussed in Section 10.1.2, we have to constrain all the local hidden dimensions to be the same, while for other criteria approaches the hidden dimensions can be assigned different.

Table 10.4. Rough description of real world datasets

Datasets	Dimensions	Classes	Training	Testing
	Datasets Description			
(O)			7,494	3,498
PEN (H)	16	10	3,747	7,245
(Q)			1,874	9,118
(O)			2,880	1,797
OPT (H)	64	10	1,440	3,237
(Q)			720	3,957
(O)			700	1,610
SEG (H)	14	7	350	1,960
(Q)			175	2,135
(O)			300	4,700
WAV (H)	21	3	150	4,850
(Q)			75	4,925

Table 10.5. Experimental results on real world datasets. The correct classification percentages are reported in the form of *average ± standard deviation* after 20 independent implementations. The best average result on each dataset is bolded.

PEN

Methods	PEN(O)	PEN(H)	PEN(Q)
AIC	94.28 ± 0.14	85.33 ± 2.61	72.16 ± 5.12
CAIC	95.18 ± 0.16	92.74 ± 1.99	80.34 ± 4.76
BIC	97.77 ± 0.13	91.35 ± 2.32	83.63 ± 4.47
SRM	96.92 ± 0.18	86.29 ± 3.10	78.44 ± 5.89
CV-5	95.39 ± 0.13	88.81 ± 2.03	78.31 ± 4.20
BYY-C	97.93 ± 0.16	**93.92** ± 2.22	**85.35** ± 4.31
IMoFA	97.90 ± 0.20	84.15 ± 4.93	67.14 ± 7.15
VBMFA	98.30 ± 0.24	86.96 ± 3.97	69.96 ± 6.30
BYY-A	**98.41** ± 0.19	87.58 ± 4.09	72.65 ± 6.88

OPT

Methods	OPT(O)	OPT(H)	OPT(Q)
AIC	92.76 ± 0.32	87.35 ± 2.38	66.36 ± 5.55
CAIC	96.98 ± 0.49	89.16 ± 2.04	74.79 ± 5.17
BIC	**97.82** ± 0.56	90.55 ± 1.87	73.33 ± 4.87
SRM	97.15 ± 0.69	88.83 ± 2.24	69.05 ± 5.26
CV-5	97.52 ± 0.46	88.34 ± 1.69	69.77 ± 4.51
BYY-C	97.61 ± 0.41	**92.96** ± 2.02	**77.98** ± 5.02
IMoFA	96.39 ± 0.49	85.41 ± 4.81	63.74 ± 7.97
VBMFA	96.42 ± 0.72	86.13 ± 4.03	65.89 ± 6.96
BYY-A	96.96 ± 0.69	86.37 ± 3.84	68.68 ± 7.03

SEG

Methods	SEG(O)	SEG(H)	SEG(Q)
AIC	77.48 ± 2.31	67.67 ± 6.35	54.45 ± 7.62
CAIC	84.08 ± 3.89	73.97 ± 5.34	58.12 ± 6.91
BIC	82.61 ± 2.54	73.81 ± 5.10	**61.70** ± 6.60
SRM	75.02 ± 3.67	69.12 ± 5.58	56.28 ± 7.51
CV-5	81.20 ± 2.91	70.43 ± 4.63	57.73 ± 6.29
BYY-C	85.33 ± 2.45	**74.11** ± 5.04	60.21 ± 6.57
IMoFA	86.11 ± 3.90	65.38 ± 7.49	49.82 ± 9.31
VBMFA	**88.32** ± 3.61	67.55 ± 6.51	53.67 ± 8.12
BYY-A	87.81 ± 2.37	67.29 ± 6.13	55.30 ± 8.35

WAV

Methods	WAV(O)	WAV(H)	WAV(Q)
AIC	71.28 ± 1.24	65.39 ± 3.86	57.62 ± 6.62
CAIC	75.85 ± 2.31	69.25 ± 3.42	60.28 ± 6.97
BIC	82.74 ± 2.04	71.34 ± 3.50	61.63 ± 6.42
SRM	81.24 ± 2.34	66.37 ± 4.11	53.34 ± 7.26
CV-5	79.61 ± 2.10	67.42 ± 3.87	57.45 ± 5.88
BYY-C	**83.11** ± 3.03	**73.56** ± 3.29	**62.40** ± 6.31
IMoFA	81.88 ± 2.99	61.33 ± 6.23	50.23 ± 8.90
VBMFA	81.43 ± 2.45	64.25 ± 5.54	53.25 ± 8.13
BYY-A	82.83 ± 1.87	65.77 ± 5.80	55.01 ± 8.27

Table 10.6. CPU time on real world datasets (in minutes). The item *Crit* refers to the average of AIC, CAIC, BIC, and SRM, due to their computational costs' similarity. Each item represents the average time on the corresponding three datasets. The experiments were carried out by MATLAB 7.4.0 (R2007a) on a P4 3.2 GHz 1 GB RAM PC.

CPU Time (in minutes)			
Methods	PEN	OPT	SEG	WAV
Crit	131	183	174	103
CV-5	596	712	689	95
BYY-C	159	195	188	117
IMoFA	20	39	35	17
VBMFA	31	50	47	23
BYY-A	23	35	32	17

We comparatively investigate the performance of each method on four real world datasets, including Pendigits (PEN), Optdigits (OPT), Segment (SEG) and Waveform (WAV), all collected from the UCI machine learning repository [3]. In order to investigate the effect of sample sizes, another two pairs of training and testing datasets are generated from each original dataset by moving samples randomly from the training set to the testing set, with the training set reduced to a half and a quarter, respectively. Thus, we have three datasets shortly denoted as O, H and Q, respectively. Dataset series under consideration are shown in Table 10.4, e.g., PEN(O) denotes the original dataset PEN, while PEN(H) denotes one with a half of samples from PEN(O).

In implementation, 20 independent simulations are conducted by each method on each of the datasets PEN, OPT, SEG and WAV. Again, for each simulation we conduct the learning procedure for 10 times, and then the results with the highest target functions are selected, as discussed in Section 10.3. The resulted classification accuracy are listed in Table 10.5, where each item is the correct classification percentage expressed in a form of *average $\pm$ standard deviation*. These results show that, BYY-C still generally outperforms all the other criteria, especially for the small-sample-size cases, while BYY-A produces better or at least similar results compared with IMoFA and VBMFA. Interestingly, when the sample sizes are large, automatic and incremental model selection methods produce performances similar to or sometimes even better than the two-phase methods. As the training sample sizes decrease, all methods' performances deteriorate considerably, where the two-phase methods' performances become better again.

Table 10.6 compares the CPU time of each method on the real world datasets, taking the average on the three training sets (O, H and Q). Those with criteria AIC, CAIC, BIC, and SRM in the two-phase procedure are merged into one item since their performances are very similar. BYY-C spends a little more computational cost due to the calculation of smoothing parameters. CV-5 requires the highest computational cost. Computational costs reduce considerably for automatic and incremental methods, where interestingly BYY-A runs faster than IMoFA and VBMFA.

---

[3] Website of UCI machine learning repository: http://archive.ics.uci.edu/ml/.

10.5   Concluding Remarks

Focusing on the model selection task on LFA, both an automatic BYY harmony learning algorithm (BYY-A) and a BYY harmony criterion (BYY-C) in the two-phase implementation are introduced. For a systematic comparison, not only several typical model selection criteria including AIC, CAIC, BIC(MDL), SRM and CV, but also two incremental learning methods named Incremental Mixture of Factor Analyzers (IMoFA) and Variational Bayesian Mixture of Factor Analyzers (VBMFA) are adopted. Comparative experiments conducted on both synthetic and real world datasets mainly show the following observations. First, BYY-C is generally best in terms of performance, while BYY-A is usually the best in terms of computational cost. Importantly, when sample size is small or dimension is high, those discussed LFA methods all face a risk of misselection. However, BYY-C and BYY-A show advantages of producing more accurate selection for small-sample-size cases compared to the remaining methods. Thus, for the LFA model selection task, we recommend using BYY-C for small-sample-size cases in a two-phase implementation, while BYY-A for practical datasets with large sample sizes.

Last but not least, as recognized in the machine learning literature, among different model selection criteria or methods, one may work better in one case while the other may work better in another. None can be said to be better than the others, and thus this chapter only compares those methods on LFA, which is widely-used and representative for model selection problem. We expect these results could give some valuable and related observations for future research.

Acknowledgement

The work described in this chapter was fully supported by a grant from the Research Grant Council of the Hong Kong SAR (Project No: CUHK4173/06E). The author would like to express many thanks to his supervisor Professor Lei Xu, who gives lots of helpful and valuable instructions for this work.

References

1. Akaike, H.: A new look at the statistical model identification. IEEE Trans. Automatic Control 19(6), 716–723 (1974)
2. Barron, A.R., Rissanen, J., Yu, B.: The Minimum Description Length Principle in Coding and Modeling. IEEE Trans. Infor. Thr. 44(6), 2743–2760 (1998)
3. Bengio, Y., Grandvalet, Y.: No Unbiased Estimator of the Variance of K-Fold Cross-Validation. JMLR 5, 1089–1105 (2004)
4. Bowman, A.W., Azzalini, A.: Applied Smoothing Techniques for Data Analysis. Oxford Statistical Science Series. Oxford Science Publications (1997)
5. Bozdogan, H.: Model selection and Akaike's information criterion (AIC): the general theory and its analytical extensions. Psychometrika 52(3), 345–370 (1987)
6. Bozdogan, H.: Akaike's Information Criterion and Recent Developments in Information Complexity. J. Math. Psychol. 44(1), 62–91 (2000)

7. Burnham, K.P., Anderson, D.: Model Selection and Multi-Model Inference. Springer, Heidelberg (2002)
8. Constantinopoulos, C., Titsias, M.K.: Bayesian Feature and Model Selection for Gaussian Mixture Models. IEEE Trans. Pattern Anal. Mach. Intell. 28(6), 1013–1018 (2006)
9. Figueiredo, M.A.F., Jain, A.K.: Unsupervised Learning of Finite Mixture Models. IEEE Trans. PAMI 24(3), 381–396 (2002)
10. Ghahramani, Z., Beal, M.J.: Variational Inference for Bayesian Mixtures of Factor Analysers. In: NIPS, pp. 449–455 (1999)
11. Hinton, G.E., Revow, M., Dayan, P.: Recognizing Handwritten Digits Using Mixtures of Linear Models. In: NIPS, pp. 1015–1022 (1994)
12. Mclachlan, G.J., Krishnan, T.: The EM Algorithm and Extensions (Wiley Series in Probability and Statistics). Wiley Series in Probability and Statistics. Wiley-Interscience, Chichester (2007)
13. Parzen, E.: On the estimation of a probability density function and mode. Annals of Mathematical Statistics 33, 1065–1076 (1962)
14. Redner, R., Walker, H.: Mixture Densities, Maximum Likelihood and the EM Algorithm. SIAM Review 26(2), 195–239 (1984)
15. Roberts, S.J., Husmeier, D., Penny, W., Rezek, I.: Bayesian Approaches to Gaussian Mixture Modeling. IEEE Trans. Pattern Anal. Mach. Intell. 20(11), 1133–1142 (1998)
16. Rubin, D., Thayer, D.: EM algorithms for ML factor analysis. Psychometrika 47(1), 69–76 (1982)
17. Salah, A.A., Alpaydin, E.: Incremental Mixtures of Factor Analysers. In: Proc. ICPR, vol. 1, pp. 276–279 (2004)
18. Schwarz, G.: Estimating the dimension of a model. Annals of Statistics 6, 461–464 (1978)
19. Stone, M.: Asymptotics For and Against Cross-Validation. Biometrika 64(1), 29–35 (1977)
20. Tikhonov, A., Arsenin, V.: Solutions of Ill-posed Problems. Winston and Sons (1977)
21. Vapnik, V.: The Nature of Statistical Learning Theory. Springer, New York (1995)
22. Vapnik, V., Sterin, A.: On structural risk minimization or overall risk in a problem of pattern recognition. Automation and Remote Control 10(3), 1495–1503 (1977)
23. Wand, M., Jones, M.: Kernel Smoothing. Monographs on Statistics and Applied Probability. Chapman and Hall, London (1995)
24. Wang, L., Feng, J.: Learning Gaussian mixture models by structural risk minimization. In: Proc. Intnl. Conf. on Machine Learning and Cybernetics 2005, Guangzhou, China, vol. 8, pp. 4858–4863 (2005)
25. Xu, L.: Bayesian Ying Yang Learning. In: Scholarpedia, vol. 2(3) (1809), http://scholarpedia.org/article/Bayesian_Ying_Yang_Learning
26. Xu, L.: Bayesian-Kullback coupled Ying-Yang machines: Unified learnings and new results on vector quantization. In: Proc. of ICONIP 1995, Beijing, China, pp. 977–988 (1995)
27. Xu, L.: Bayesian Ying-Yang system and theory as a unified statistical learning approach (II) From unsupervised learning to supervised learning and temporal modeling. In: Wong, K.W., King, I., Leung, D. (eds.) Theoretical aspects of neural computation: A multidisciplinary perspective, pp. 25–42. Springer, Berlin (1997)
28. Xu, L.: RBF nets, mixture experts, and Bayesian Ying-Yang learning. Neurocomputing 19(1-3), 223–257 (1998)
29. Xu, L.: Best harmony, unified RPCL and automated model selection for unsupervised and supervised learning on Gaussian mixtures, ME-RBF models and three-layer nets. International Journal of Neural Systems 11(1), 3–69 (2001)
30. Xu, L.: BYY harmony learning, independent state space, and generalized APT financial analyses. IEEE Trans. Neural Networks 12, 822–849 (2001)

31. Xu, L.: BYY harmony learning, structural RPCL, and topological self-organizing on mixture models. Neural Networks 15(8-9), 1125–1151 (2002)
32. Xu, L.: Data smoothing regularization, multi-sets-learning, and problem solving strategies. Neural Networks 16(5-6), 817–825 (2003)
33. Xu, L.: Advances on BYY harmony learning: information theoretic perspective, generalized projection geometry, and independent factor auto-determination. IEEE Trans. Neural Networks 15(5), 885–902 (2004)
34. Xu, L.: Bayesian Ying-Yang Learning(I): A unified perspective for statistical modeling. In: Zhong, N., Liu, J. (eds.) Intelligent Technologies for Information Analysis, pp. 615–659. Springer, Heidelberg (2004)
35. Xu, L.: Bayesian Ying-Yang Learning(II): A new mechanism for model selection and regularization. In: Zhong, N., Liu, J. (eds.) Intelligent Technologies for Information Analysis, pp. 661–706. Springer, Heidelberg (2004)
36. Xu, L.: Fundamentals, Challenges, and Advances of Statistical Learning for Knowledge Discovery and Problem Solving: A BYY Harmony Perspective. In: Proc. of Int. Conf. on Neural Networks and Brain, Beijing, China, pp. 24–55 (2005)
37. Xu, L.: Trends on Regularization and Model Selection in Statistical Learning: A Perspective from Bayesian Ying Yang Learning. In: Duch, W., Mandziuk, J., Zurada, J.M. (eds.) Studies in Computational Intelligence, pp. 365–406. Springer, Heidelberg (2007)
38. Xu, L.: A unified perspective and new results on RHT computing, mixture based learning, and multi-learner based problem solving. Pattern Recognition, 2129–2153 (2007)

11

The Concept of Opposition and Its Use in Q-Learning and $Q(\lambda)$ Techniques

Maryam Shokri, H.R. Tizhoosh, and Mohamed S. Kamel

Pattern Analysis and Machine Intelligence Laboratory, University of Waterloo, Canada
mshokri@engmail.uwaterloo.ca, tizhoosh@uwaterloo.ca,
mkamel@uwaterloo.ca

Summary. Reinforcement learning (RL) is a goal-directed method for solving problems in uncertain and dynamic environments. RL agents explore the states of the environment in order to find an optimal policy which maps states to reward-bearing actions. This chapter discusses recently introduced techniques to expedite some of the tabular RL methods for off-policy, step-by-step, incremental and model-free reinforcement learning with discrete state and action space. The concept of opposition-based reinforcement learning has been introduced for Q-value updating. Based on this concept, the Q-values can be simultaneously updated for action and opposite action in a given state. Hence, the learning process in general will be accelerated. Several algorithms are outlined in this chapter. The $OQ(\lambda)$ has been introduced to accelerate $Q(\lambda)$ algorithm in discrete state spaces. The $NOQ(\lambda)$ method is an extension of $OQ(\lambda)$ to operate in a broader range of non-deterministic environments. The update of the opposition trace in $OQ(\lambda)$ depends on the next state of the opposite action (which generally is not taken by the agent). This limits the usability of this technique to the deterministic environments because the next state should be known to the agent. $NOQ(\lambda)$ is presented to update the opposition trace independent of knowing the next state for the opposite action. The primary results show that $NOQ(\lambda)$ can be employed in non-deterministic environments and performs even faster than $OQ(\lambda)$.

11.1 Introduction

The ultimate goal of machine intelligence is to develop autonomous artificial agents that are able to think and act rationally. Currently, there exist numerous machine intelligence methods that are involved in learning, understanding, adapting, interacting, achieving goals or objectives, reasoning, predicting, recognizing, or acting rationally.

Reinforcement learning (RL) is one of the techniques in machine intelligence that can be considered a goal-directed method for solving problems in uncertain and dynamic environments. The RL agent learns by receiving reinforcement signals (reward or punishment) from its environment. One of the advantages of using reinforcement learning is its independence from a-priori knowledge in terms of required training data. The learning is rather performed based on trial and error. This behavior is useful for all user-dependent applications where it is generally difficult to obtain sufficiently large training data.

One of reinforcement learning schemes is the temporal differencing (TD) which generally does not require a model of the environment and is based on step-by-step,

H.R. Tizhoosh, M. Ventresca (Eds.): Oppos. Concepts in Comp. Intel., SCI 155, pp. 233–253, 2008.
springerlink.com

incremental computation. Q-learning is off-policy TD control and one of the most popular methods in reinforcement literature. In off-policy techniques, agent learns a greedy policy and also applies an exploratory policy for action selection. One of the characteristics of Q-learning is model-freedom which makes it suitable for many real-world applications.

Dynamic programming, Monte Carlo algorithms, and temporal-difference learning (TD) are three classes of techniques for solving problems using reinforcement idea. Some of the RL methods are based on the concept of an eligibility trace which provides a bridge between TD techniques and Monte Carlo methods. The underlaying idea is that only eligible states or actions will be assigned a credit or blamed for an error.

Most real-world applications constitute large environments which are dynamic, stochastic, nondeterministic, hidden, and/or only partially observable. One of the theoretical conditions for convergence of reinforcement learning methods such as Q-learning [37] requires that each state-action pair must be visited infinite times (in practice, of course, this reduces to 'multiple times'). Hence, tabular and naive RL algorithms (e.g. Q-learning) could benefit from any extension capable of decreasing their computation time in the case of large state spaces where multiple visits of all states can easily become infeasible.

This chapter outlines the newly proposed opposition-based RL algorithm for accelerating the learning process in off-policy, step by step, incremental and model-free reinforcement learning with discrete state-action space. We integrate the concept of opposition [19, 20, 21, 29, 35] within a new scheme to make some tabular RL algorithms perform faster. The general idea behind the algorithms is that in Q-value updating, the agent updates the value of an action in a given state. If the agent knows the value of the opposite state or the opposite action, then instead of one value, the agent can update two Q-values at the same time without taking the corresponding opposite action. This accelerates the learning process, particularly the exploration phase. The most challenging part of this scheme, however, is defining the opposite state and/or the opposite action. For some applications, the concept of opposition can be established heuristically. A variety of algorithms can be generated based on the concept of opposition to improve learning and casue a faster convergence [28, 31, 32, 33]. For other cases where oppositional relationships are not obvious (hidden), incremental opposition mining algorithms are required to exploit the benefit of simultaneous multiple-updates.

The chapter consists of four sections. Section 11.2 introduces the reinforcement learning techniques as well as a survey of the techniques for accelerating reinforcement learning. In Section 11.3 a class of new opposition-based RL algorithms are introduced and major techniques based on opposition are described. Experimental results are presented in Section 11.4. Section 11.5 draws some conclusions.

11.2   Reinforcement Learning

Reinforcement learning (RL) can be considered as a class of goal-directed intelligent techniques for solving problems in uncertain and dynamic environments. An agent is a function or software component that maps the states of an environment to a set of actions to establish an steady (stable) state of the environment. The agent attempts to maximize the profit (performance/reward) in order to reach this goal. Intelligent robots and softbots (software agents) are examples of machine intelligent agents. A reinforcement learning agent is autonomous [23] meaning that its behavior is determined by its own experience. What is outside the agent is considered the environment. The states are parameters (features) describing the environment. An RL agent has the ability to sense the environment and learn the optimal policy (or a good policy) for taking optimal/good actions in every state of the environment to achieve its goal. As a result of taking an action, the agent has the ability to influence the state of the environment (and can map the states to appropriate actions). The agent must be aware of the states of the environment by interacting with the environment, and learns from receiving reinforcement feedback through a reward or punishment function. The reward function represents, directly or indirectly, the goal of the reinforcement learning problem. RL agents try to maximize the reward or minimize the punishment [2, 4, 30]. The reward values could be objective or subjective. In the subjective case, the agent will receive reward and punishment directly from the interactive user (in some cases by using a software interface). In the objective case, the reward is defined based on some optimality measures or desired properties of the results [25, 27, 30]. Actions could affect the next state and subsequent rewards and have the ability to optimize the environment's state [30].

As mentioned earlier, one of the advantages of some reinforcement learning techniques is their independency from a priori knowledge by learning based on trial and error. Some RL agents learn from their own experience without relying on a teacher or training data. This kind of learning is not supervised, but because of using a reward function and exploitation of rewarding actions a *weak supervision* can be assumed. The agent can learn online by adapting through interaction with the environment while performing the required task and improving its behavior in real time. This behavior is useful for all user-dependent applications where a sufficiently large training data is difficult or impossible to obtain.

Learning is the core characteristic of any intelligent system and can be described as "modification of a behavioral tendency by experience" [14]. For an RL agent, learning by trial and error has mainly two stages, namely exploration and exploitation. Exploration means that the agent tries to discover which actions yield the maximum reward by taking different actions repeatedly. Exploitation, on the other hand, means that the agent takes those actions that yield more reward. The transition between exploration and exploitation generally occurs in a gradual manner depending on the selected action policy.

The history of RL has two major parts: the study of animal learning, and the solution of optimal control problems using value functions and dynamic programming [30]. Value functions are functions of states or of state-action pairs that quantify the performance of an action in a given state. Watkins developed the Q-learning algorithm in 1989 [36, 37] in such a way that the agent maintains a value for both state and action representing a prediction of the value of taking that action from the state.

11.2.1   General Framework

The design of an RL agent is based on the characteristics of the problem at hand. First of all the problem must be clearly defined and analyzed and the purpose of designing

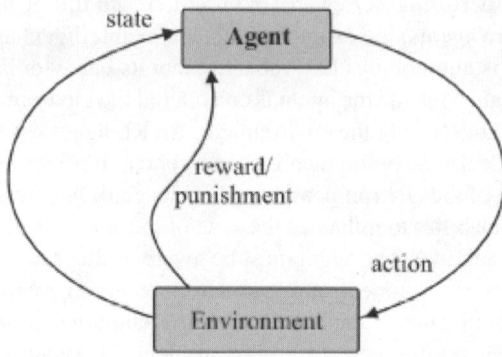

Fig. 11.1. Basic components of reinforcement learning

the agent must be determined. Figure 11.1 illustrates the components, which constitute the general idea behind reinforcement learning.

The RL agent takes an action that influences the environment. The agent acquires knowledge of the actions that generate rewards and punishments and eventually learns to perform the actions that are most rewarding in order to attain a certain goal. In the RL model presented in Figure 11.1 the learning process is as follows [23, 30]:

- Observe the state of the environment
- Take an action
- Observe reward/punishment
- Observe the new state

11.2.2   Action Policy

Another key element of reinforcement learning is the *action policy* which defines the agent's behavior at a given time. It maps the perceived states to the actions to be taken [30]. There are three common policies, softmax, ϵ-greedy, and greedy policy. The ϵ-greedy or near greedy policy is based on the idea of selecting greedy actions most of the time but a few times selecting a random action with probability of ϵ. Based on the near greedy policy an agent has a chance to explore more and balance exploration with exploitation, in contrast to greedy policy that always selects rewarding actions. In this situation all actions may not be explored.

Generally, choosing the appropriate policy depends on the application but it must be considered that the greedy policy sometimes leads to sub-optimal solution which is a common problem in many intelligent systems. On the other hand, one of the disadvantages of using ϵ-greedy is that this technique uses random action selection by considering equal probability for each action based on a uniform distribution for the exploration. An alternative solution is considering the estimated value of each action to vary the probability of action selection. This technique is called softmax action selection [30]. The Boltzmann policy is the most common softmax method which uses a Gibbs, or Boltzmann distribution for defining the policy. The probability of taking each action is presented based on Boltzmann policy in Equation 11.1, where Q(s, a) is a Q-matrix [36, 37] representing the state-action values, τ is a parameter called *temperature*, s is state, and a is action:

$$P(a) = \frac{e^{\frac{Q(s,a)}{\tau}}}{\sum e^{\frac{Q(s,:)}{\tau}}}, \tag{11.1}$$

where the parameter τ is used to control the exploration. At the beginning of learning τ has higher values which enforces more exploration. As time increases τ decreases and consequently exploitation intensifies [6].

11.2.3 Markov Decision Process (MDP)

Reinforcement learning is learning via interaction of the agent with the environment to achieve a certain goal, i.e. maximizing the accumulated rewards over the long run [10]. The other concept that should be addressed here is the Markov property for reinforcement learning. If an environment is Markovian then its states at the time step $t+1$ depend on the state and action at time t. If the reinforcement learning task satisfies the Markov property it is called a finite Markov decision process (MDP) [30]. Regarding RL problems as MDPs, it is assumed that the next state depends on the finite history of previous states [25]. Hence, the environment can be modeled as a Markov decision processe (MDP). The first order Markov property used to predict the probability of a possible next state as follows:

$$P_{ss'}^a = Pr\{s_{t+1}|s_t = s, a_t = a\}, \tag{11.2}$$

where $P_{ss'}^a$ is the transition probability to next state s' given any current state s and action a [30].

11.2.4 Temporal-Difference Learning

Temporal-difference learning is a combination of Monte Carlo and dynamic programming ideas. If at time t a non-terminal state s_t is visited, TD methods estimate the value of that state, $V(s_t)$, based on what happens after that visit. TD methods wait until the next step $(t + 1)$ to determine the increment to $V(s_t)$ as opposed to Monte Carlo that must wait until the end of the learning episode. The simplest TD update, called TD(0), is presented in Equation 11.3 where α is a step-size parameter $(0 < \alpha \leq 1)$, r is result of taking an action[1], γ is a discount-rate parameter $(0 \leq \gamma \leq 1)$ and $V(s_t)$ is value function in each time step t:

$$V(s_t) \leftarrow V(s_t) + \alpha[r_{t+1} + \gamma V(s_{t+1}) - V(s_t)]. \tag{11.3}$$

The tabular TD(0) for estimating the value function is presented in Algorithm 1 where V is value function and π is the action policy to be evaluated.

---

[1] Generally, in all RL algorithms r represents the result of the action which can be reward or punishment. The variable r represents the reward and a variable p represents the punishment.

Algorithm 1. Tabular TD(0) for estimating value function [30]

1: Initialize $V(s)$ arbitrarily, π the policy to be evaluated
2: **for** each episode **do**
3:     Initialize s
4:     **for** each step of episode **do**
5:         $a \leftarrow$ action given by π for s
6:         Take action a: observe reward, r, and next state, s'
7:         $V(s) \leftarrow V(s) + \alpha[r + \gamma V(s') - V(s)]$
8:         $s \leftarrow s'$
9:     **end for**
10: **end for**

11.2.5   Sarsa

Sarsa is on-policy TD control [30]. For an on-policy method the state-action value $Q^\pi(s, a)$ must be estimated for the current policy π, and all states s and actions a. In the on-policy Sarsa, the learned policy for the action-value function is the same as the policy which is applied for action selection. The general algorithm for Sarsa is presented in Algorithm 2.

Algorithm 2. Sarsa: An on-policy TD control algorithm [30]

1: Initialize $Q(s, a)$ arbitrarily
2: **for** each episode **do**
3:     Initialize s
4:     Choose a from s using policy derived from Q (e.g.,ϵ-greedy)
5:     **for** each step of episode **do**
6:         Take action a, observe r, s'
7:         Choose a' from s' using policy derived from Q (e.g., ϵ-greedy)
8:         $Q(s, a) \leftarrow Q(s, a) + \alpha[r + \gamma Q(s', a') - Q(s, a)]$
9:         $s \leftarrow s', a \leftarrow a'$
10:     **end for**
11: **end for**

11.2.6   Q-Learning

Q-learning is off-policy TD control and is one of the most popular methods in reinforcement learning. In an off-policy technique the learned action-value function, $Q(s, a)$, directly approximates the optimal action-value function Q^*, independent of the policy being followed. In other words, the agent learns a greedy policy as well as applying exploratory policy for action selection [30]. This simplifies the algorithm and facilitates convergence. The agent learns to act optimally in Markovian domains by experiencing sequences of actions. The agent takes an action at a particular state and uses immediate reward and punishment and estimates the state value. It evaluates the consequences of taking different actions. By trying all actions in all states multiple times, the agent learns which action is best overall for each visited state [37]. Therefore, the agent must determine an optimal policy and maximize the total discounted expected reward. Equation 11.4 must be employed for updating the action-value function, Q, where s is state, a is action, α is learning step, γ is discount factor, r is the reward or punishment, a' is the next action, and s' is the next state:

$$Q(s,a) \leftarrow Q(s,a) + \alpha[r + \gamma \max_{a'} Q(s',a') - Q(s,a)]. \tag{11.4}$$

The task of Q-learning is determining an optimal policy π^*. The values of the Q matrix are the expected discounted reward for executing action a at state s using policy π [37]. The (theoretical) condition for convergence of the Q-learning algorithm is that the sequence of episodes which forms the basis of learning must visit all states infinitely. It must be mentioned that based on Watkins and Dayan's Theorem in [37] the rewards and learning rate are bounded ($|r_n| \leq R_{max}, 0 \leq \alpha_n < 1$). The Q-learning algorithm is presented in Algorithm 3.

Algorithm 3. Q-Learning: An off-policy TD control algorithm [30]

1: Initialize $Q(s,a)$ arbitrarily
2: **for** each episode **do**
3:     Initialize s
4:     **for** each step of episode **do**
5:         Choose a from s using policy derived from Q {e.g., ϵ-greedy}
6:         Take action a, observe r, s'
7:         $Q(s,a) \leftarrow Q(s,a) + \alpha[r + \gamma \max_{a'} Q(s',a') - Q(s,a)]$
8:         $s \leftarrow s'$;
9:     **end for**
10: **end for**

Q-learning is model-free and uses bootstrapping. The bootstrapping property helps the algorithm to use step-by-step reward, which is the result of the taking an action in each step. This reward is a useful source of information which guides the agent through the environment. Q-learning is considered a general and simple technique for learning through interaction in unknown environments.

Initialization of parameters can be considered as using a priori knowledge for reinforcement learning. Choosing the appropriate parameters may influence the convergence rate [18]. The focus of opposition-based reinforcement learning is on accelerating the update process for Q-learning which is a major obstacle for its application and is mainly a result of the hyper-dimensionality of states and actions in real-world problems.

11.2.7 $Q(\lambda)$ – A Bridge between Monte Carlo and Q-Learning

Sutton et al. [30] introduce the eligibility trace as a bridge between TD techniques and Monte Carlo methods. The idea behind the eligibility traces is that only eligible states or actions will be assigned a credit or blame for the error. TD(λ) is based on a weighted averaging of n-step backups, λ^{n-1} ($0 \leq \lambda \leq 1$). This weighted averaging is presented in Equation 11.5 and Equation 11.6 (λ-return) where R_t^n is the n-step target at time t [30]:

$$R_t^n = r_{t+1} + \gamma r_{t+2} + \gamma^2 r_{t+3} + ... + \gamma^{n-1} r_{t+n} + \gamma^n V_t(s_{t+n}), \tag{11.5}$$

$$R_t^\lambda = (1 - \lambda) \sum_{n=1}^{\infty} \lambda^{n-1} R_t^n. \tag{11.6}$$

The alternative choice for non-Markovian tasks with delayed rewards is using the eligibility traces. "Eligibility trace is a temporary record of occurance of an event" and "marks the memory parameter associated with the event as eligible for undergoing learning changes" [30]. Eligibility traces require more computation but yield faster learning especially for applications with delayed rewards [30]. The parameter λ ($0 \leq \lambda \leq 1$) must be adjusted to place the eligibility somewhere between TD and Monte Carlo. For $\lambda = 1$ the algorithm behaves like the Monte Carlo technique. In contrast, if the $\lambda = 0$, then the overall backups behave like one-step TD backup. Sutton et al. suggest that there is not sufficient theoretical investigations on determining the suitable location for placing the eligibility [30].

The idea of eligibility traces can be applied to TD techniques such as Sarsa and Q-learning. Watkins's $Q(\lambda)$ unlike the $TD(\lambda)$ and Sarsa(λ) looks ahead until the next exploratory action. TD(λ) and Sarsa(λ) need to look ahead until the end of the episode. In contrast, in $Q(\lambda)$, if an agent takes an exploratory action then the eligibility traces will become zero. During the exploration process in early learning the traces will be cut off [30]. Peng et al. [16] introduced a solution for this problem by using a mixture of backups. The implementation of Peng's technique is more difficult than the $Q(\lambda)$ [16, 30]. This technique is "experimentation-sensitive" when $\lambda > 0$. Rewards associated with "non-greedy action" will not be used for evaluating "greedy policy" [16].

Sutton et al. proposed a third variation of $Q(\lambda)$ [30], namely Naive $Q(\lambda)$, which is similar to Watkins's $Q(\lambda)$ with the difference that in this version the traces related to exploratory actions are not set to zero. In this chapter the Watkins's $Q(\lambda)$ has been implemented (presented in Algorithm 4). The reason for this is that in Watkins's $Q(\lambda)$ the representation of the update for the value function is based on using $Q(s, a)$, contrary to Peng's technique which is based on a combination of backups based on both $V(s)$ and $Q(s, a)$. The opposition-based RL technique benefits from the concept of opposite actions, therefore the updates should be presented for $Q(s, a)$ and $Q(s, \breve{a})$ where $\breve{a}$ is opposite action (see section 11.3).

11.2.8   Benefits and Shortcomings of Reinforcement Learning

Reinforcement learning has the ability of learning through interaction with a dynamic environment and using reward and punishment, generally independent of any training data set. RL agents learn from their own experience without relying on teachers or training data. The agent does not need a set of training examples. Instead, it learns online by continuously learning and adapting through interaction with the environment. In some of the RL techniques the model of the environment can be applied to predict the next state and next reward by using the given state and action. Model-based reinforcement

---

Algorithm 4. Tabular Watkins's $Q(\lambda)$ algorithm [30]

1: Initialize $Q(s,a)$ with arbitrary numbers and initialize $e(s,a) = 0$ for all s and a
2: **for** each episode **do**
3:    Initialize s and a
4:    **for** each step of episode **do**
5:       Take action a, observe r and next state s'
6:       Choose next action a' from s' using policy
7:       $a^* \longleftarrow argmax_b\ Q(s',b)$ {if a' ties for max, then $a^* \leftarrow a'$}
8:       $\delta \longleftarrow r + \gamma\ Q(s',a^*) - Q(s,a)$
9:       $e(s,a) \longleftarrow e(s,a) + 1$
10:      **for** all s, a **do**
11:         $Q(s,a) \longleftarrow Q(s,a) + \alpha\delta e(s,a)$
12:         **if** $a' = a^*$ **then**
13:            $e(s,a) \longleftarrow \gamma\lambda e(s,a)$
14:         **else**
15:            $e(s,a) \longleftarrow 0$
16:         **end if**
17:      **end for**
18:      $s \longleftarrow s'; a \longleftarrow a'$
19:   **end for**
20: **end for**

---

learning methods have the advantage of yielding more accurate value estimation but they usually need more memory and may suffer from the curse of dimensionality. The other major problem is that the model of the environment is not always available.

The strengths and shortcomings of RL can be summarized as follows:

Strengths:

- Learning through interaction with environment (Online learning)
- No training data is required for some of the RL techniques
- No model is required for some of the RL techniques
- Independency of a priori knowledge for some of the RL techniques

Shortcomings:

- Exploring the state space is computationally expensive
- The large number of actions also makes the RL techniques more computationally expensive
- Design of RL agents is not straightforward for all applications [7, 9, 17]

Existing Literature

The condition for convergence of Q-learning requires that each state-action pair must be visited infinite times. This requirement is impossible to fulfill in the real world. In practice, nearly optimal action policies can usually be achieved when all actions have been taken multiple times for all states [22]. In real world applications generally the agent faces a very large state or action space which limits the usability of reinforcement learning. There are several solutions proposed in literature. Some general solutions to this problem are briefly reviewed in the following.

Using a priori knowledge – It is argued that RL is not always a *tabula rasa* and additional knowledge can be provided for the agent [12, 22]. The knowledge can be gained through imitation of other agents or transformation of knowledge from previously solved problems [12]. Ribeiro proposed a technique of using knowledge about rates of variation for action values in Q-learning and updating temporal information for the visited states and neighboring states which have similarities with the visited states [22]. Knowledge can be transfered to the agent by a human user through giving advice to the agent. Maclin et al. [11] propose the Preference Knowledge-Based Kernel Regression algorithm (Pref-KBKR). Their proposed algorithm uses human advice as a policy based on if-then rules.

Using hierarchical models and task decomposition – The idea of a hierarchical model is presented by Mahadevan and Kaelbling in [12] as a way of increasing the speed of learning by decomposing the task into a collection of simpler subtasks. Goel provides a technique of sub-goal discovery based on learned policy for hierarchical reinforcement learning [5]. In this technique, hierarchies of actions are produced by using sub-goals in a learned policy model. Hierarchies of actions can then be applied for more effective exploration and acceleration of the process [12]. Goel emphasizes that for finding the sub-goals the states with certain structural properties must be searched [12]. Kaelbling [8] introduces hierarchical distance to goal learning (HDG) by using decomposition of the state space. She argues that this modified version achieves the goal more efficiently. Shapiro et al. [26] combine the hierarchical RL with background knowledge. They propose *Icarus*, an agent architecture that embeds hierarchical RL within a programming language representing an agent's behavior, where the programmer writes an approximately correct plan including options in different level of details to show how to behave with different options. The agent then learns the best options from experience and the reward function given by the user.

Parameters optimization – Potapov and Ali mention that choosing the appropriate parameters may influence convergence rate of reinforcement learning [18]. Specifically, they investigate the problem of selecting parameters for Q-learning method.

Function approximation – Function approximation techniques such as neural networks can also be considered as a technique for reducing the large stochastic state space, especially in continuous domains [12, 24]. These techniques provide methods for approximating the value function [12].

Offline training – The RL agent could be trained offline. The demands of the user can be learned offline in order to minimize the online learning time [34].

Generalization – Agent generalizes by learning similar or closely related tasks. The generalization techniques "allow compact storage of learned information and transfer of knowledge between "similar" states and actions" [10].

Hybrid techniques – As mentioned earlier, neural networks can be applied in the framework of the RL as function approximators [12, 24]. In the case of model learning, the Bayesian network can represent the state transition and reward function [12].

Using macro actions – Mc Govern and Sutton [13] introduce the macro actions as temporally extended actions by combining smaller actions. They argue that macro actions can affect the speed of learning based on the task at hand.

Relational state abstraction – Morales proposes relational state abstraction for reinforcement learning [15]. In this technique Morales describes that the states can be presented as a set of relational properties which yields to abstraction and simplification of the state space. Then, the agent can learn over abstracted space to produce reusable policies which can be transferred and applied to problems with the same relational properties [15].

11.3   Opposition-Based Reinforcement Learning

The goal of this chapter is to introduce opposition-based RL techniques which expedite the learning process for Q-learning and $Q(\lambda)$. These methods are based on multiple and concurrent Q-value updates. In conventional Q-learning, an agent considers one state-action pair and updates one Q-value per iteration. In opposition-based Q-learning we assume that for each action there is an *opposite action*. If the agent takes an action and receives a reward (or punishment) for that action, it would receive punishment (reward) for the opposite action. Therefore, instead of taking the opposite action in a given state, the agent will simply update its associated Q-value. It means that for a given state the Q-values can be updated for both the action (which is taken) and its corresponding opposite action (which is not taken) at the same time. This strategy saves time and accelerates the learning process [31, 32, 33]. In this section we introduce two oppositional RL techniques, opposition-based $Q(\lambda)$ ($OQ(\lambda)$) and opposition-based $Q(\lambda)$ with Non-Markovian Update ($NOQ(\lambda)$) [28, 29].

11.3.1 $OQ(\lambda)$ – Opposition-Based $Q(\lambda)$

The relationship between the idea of opposition and RL has been explored in the framework of the opposition-based $Q(\lambda)$ technique, $OQ(\lambda)$ [28, 29]. If an $OQ(\lambda)$ agent at each time t receives a reward for taking the action a in a given state s, then at that time the agent may also receive punishment for opposite action $\breve{a}$ in the same state s without taking that opposite action. It means that the value function, Q, (e.g. Q-matrix in tabular Q-learning) can be updated for two values $Q(s, a)$ and $Q(s, \breve{a})$ instead of only one value $Q(s, a)$. Therefore, an agent can simultaneously explore actions and integrate knowledge about the opposite actions. Consequently, updating the Q-values for two actions in a given state for each time step can lead to faster convergence since the Q-matrix can be filled in a shorter time [31, 32, 33].

Figure 11.2 demonstrates the difference between Q-matrix updating using reinforcement learning (left image) and Q-matrix updating using opposition-based reinforcement learning (right image). In the left image the states are s_i and actions are a_j where $1 \leq i \leq 5$ and $1 \leq j \leq 4$. The agent in state s_2 takes action a_2 and receives reward r. Then by using reward r the value $v_1 = Q(s_2, a_2)$ of action a_2 in the state s_2 is calculated using the updating formula of Q-learning algorithm as follows:

$$Q(s_2, a_2) \leftarrow Q(s_2, a_2) + \alpha[r + \gamma Q(s', a') - Q(s_2, a_2)]. \tag{11.7}$$

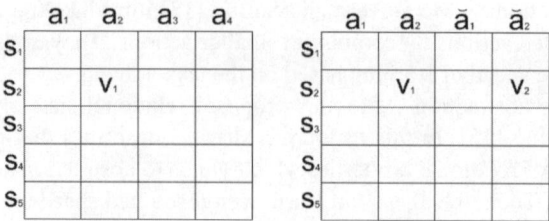

Fig. 11.2. Q-matrix updating; Left: Q-matrix updating using reinforcement learning, right: Q-matrix updating using opposition-based reinforcement learning where an additional update (v_2) can be performed for the opposite action $\breve{a}_2$

In the right image of Figure 11.2, there are two actions a_1 and a_2 with their associated opposite actions $\breve{a}_1$ and $\breve{a}_2$. The agent in state s_2 takes action a_2 and receives reward r. By using reward r the value v_1 of action a_2 in the state s_2 is calculated as before. In the opposition-based technique we assume that the agent will receive an opposite reward by taking opposite action. Hence, by assuming that the agent will receive punishment p (opposite reward) by taking the opposite action $\breve{a}_2$ in state s_2, the value $v_2 = Q(s_2, \breve{a}_2)$ in Q-matrix is also updated as follows [31, 32, 33]:

$$Q(s_2, \breve{a}_2) \leftarrow Q(s_2, \breve{a}_2) + \alpha[p + \gamma Q(s', a') - Q(s_2, \breve{a}_2)]. \tag{11.8}$$

It means that the value function Q can be updated for two values instead of only one value. Therefore, an agent can simultaneously explore actions and update its knowledge about the opposite actions. Hence, the additional update should accelerate the learning.

The $OQ(\lambda)$ algorithm is constructed based on opposition traces which represent eligibility traces for opposite actions. Assume that $e(s, a)$ is the eligibility trace for action a in state s, then the opposition trace is $\breve{e} = e(s, \breve{a})$. For updating the Q-matrix in a given state s, agent takes action a and receives reward r. Then by using reward r the Q-matrix will be updated for all states and actions:

$$\delta_1 \longleftarrow r + \gamma Q(s', a^*) - Q(s, a), \tag{11.9}$$

$$Q(s, a) \longleftarrow Q(s, a) + \alpha \delta_1 e(s, a), \tag{11.10}$$

By assuming that the agent will receive punishment p by taking opposite action $\breve{a}$, the Q-matrix will be updated for all states s and opposite actions $\breve{a}$:

$$\delta_2 \longleftarrow \breve{r} + \gamma Q(s'', a^{**}) - Q(s, \breve{a}), \tag{11.11}$$

$$Q(s, \breve{a}) \longleftarrow Q(s, \breve{a}) + \alpha \delta_2 e(s, \breve{a}), \tag{11.12}$$

The $OQ(\lambda)$ algorithm [28] is presented in Table 5.

$OQ(\lambda)$ differs from $Q(\lambda)$ algorithm in the Q-value updating. In $OQ(\lambda)$ the opposition trace facilitates updating of the Q-values for opposite actions and instead of

---

Algorithm 5. $OQ(\lambda)$ [28]. If the agent receives punishment p for taking an action then the opposite action receives reward r

---

1: For all s and a initialize $Q(s,a)$ with arbitrary numbers and initialize $e(s,a) = 0$
2: **for** each episode **do**
3:    Initialize s and a
4:    **for** each step of episode **do**
5:       Take action a, observe r and next state s'
6:       Determine opposite action $\breve{a}$ and next state s''
7:       Calculate opposite reward (punishment) $\breve{r} = p$
8:       Choose next action a' from s' using policy
9:       Determine next opposite action $\breve{a}'$ from s''
10:      $a^* \longleftarrow argmax_b\, Q(s', b)$ {if a' ties for max, then $a^* \leftarrow a'$}
11:      $a^{**} \longleftarrow argmax_b\, Q(s'', b)$ {if $\breve{a}'$ ties for max, then $a^{**} \leftarrow \breve{a}'$}
12:      $\delta_1 \longleftarrow r + \gamma\, Q(s', a^*) - Q(s, a)$
13:      $\delta_2 \longleftarrow \breve{r} + \gamma\, Q(s'', a^{**}) - Q(s, \breve{a})$
14:      $e(s, a) \longleftarrow e(s, a) + 1$
15:      $e(s, \breve{a}) \longleftarrow e(s, \breve{a}) + 1$
16:      **for** all s, a in action set A **do**
17:         $Q(s, a) \longleftarrow Q(s, a) + \alpha\delta_1 e(s, a)$
18:         **if** $a' = a^*$ **then**
19:            $e(s, a) \longleftarrow \gamma\lambda e(s, a)$
20:         **else**
21:            $e(s, a) \longleftarrow 0$
22:         **end if**
23:      **end for**
24:      $s \longleftarrow s'; a \longleftarrow a'$
25:      **for** all $s, \breve{a}$ in the opposite action set $\breve{A}$ where $\breve{A}$ is not $\subset A$ **do**
26:         $Q(s, \breve{a}) \longleftarrow Q(s, \breve{a}) + \alpha\delta_2 e(s, \breve{a})$
27:         **if** $\breve{a}' = a^{**}$ **then**
28:            $e(s, \breve{a}) \longleftarrow \gamma\lambda e(s, \breve{a})$
29:         **else**
30:            $e(s, \breve{a}) \longleftarrow 0$
31:         **end if**
32:      **end for**
33:   **end for**
34: **end for**

---

punishing/rewarding the action and opposite action[2] we punish/reward the eligible trace and opposite trace.

Reward/Punishment Confusion

An issue that must be addressed here is that in some situations an action and its opposite action may yield the same result (=reward/punishment) in a given state. The action and

---

[2] It is assumed that when the agent receives reward (r)/punishment ($\breve{r}$) for taking an action, it will receive punishment ($\breve{r}$)/reward (r) for taking the opposite action.

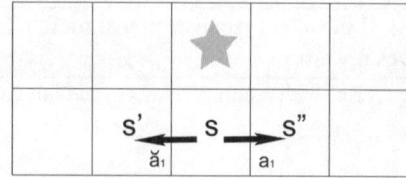

Fig. 11.3. Both action and opposite action should be punished for going away from the target (star)

opposite action may both lead to reward, or both lead to punishment. The example in Figure 11.3 illustrates this situation using the grid world problem [28]. The goal is presented by a star and the present state is s. For both action a_1, and its opposite $\breve{a}_1$ the result is punishment because they both increase the distance of the agent from the goal. Hence, both of them should be punished. Rewarding one of them will falsify the value function and affect the convergence.

Opposition traces are possible solutions for this problem since the Q-matrix updating is not limited to one action and one opposite action at a given state, but also depends on updating more Q-values by using eligibility and opposition traces. Therefore, all actions in the trace and all opposite actions in the opposition trace will affect the learning process and help to avoid/reduce the effect of reward/punishment confusion. It can also be stated that for grid-based problems with one target the influence of confusing cases becomes completely negligible with increase in dimensionality. However, the confusing cases can cause algorithm oscillations around the corresponding state if mechanisms such as opposition traces are not employed. This has been reported for standard Q-learning [33].

11.3.2   Opposition-Based Q(λ) with Non-markovian Update

It has been assumed that the agent has knowledge of opposite state. This may limit the usability of the technique to deterministic environments because the next state of the environment (for the opposite action) should be provided, considering that the agent will not actually take the opposite action. In order to relax this assumption, the Non-Markovian Opposition-Based $Q(\lambda)$ ($NOQ(\lambda)$) algorithm is introduced in this section.

The new method is a hybrid of Markovian update for eligibility traces and non-Markovian-based update for opposition traces. The $NOQ(\lambda)$ method specifically focuses on investigating the possible Non-Markovian updating of opposition traces where the next state for the opposite action may not be available. This extends the usability of $OQ(\lambda)$ to a broader range of applications where the model of environment is not provided for the agent.

One issue regarding the $OQ(\lambda)$ algorithm is the problem of Markovian-based updating of the opposite trace which will be addressed here. As it is presented in Algorithm 5, line 6, the agent should determine the next state s'' after defining the opposite action $\breve{a}$. In this algorithm the agent does not actually take $\breve{a}$. For instance, if the action is going to the left, then the opposite action is determined (not taken) as going to the right. In a deterministic environment the agent can figure out the next state by using the model of the environment. In the case of the Gridworld problem the agent assumes that the next state s'' for opposite action $\breve{a}$ is the opposite of next state $\breve{s}'$ with respect to initial states. In this case the Q values can be updated for the opposition trace as follows:

$$a^{**} \longleftarrow \mathrm{argmax}_b Q(s'', b), \tag{11.13}$$

$$\delta_2 \longleftarrow \breve{r} + \gamma Q(s'', a^{**}) - Q(s, \breve{a}), \tag{11.14}$$

$$Q(s, \breve{a}) \longleftarrow Q(s, \breve{a}) + \alpha \delta_2 e(s, \breve{a}), \tag{11.15}$$

where α is step-size parameter and γ is a discount-rate parameter. Equations 11.13, 11.14, and 11.15 present the Markovian-based updating by considering the next state of s'' [29]. In $NOQ(\lambda)$ we address this problem by introducing the non-Markovian updating for opposition traces.

The $OQ(\lambda)$ method updates opposite traces without taking opposite actions (which intuitively would not make sense in any event). For this reason, the opposition update (the Markovian update) depends on the next state of the environment that should be known by the agent. This limits the applicability of the $OQ(\lambda)$ to deterministic environments. We relax the constraint of Markovian updating by introducing a new update for opposition traces. Equation 11.16 presents the new update formula for opposition traces where $\breve{r}$ is the opposite reward, and $e(s, \breve{a})$ is the opposite trace:

$$Q(s, \breve{a}) \longleftarrow Q(s, \breve{a}) + W \times \breve{r} \times e(s, \breve{a}). \tag{11.16}$$

The parameter $W \in [0, 1]$ is introduced to impose a weight on opposition update. If in some application the definition of opposite action is not straightforward, then it should be defined as a function of the number of iterations. Hence, we assume that at the beginning of learning the weight of the update is low and increases gradually as the agent explores the actions and the opposite actions. In the case of the Gridworld problem, the definition of opposite actions are known and hence $W = 1$.

As it is presented in Equation 11.16, $Q(s, \breve{a})$ does not depend on the next state, in contrast to the $OQ(\lambda)$ technique which depends on s'' (see Equations 11.15 and 11.14). The algorithm of the $NOQ(\lambda)$ technique is presented in the Algorithm 6.

11.4 Experimental Results

The Gridworld [29] problem with three sizes (20×20, 50×50, and 100×100) is chosen as a test-case problem. The grid represents the learning environment and each cell of the grid represents a state of the environment. A sample Gridworld is presented in Figure 11.4. The agent can move in 8 possible directions indicated by arrows in the figure. The goal of the agent is to reach the defined target in the grid which is marked by a star.

Four actions with their corresponding four opposite actions (if $a = Up$ then $\breve{a} = Down$) are defined. By taking an action, the agent has the ability to move to

---

Algorithm 6. $NOQ(\lambda)$ Algorithm [29]. If the Agent Receives Punishment p for Taking an Action then the Opposite Action receives Reward r

---

1: For all s and a initialize $Q(s,a)$ with arbitrary numbers and initialize $e(s,a) = 0$
2: **for** each episode **do**
3:     Initialize s and a
4:     **for** each step of episode **do**
5:         Take action a, observe r and next state s'
6:         Determine opposite action $\breve{a}$
7:         Choose next action a' from s' using policy
8:         Determine next opposite action $\breve{a}'$
9:         $a^* \longleftarrow argmax_b\, Q(s', b)$ {if a' ties for max, then $a^* \leftarrow a'$}
10:         $\delta_1 \longleftarrow r + \gamma\, Q(s', a^*) - Q(s, a)$
11:         $e(s,a) \longleftarrow e(s,a) + 1$
12:         $e(s,\breve{a}) \longleftarrow e(s,\breve{a}) + 1$
13:         **for** all s, a in the action set A **do**
14:             $Q(s,a) \longleftarrow Q(s,a) + \alpha\delta_1 e(s,a)$
15:             **if** $a' = a^*$ **then**
16:                 $e(s,a) \longleftarrow \gamma\lambda e(s,a)$
17:             **else**
18:                 $e(s,a) \longleftarrow 0$
19:             **end if**
20:         **end for**
21:         **for** all $s, \breve{a}$ in the opposite action set $\breve{A}$ where $\breve{A}$ is not $\subset A$ **do**
22:             $Q(s,\breve{a}) \longleftarrow Q(s,\breve{a}) + W * \breve{r} * e(s,\breve{a})$
23:             **if** $a' = a^*$ **then**
24:                 $e(s,\breve{a}) \longleftarrow \gamma\lambda e(s,\breve{a})$
25:             **else**
26:                 $e(s,\breve{a}) \longleftarrow 0$
27:             **end if**
28:         **end for**
29:         $s \longleftarrow s';\ a \longleftarrow a'$
30:     **end for**
31: **end for**

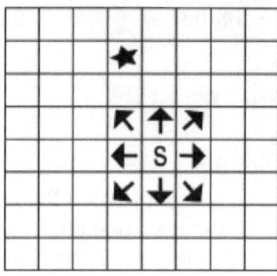

Fig. 11.4. Sample Gridworld [29]. There are eight possible actions presented by arrows. The letter S represents the state and the star is the target.

Table 11.1. The Initial Parameters for all experiments [29]

n_E	I_{max}	α	γ	λ
100	1000	0.3	0.2	0.5

Table 11.2. The results for the four measures of $\overline{\overline{I}}$, $\overline{T}$, χ, and ζ for algorithms ($Q(\lambda)$, $OQ(\lambda)$, and $NOQ(\lambda)$) [29]. The results are based on 100 runs for each algorithm [29].

	$Q(\lambda)$			$OQ(\lambda)$			$NOQ(\lambda)$		
$X \times Y$	20 × 20	50 × 50	100 × 100	20 × 20	50 × 50	100 × 100	20 × 20	50 × 50	100 × 100
$\overline{\overline{I}}$	255 ± 151	528 ± 264	431 ± 238	20 ± 8	57 ± 26	103 ± 50	19 ± 7	48 ± 23	100 ± 46
$\overline{T}$	3 ± 2	17 ± 6.6	102 ± 25	0.3 ± 0.1	5 ± 1	80 ± 7	0.2 ± 0.1	5 ± 0.7	58 ± 6
χ	0	11	9	0	0	0	0	0	0
ζ		93.3%			100%			100%	

one of the neighboring states. The actions/opposite actions are left/right, up/down, up-right/down-left, and up-left/down-right. The initial state is selected randomly for all experiments. If the size of a grid is (X_{max}, Y_{max}), then the coordinates of the target are fixed at $(\frac{X_{max}}{2}, \frac{Y_{max}}{3})$. The value of immediate reward is 10, and punishment is -10. After the agent takes an action, if the distance of the agent from the goal is decreased, then agent will receive a reward. If the distance is increased or not changed, the agent receives punishment. The Boltzmann policy is applied for all implementations. The $Q(\lambda)$, $OQ(\lambda)$, and $NOQ(\lambda)$ algorithms are implemented. The initial parameters for the algorithms are presented in Table 11.1.

The following measurements are considered for comparing the results of $Q(\lambda)$, $OQ(\lambda)$, and $NOQ(\lambda)$ algorithms:

- overall average iterations $\overline{\overline{I}}$: average of iterations over 100 runs
- average time $\overline{T}$: average running time (seconds) over 100 runs
- number of failures χ: number of failures over 100 runs

The agent performs the next episode if it reaches the target represented by the star. Learning stops when the accumulated reward of the last 15 iterations has a standard deviation below 0.5. The results are presented in Table 11.2.

The results presented in Table 11.2 are also plotted in Figures 11.5 and 11.6 for visual comparisons. Figure 11.5 presents changes in the overall average number of iterations for $Q(\lambda)$, $OQ(\lambda)$, and $NOQ(\lambda)$ algorithms for the three grid sizes. We observe that the total number of iterations for convergence of $Q(\lambda)$ is far higher than $OQ(\lambda)$ and $NOQ(\lambda)$ algorithms. The $NOQ(\lambda)$ takes slightly less iterations than $OQ(\lambda)$. The reason for this may be related to the independency of $NOQ(\lambda)$ of next state for the opposite action.

Figure 11.6 presents the average time for $Q(\lambda)$, $OQ(\lambda)$, and $NOQ(\lambda)$ algorithms for the three grid sizes. Even though the number of iterations for $OQ(\lambda)$ and $NOQ(\lambda)$

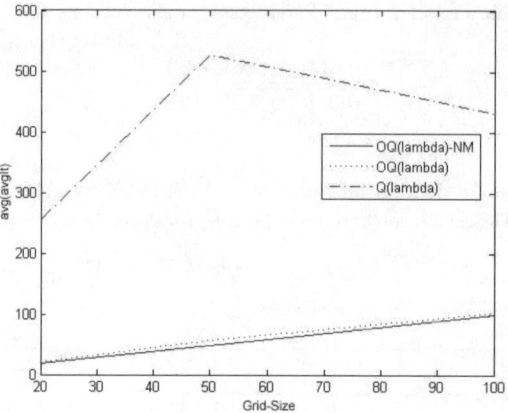

Fig. 11.5. The average of average iterations $\overline{I}$ for $Q(\lambda)$, $OQ(\lambda)$, and $NOQ(\lambda)$;

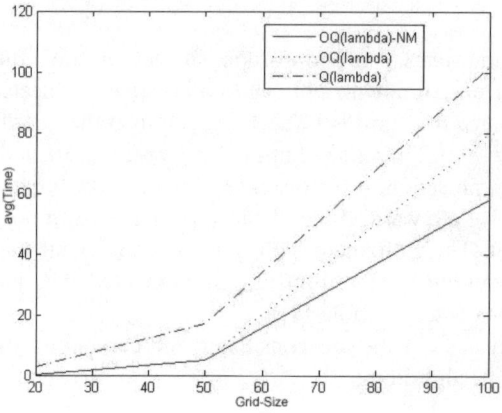

Fig. 11.6. The average time T for $Q(\lambda)$, $OQ(\lambda)$, and $NOQ(\lambda)$ [29]

are almost the same, the average computation time of $NOQ(\lambda)$ is much less than the average time of the $OQ(\lambda)$ for 100×100 grid. The reason is that $NOQ(\lambda)$ algorithm is more efficient than $OQ(\lambda)$ due to decrease in the computational overhead associated with updating the opposition traces.

In order to compare $Q(\lambda)$, $OQ(\lambda)$, and $NOQ(\lambda)$ algorithms we also need to consider that $Q(\lambda)$ failed to reach the goal or target 20 times[3]. To reflect this failure in the performance measure, the success rate $\zeta_{overall}$ for the algorithms (presented in [28]) is calculated:

---

[3] If the fixed numbers of iterations and episodes are not enough for the algorithm to reach the target (in one run) then we consider this as one failure.

$$\zeta_{overall} = \left(1 - \frac{\sum\limits_{i=1}^{k} \chi_i}{\sum\limits_{k} H}\right) \times 100, \tag{11.17}$$

where k is the number of grids tested (in this case $k = 3$), χ is the number of failures, and H is the number of times the code is run for each grid. Considering the convergence conditions, for the $Q(\lambda)$ algorithm, the overall success rate is $\zeta_{overall} = 93.3\%$ because the agent failed to reach the goal 20 times. For the proposed algorithm $NOQ(\lambda)$, and the oppsotion-based algorithm $OQ(\lambda)$, the overall success rate is $\zeta_{overall} = 100\%$; Both always successfully find the target.

11.5 Conclusions

The goal of this chapter was to review new techniques for expediting some of the tabular RL methods for off-policy, step-by-step, incremental and model-free reinforcement learning with discrete state and action space. To solve this problem the concept of opposition-based reinforcement learning has been used. The general idea is that in Q-value updating, the agent updates the value of an action in a given state. If the agent knows the value of the opposite state or the opposite action, then instead of one value, the agent can update multiple Q-values at the same time without taking the associated opposite action.

A variety of algorithms can be generated based on the concept of opposition to improve the learning and design faster RL techniques. Opposition is applied to create several algorithms based on using Q-learning. The $OQ(\lambda)$ has been introduced to accelerate $Q(\lambda)$ algorithm with discrete state and action space. The $NOQ(\lambda)$ method is an extension of $OQ(\lambda)$ to cover a broader range of non-deterministic environments. The update of the opposition trace in $OQ(\lambda)$ depends on the next state of the opposite action (which cannot be taken). This limits the usability of this technique to the deterministic environments because the next state should be detected or known by the agent. $NOQ(\lambda)$ was presented to update the opposition traces independent of knowing the next state for the opposite action. The primary results show that $NOQ(\lambda)$ can be employed in non-deterministic environments and performs even faster (see Figure 11.6) than $OQ(\lambda)$.

The future work will focus on the extending the opposition-based technique to other reinforcement learning algorithms. We will also study the effects of the opposite states on the performance of RL algorithms with respect to the number of iterations and running time of the codes. The investigation of the effects of opposition-based RL methods in hybrid techniques is an interesting research topic which should be considered as well.

References

1. Alpaydin, E.: Introduction to Machine Learning. MIT Press, Cambridge (2004)
2. Ayesh, A.: Emotionally Motivated Reinforcement Learning Based Controller. IEEE SMC, The Hague (2004)
3. Collins Cobuild English Dictionary, pp. 77–85. HarperCollins Publishers, Fulham Palace Road, London, England (2000)

4. Gadanho, S.: Reinforcement Learning in Autonomous Robots: An Empirical Investigation of the Role of Emotions, PhD Thesis, University of Edinburgh, Edinburgh (1999)
5. Goel, S.K.: Subgoal Discovery for Hierarchical Reinforcement learning Using Learned Policies, Department of Computer Science and Engineering, University of Texas at Arlington, TX, USA, Master of Science in Computer Science and Engineering (2003)
6. Humphrys, M.: Action Selection Methods Using Reinforcement Learning, PhD Theses, University of Cambridge (1997)
7. Jaakkola, T., Singh, S.P., Jordan, M.I.: Reinforcement Learning Algorithm for Partially Observable Markov Decision Problems. In: Advances in Neural Information Processing Systems, vol. 7 (1994)
8. Kaelbling, L.P.: Hierarchical Reinforcement Learning: Preliminary Results. In: Proceedings of the Tenth International Conference on Machine Learning (1993)
9. Kaelbling, L.P., Littman, M.L., Cassandra, A.R.: Planning and Acting in Partially Observable Stochastic Domains. Artificial Intelligence 101, 99–134 (1998)
10. Kaelbling, L.P., Littman, M.L., Moore, A.W.: Reinforcement Learning: A Survey. Journal of Artificial Intelligence Research 4, 237–285 (1996)
11. Maclin, R., Shavlik, J., Torrey, L., Walker, T., Wild, E.: Giving Advice about Preferred Actions to Reinforcement Learners Via Knowledge-Based Kernel Regression. American Association for Artificial Intelligence (2005)
12. Mahadevan, S., Kaelbling, L.P.: The NSF Workshop on Reinforcement Learning: Summary and Observations. In: AI Magazine (1996)
13. Mc Govern, A., Sutton, R.S.: Macro-Actions in Reinforcement Learning: An Empirical Analysis, University of Massachusetts, Amherst, Technical Report Number 98-70 (1998)
14. Merriam-Webster Online English Dictioinary, http://www.m-w.com
15. Morales, E.F.: Relational State Abstractions for Reinforcement Learning. In: Proceedings of the ICML 2004 workshop on Relational Reinforcement Learning, Banff, Canada (2004)
16. Peng, J., Williams, R.J.: Incremental Multi-Step Q-Learning. Machine Learning 22 (1996)
17. Pham, T.D.: Perception-Based Hidden Markov Models: A Theoretical Framework for Data Mining and Knowledge Discovery. Soft Computing 6, 400–405 (2002)
18. Potapov, A., Ali, M.K.: Convergence of Reinforcement Learning Algorithms and Acceleration of Learning. Physical Review E 67, 026706 (2003)
19. Rahnamayan, S., Tizhoosh, H.R., Salama, M.: Opposition-Based Differential Evolution Algorithms. In: IEEE International Joint Conference on Neural Networks (IJCNN), IEEE World Congress on Computational Intelligence, Vancouver, July 16-21 (2006)
20. Rahnamayan, S., Tizhoosh, H.R., Salama, M.: Opposition-Based Differential Evolution for Optimization of Noisy Problems. In: IEEE Congress on Evolutionary Computation, IEEE World Congress on Computational Intelligence, Vancouver, July 16-21 (2006)
21. Rahnamayn, S., Tizhoosh, H.R., Salama, M.: A Novel Population Initialization Method for Accelerating Evolutionary Algorithms. In: Computers and Mathematics with Applications (2006)
22. Ribeiro, C.H.: Embedding a Priori Knowledge in Reinforcement Learning. Journal of Intelligent and Robotic Systems 21, 51–71 (1998)
23. Ribeiro, C.: Reinforcement Learning Agent. Artificial Intelligence Review 17, 223–250 (2002)
24. Sutton, R.S.: Open theoretical questions in reinforcement learning. In: Fischer, P., Simon, H.U. (eds.) EuroCOLT 1999. LNCS (LNAI), vol. 1572, pp. 11–17. Springer, Heidelberg (1999)
25. Russell, S.J., Norvig, P.: Artificial Intelligence: A Modern Approach. Pearson Education Inc., New Jersey (2003)

26. Shapiro, D., Langley, P., Shachter, R.: Using Background Knowledge to Speed Reinforcement Learning in Physical Agents. In: AGENTS 2001, Montréal, Quebec, Canada (2001)
27. Shokri, M., Tizhoosh, H.R.: Using Reinforcement Learning for Image Thresholding. In: Canadian Conference on Electrical and Computer Engineering, vol. 1, pp. 1231–1234 (2003)
28. Shokri, M., Tizhoosh, H.R., Kamel, M.: Opposition-Based $Q(\lambda)$ Algorithm. In: International Joint Conference on Neural Networks, IJCNN, pp. 646–653 (2006)
29. Shokri, M., Tizhoosh, H.R., Kamel, M.S.: Opposition-Based $Q(\lambda)$ with Non-Markovian Update. In: IEEE International Symposium on Approximate Dynamic Programming and Reinforcement Learning, Hawaii, USA (accepted, 2007)
30. Sutton, R.S., Barto, A.G.: Reinforcement learning: An Introduction. MIT Press, Cambridge (1998)
31. Tizhoosh, H.R.: Reinforcement Learning Based on Actions and Opposite Actions. In: ICGST International Conference on Artificial Intelligence and Machine Learning (AIML 2005), Cairo, Egypt (2005)
32. Tizhoosh, H.R.: Opposition-Based Learning: A New Scheme for Machine Intelligence. In: International Conference on Computational Intelligence for Modeling Control and Automation - CIMCA 2005, Vienna, Austria, vol. I, pp. 695–701 (2005)
33. Tizhoosh, H.R.: Opposition-Based Reinforcement learning. Journal of Advanced Computational Intelligence and Intelligent Informatics 10(4), 578–585 (2006)
34. Tizhoosh, H.R., Shokri, M., Kamel, M.: The Outline of a Reinforcement-Learning Agents for E-Learning Applications. In: Pierre, S. (ed.) E-Learning Networked Environments and Architectures: A Knowledge Processing Perspective. Springer, Heidelberg (2005)
35. Ventresca, M., Tizhoosh, H.R.: Improving the Convergence of Backpropagation by Opposite Transfer Functions. In: IEEE Congress on Evolutionary Computation, IEEE World Congress on Computational Intelligence, Vancouver, July 16-21 (2006)
36. Watkins, C.J.C.H.: Learning from Delayed Rewards. Cambridge University, Cambridge (1989)
37. Watkins, C.J.H., Dayan, P.: Technical Note, Q-Learning. Machine Learning 8, 279–292 (1992)

12

Two Frameworks for Improving Gradient-Based Learning Algorithms

Mario Ventresca and H.R. Tizhoosh

Department of Systems Design Engineering, University of Waterloo, Canada
{mventres,tizhoosh}@pami.uwaterloo.ca

Summary. Backpropagation is the most popular algorithm for training neural networks. However, this gradient-based training method is known to have a tendency towards very long training times and convergence to local optima. Various methods have been proposed to alleviate these issues including, but not limited to, different training algorithms, automatic architecture design and different transfer functions. In this chapter we continue the exploration into improving gradient-based learning algorithms through dynamic transfer function modification. We propose opposite transfer functions as a means to improve the numerical conditioning of neural networks and extrapolate two backpropagation-based learning algorithms. Our experimental results show an improvement in accuracy and generalization ability on common benchmark functions. The experiments involve examining the sensitivity of the approach to learning parameters, type of transfer function and number of neurons in the network.

12.1 Introduction

Neural network learning algorithms are extremely capable of discovering solutions in quadratic or simply-shaped, non-complex error surfaces [4, 23]. Some popular learning algorithms such as backpropagation [39, 52, 53], Levenbeg-Marquardt [21], quick propagation [18], resilient propagation [38] and various conjugate gradient-based approaches [7, 34] have been developed which are capable of learning in more complex scenarios. However, no learning algorithm is perfect, and countless heuristics and other techniques to improve on these and other algorithms have been developed.

In this chapter we propose two frameworks for hybridizing gradient-based learning algorithms with notions of opposition. Our global decision-based framework allows a network to decide which neurons are candidates for a transfer function change. Whereas our second approach allows specific neurons to decide whether to change their transfer function based only on local information. These frameworks show improvements in network accuracy at the expense of a small amount of computational overhead.

Both frameworks are based on employing opposite networks which utilize opposite transfer functions in different manners [48, 49, 50]. An opposite transfer function is essentially a transformation on a neuron transfer function which alters the input/output mapping of the network. While other structural transformations do not guarantee a change in the network input/mapping, opposite transfer function guarantee this for minimal networks. Each combination of these functions represents a unique location on

H.R. Tizhoosh, M. Ventresca (Eds.): Oppos. Concepts in Comp. Intel., SCI 155, pp. 255–284, 2008.
springerlink.com

the training error surface and transformations between all networks is accomplished in $\mathcal{O}(1)$ computational complexity. This property allows for the efficient consideration of opposite networks.

The remainder of this chapter is organized as follows. Section 12.2 will discuss the backpropagation algorithm and previous works concerning the influence of adaptive transfer functions on its performance. We also point out properties of symmetric transformations and their effect on the input/output mapping and error surface. In Section 12.3 we introduce the concept of an opposite network, which has its foundation on the notion of an opposite transfer function. We also discuss properties of these types of networks. Then, in Section 12.4 we propose two frameworks for using opposite networks, based on global and local decision rules. The backpropagation and backpropagation through time algorithms are extended as examples of each framework. Our experimental results on both modified algorithms are presented in Section 12.5 and our conclusions and some possible directions for future works are provided in Section 12.6.

12.2   Background

Many neural network learning algorithms are based on utilizing gradient information $\nabla E(\mathbf{w})$ to guide the weight optimization procedure, where $\mathbf{w} \in \mathcal{W}$ is a configuration of weights and biases (hereafter simply referred to as weights) of the space of possible weight configurations, $\mathcal{W}$. The gradient is calculated with respect to the error function E which summarizes the network error over all training patterns $\{(\mathbf{x}, \mathbf{d})_p\}_{p=1}^P$. For each input vector $\mathbf{x}$ there exists a target vector $\mathbf{d}$, where commonly $|\mathbf{x}| \gg |\mathbf{d}|$. The input and output layer sizes of the network equal the size of the input and target vectors, respectively.

12.2.1   Backpropagation

Backpropagation is the most popular gradient-based algorithm for training feedforward and recurrent neural networks [39, 52, 53]. Many other learning approaches either employ backpropagation (BP) directly, or are based on its principles. Therefore, using this algorithm in our study potentially has a wider influence than other learning algorithms. The general idea of BP is to update $\mathbf{w}$ in the direction of the greatest rate of decrease of the error function.

We will assume a feedforward network with L layers indexed by $l = 0..L$ where $|l|$ is the size of the lth layer. It must be that $|l = 0| = |\mathbf{x}|$ and $|l = L| = |\mathbf{d}|$. Further assuming that network performance E is measured by the mean-squared error (MSE) function where y_k is the kth output of the network then,

$$E = \frac{1}{2P} \sum_{p=1}^{P} \sum_{k=1}^{K} (d_k - y_k)^2 \qquad (12.1)$$

and we can quickly derive the backpropagation algorithm. Since E is a function of a neuron's output error (e_j), which is a function of the neuron output (y_j), which is itself a function of the neuron input (v_j) it follows from the Chain Rule of Calculus that

$$\frac{\partial E}{\partial w_{ij}} = \frac{\partial E}{\partial e_j} \frac{\partial e_j}{\partial y_j} \frac{\partial y_j}{\partial v_j} \frac{\partial v_j}{\partial w_{ij}} \qquad (12.2)$$

represents the derivative of E with respect to each weight w_{ij} in the network between every neuron pair (i, j). Separately calculating the derivative of each value in (12.2) yields

$$\frac{\partial E}{\partial e_j} = e_j \text{ and } \frac{\partial e_j}{\partial y_j} = -1 \text{ and } \frac{\partial y_j}{\partial v_j} = \phi'(v_j) \text{ and } \frac{\partial v_j}{\partial w_{ij}} = y_i, \qquad (12.3)$$

where ϕ' is the derivative of the neuron transfer function, ϕ. It is common to group the local gradient values of neuron j as

$$\delta_j = e_j \phi'(v_j). \qquad (12.4)$$

Algorithm 1. The Backpropagation Algorithm

1: randomly select $\mathbf{w} \in \mathcal{W}$
2: **while** termination criteria not satisfied **do**
3:   **for** each training pattern $\{(\mathbf{x}, \mathbf{d})\}$ **do**
     {set input pattern into network}
4:     $\mathbf{y}^0 = \mathbf{x}$

     {forward propagation}
5:     **for** each layer, l=1 to L **do**
6:       **for** each neuron, $j = 1 \dots |l|$ **do**
7:         $v_j^l = \sum_{i=0}^{|l-1|} w_{ij}^l y_i^{l-1}$
8:         $y_j^l = \phi(v_j^l)$
9:       **end for**
10:    **end for**

     {calculate error of each output neuron}
11:    $\mathbf{e}^L = \mathbf{d} - \mathbf{y}^L$

     {backward propagation}
12:    **for** $l = L \dots 1$ **do**
13:      **for** $j = 1 \dots |l|$ **do**
14:        $\delta_j^l = \begin{cases} e_j^L \phi_j'(v_j^L) & \text{if } l\text{=L} \\ \phi_j'(v_j^l) \sum_{k=1}^{|l+1|} \delta_k^{l+1} w_{kj}^{l+1} & \text{otherwise} \end{cases}$
15:        $w_{ij}^l = w_{ij}^l + \alpha \delta_j^l y_i^{l-1}$, for $i = 0 \dots |l-1|$
16:      **end for**
17:    **end for**
18: **end for**
19: **end while**

Weight update in backpropagation then follows the Delta Rule,

$$\Delta \mathbf{w} = \alpha \frac{\partial E}{\partial w_{ij}}. \tag{12.5}$$

where $0 < \alpha \leq 1$ represents a learning rate parameter and controls how $\mathbf{w}$ is updated. Learning typically begins at an initially random $\mathbf{w}$ and proceeds until a maximum number of iterations is met or $E < \varepsilon$, where $\varepsilon \geq 0$ is an acceptable performance threshold.

The pseudocode for backpropagation is presented in Algorithm 1. For readability we use index l to refer to a specific layer, i.e. w_{ij}^l represents the weight connecting neuron i in layer $l - 1$ to neuron j in layer l. The forward propagation steps (lines 5-10) determine the input and output value of each neuron in the network. The network error on the current training pattern is calculated in line 11. In lines 12-17 the δ values of each neuron in the output and hidden layers are computed and each w_{ij}^l is updated accordingly. This online version of backpropagation will continue until the termination criteria are satisfied.

Extending backpropagation to recurrent neural networks is possible through the backpropagation through time algorithm [52]. In this approach, the recurrent connections are unfolded such that the network is essentially feedforward. Then, we can use Algorithm 1 to train the unfolded network.

As with other learning algorithms, backpropagation is not perfect. It is very sensitive to its initial conditions, i.e. the random weight settings or transfer function type, amongst others [46]. A poor initial network, an ill-conditioned one, typically will exhibit premature convergence towards a poor input/output mapping. To aid in alleviating this problem other learning algorithms utilize second-order information regarding the curvature of the error surface. Other approaches include adaptive learning rate and momentum values [12, 27, 33] to allow the learning algorithm to make larger weight updates to avoid flat areas and better explore the error surface. For more information regarding these strategies the interested reader is referred to [3, 4, 23].

12.2.2   Adaptive Transfer Functions

Typically the transfer functions used by a neural network are arbitrarily chosen and fixed before training begins. However, the Universal Approximation Theorem (see Appendix A) does not particularly favor any specific sigmoid function [19, 25]. One approach is to employ a global search algorithm to determine a better set of transfer functions before learning, however this approach tends to be very time consuming [32]. Another approach is to use additional parameters in the transfer functions which can be adapted during the learning process.

Figure 12.1 shows an example of different slope parameterizations of the logistic function, which is commonly utilized in neural network learning. Specifically, the function has the form

$$\phi(x) = \frac{1}{1 + ae^{-mx}} \tag{12.6}$$

where m is a parameter controlling the slope of the function and a controls the gain. By learning the slope and gain for each transfer function in the network it is possible to

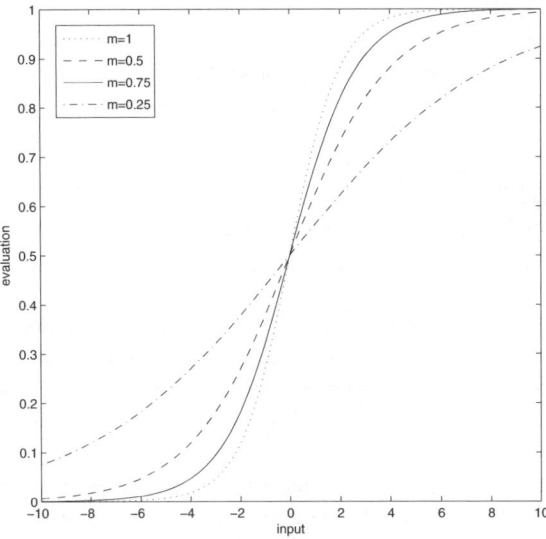

Fig. 12.1. Different slope parameterizations of the logistic function

improve on the accuracy and learning rate of backpropagation-based learning algorithms [26, 44].

Various researchers have investigated justifications for the use adaptive transfer functions. Chandra and Singh [6] hypothesize that for a specific data set there may be a preferred type of transfer function. However, deciding this specific function is too complex a task for large real world problems and therefore they raise a case for the use of transfer functions which can adapt during learning to better suit the problem. This hypothesis was also echoed by Xu and Ming [56] and by Hu and Shao [26], who also report a decrease in the complexity of the network (i.e the number of hidden neurons required to perform the given task).

Adaptive sigmoid functions have been investigated in [5]. Their proposed algorithm was shown to outperform traditional backpropagation by up to an order of magnitude. The addition of a gain parameter to adjust sigmoid networks was proposed in [29]. The method showed an increase in network performace. And in [43] another adaptive transfer function network is proposed which yields a faster convergence rate and accuracy for classification of echocardiogram data than networks with fixed sigmoid functions.

A comparison was performed in [10] between single hidden layer networks with and without adaptive transfer functions for static and dynamic patterns. It was found that utilizing adaptive transfer functions was able to model the data more accurately.Improved generalization and robustness were also observed using adaptive functions in multiresolution learning [31].

The special case of learning the amplitude of transfer functions was considered in [45]. Using a variety of function approximation, classification and regression tasks it was shown that the method yielded an increase in convergence rate and a possible improvement in generalization ability by finding better areas of the error surface. This work was extended to recurrent networks using a real-time recurrent learning algorithm and showed similar improvements [20]. Another work on adaptive amplitude transfer functions focused on complex-valued nonlinear filters [22].

Higher-order adaptive transfer functions were considered in [55]. They reported an increased flexibility, reduced network size, faster learning and lower approximation errors.

Vecci et al. investigate a special class of neural networks which are based on cubic splines [47]. The method adapts the control points of a Catmull-Rom spline. It is shown that this model has several advantages including lower training times and reduced complexity.

Hoffmann [24] proposes universal basis functions (UBFs) which are parameterized to allow for a smooth transition between different functional forms. The UBFs can be utilized for both bounded and unbounded subspaces. Hoffmann's experimental results show an improvement over radial basis functions for some common benchmark data sets.

These works show that adaptive transfer functions can lead to improvements in accuracy, robustness and generalization of neural networks. Adjusting the parameters of the neural network structure represents a structural transformation. It is possible that some transformations do not alter the network input/output mapping, making them irrelevant.

12.2.3   Symmetric Structural Transformations

Symmetry refers to a physical system's ability to remain unaffected by some transformation. A neural network represents a mapping $\Psi : \Re^n \mapsto \Re^q$ from an n-dimensional space of input features to a q-dimensional space of target values or concepts. Symmetry in neural networks typically refers to a transformation T in the network structure or free parameters which does not affect the input-output mapping. That is, Ψ is invariant to the transformation such that $T(\Psi) = \Psi$. Two networks $\mathcal{N}_1$ and $\mathcal{N}_2$ representing the same mapping are denoted as $\mathcal{N}_1 \sim \mathcal{N}_2$ ($\nsim$ is used for different mappings). In this work we concentrate on structural symmetry as it is concerned with transfer functions, but a more thorough examination can be found in [2].

Structural symmetries can arise as a result of permuting neurons within a hidden layer or by a sign inversion of a transfer function. Permutation is possible by exchanging all the input and output connections within a set of neurons from a specific layer. This permutation transformation does not affect the output of the network, and thus is a symmetric transformation. Given n neurons, a total of $n!$ permutations are possible [8].

The other coherent transformation operates directly on the transfer function and is known as a sign transformation. Given some transfer function having odd symmetry (i.e. $\phi(x) = -\phi(-x)$), multiplying all input and output weights by -1 will result in an invariant input/output mapping [8]. It has been shown that this specific symmetry is valid for any infinitely differentiable function where each successively differentiated function evaluates to zero [1].

On the other hand, if the transfer function exhibits even symmetry (i.e. $\phi(x) = \phi(-x)$) then multiplying all input connections by -1 also leaves $\Psi(\mathcal{N})$ unchanged. This symmetry is also valid for an infinitely differentiable function [1], of which the most common is the radial-basis transfer function. For either even or odd transfer functions, given a layer of n non-input neurons there exists 2^n possible sign-symmetric transformations.

The following remark summarizes the aforementioned symmetries [8]:

Remark 1. The set of all equi-output transformations on the weight space $\mathcal{W}$ forms a non-Abelian group G of order $\#G$, where

$$\#G = \prod_{l=2}^{L-1}(m_l!)(2^{m_l}) \tag{12.7}$$

where L is the number of non-input layers and m_l is the number of neurons in layer l.

Each of these strutural transformations defines a symmetry in weight space consisting of equivalent parts. By taking these symmetries into consideration it is possible to reduce the size of the weight space [1, 2, 8, 9, 28, 42]. Considering non-symmetric transformations may also be beneficial to neural network learning. Specifically, we are concerned with even-signed transformations on odd-signed transfer functions.

12.3 Opposite Networks

Before defining an opposite network we discuss the concept of an opposite transfer function (OTF). The underlying idea behind OTFs is to provide a means for altering the network structure such that knowledge stored in connection weights is retained but the input/output mapping differs. That is, viewing the purpose of each non-input neuron as a "decision maker" based on the given signal, what effect does reversing the decision have on the input-output mapping? We will show that this specific form of transfer function guarantees a different mapping when the networks are minimal.

12.3.1 Opposite Transfer Functions

Dynamically adjusting transfer function parameters, and thus modifying the error surface and input/output mapping has, of course, been investigated. Similarly, many alternative transfer functions have also been proposed, refer to [51] for a survey. However, most of these methods or alternative transfer functions increase the search space size by (a) defining a set of transfer functions which can be used instead of a single function or, (b) increasing the parameterizations of the functions or, (c) infusing more parameters into the learning algorithm. As we will see below opposite transfer functions do not imply that the size of search space increases. Although, in practice this may not necessarily be the case. Nevertheless, OTFs will be shown to be useful and warrant further investigation.

Essentially, an OTF is an even transformation of an odd transfer function and can be defined as follows [48]:

Definition 1 (Opposite Transfer Function). *Given some odd-symmetric transfer function $\varphi\colon \Re \mapsto \Re$, its corresponding opposite transfer function is $\breve{\varphi}(x) = \varphi(-x)$, where the breve notation indicates the function is an opposite and $x \in \Re$.*

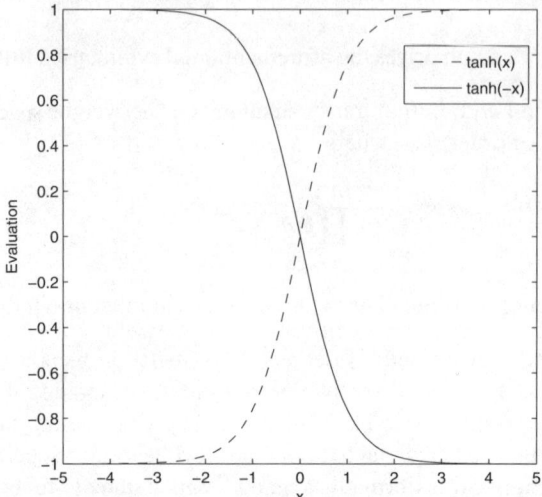

Fig. 12.2. Comparing the tanh(x) and its opposite tanh(-x)

According to this definition, the relationship between a transfer function and its corresponding opposite is not odd (i.e. $\varphi(-x) \neq -\breve{\varphi}(x)$). This will be a key feature which many useful properties of OTFs are discussed in the next section. An example of a tanh function and its opposite are presented in Figure 12.2.

From Definition 1 we notice that the transformation is equivalent to multiplying all input weights to a neuron by -1, but not the output signal, as is the case in a sign transformation. This new weight configuration lies in the same weight space as the original network, and thus does not increase the size of the search space. Actually, the OTF is simply a means to consider a different, yet related location in the existing search space.

In order to be useful in backpropagation-like learning algorithms the following characteristics of an OTF should be present:

1. Both $\varphi(x)$ and $\breve{\varphi}(x)$ are continuous and differentiable.
2. For derivatives we have $\frac{d\breve{\varphi}(x)}{dx} = -\frac{d\varphi(x)}{dx}$.

Extrapolating on Definition 1 we can now define the concept of an opposite network $\breve{\mathcal{N}}$ and the set of opposite networks $\Gamma(\mathcal{N})$ [48]:

Definition 2 (Opposite Network). *Given some minimal neural network $\mathcal{N}$ the corresponding set of opposite network(s) $\Gamma(\mathcal{N})$ is defined as all networks having identical connection structure and weight values, but differing in that at least one transfer function is in an opposite state. Any opposite network is referenced as $\breve{\mathcal{N}}$ where it is understood that $\breve{\mathcal{N}} \in \Gamma(\mathcal{N})$.*

Inherent in this definition is a notion of "degree of opposition". That is, if two networks differ by only one transfer function then they are less opposite than two networks differring in more than two functions. In this scenario, true opposite networks differ in all transfer functions.

12.3.2 Properties of Opposite Networks

While almost naive in concept, opposite networks have useful properties. The fact that given any irreducible neural network we can easily translate between any opposite network (each having a unique input-output mapping) forms the basis of the benefits of opposite networks. Furthermore, the transformation requires constant computation time (i.e. $\mathcal{O}(1)$) since only a single multiplication operation is required per transfer function.

Uniqueness of Input-Output Mapping

Aside from symmetrical transformations, it is possible for $\mathcal{N}_1 \sim \mathcal{N}_2$ if one network can be reduced to the other [42]. For example, if there exists some neuron $\eta \in \mathcal{N}$ which has all outgoing weights equal to zero. Then, the removal of η does not affect the input-output mapping of $\mathcal{N}$.

A formal definition of minimality, or equivalently, irreducibility has been given in [42]:

Definition 3 (Irreducibility). *A feedforward neural network with m input nodes and one hidden layer of n hidden neurons can be called irreducible if none of the following is true:*

1. *One of the $v_{j,k}$ vanishes.*
2. *There exists two indices $j_1, j_2 \in \{1, ..., n\}$ where $j_1 \neq j_2$ such that the functionals ψ_{j_1}, ψ_{j_2} are sign-equivalent[1].*
3. *One of the functionals ψ_j is constant.*

where ψ_j is the total input signal to neuron j. An important consequence of minimality is that every minimal network represents a unique input/output mapping [1, 42].

Using the fact that OTFs represent a non-symmetric transformation it is possible to prove the following theorem [48]:

Theorem 1 (Opposite Network Irreducibility). *Let $\mathcal{N}$ be a minimal network having $|\Gamma(\mathcal{N})| \geq 1$. Then, for $S = \{\mathcal{N}\} \cup \Gamma(\mathcal{N})$, $s_i \nsim s_j \ \forall \ s_i, s_j \in S$ where $i \neq j$ (i.e. all the networks are minimal and represent a unique input-output mapping).*

While the mappings for each network are unique, it must also be shown that there does not exist a priori bias or tendency for a certain set of transfer function states to yield a lower error, over all possible unknown data sets. Otherwise, there is no use for considering OTFs, or any other transfer function for that matter. We will consider only the case for OTFs versus similar fixed functions.

In order to prove the equi-probably mapping property we assume the underlying distribution of the input data $\mathbf{X}$ is unknown but bounded to $[-1, 1]$. We also make the

---

[1] Two functions $f_1(x), f_2(x)$ are sign-equivalent if $f_1(x) = f_2(x)$ or $f_1(x) = -f_2(x) \ \forall x \in \Re^d$ where d is the dimensionality of the space.

conjecture that for any unknown $\mathbf{X}$, and any unknown $\mathcal{N}_1$ and $\mathcal{N}_2$, both minimal, but having exactly the same architecture such that both networks lie in the same search space then we have

$$P(E_{\mathcal{N}_1}(\mathbf{X}) \leq E_{\mathcal{N}_2}(\mathbf{X})) = P(E_{\mathcal{N}_2}(\mathbf{X}) \leq E_{\mathcal{N}_1}(\mathbf{X})) = 0.5. \qquad (12.8)$$

That is, over all possible data sets there is no predisposition for one mapping to yield a more favorable outcome, with respect to all error functions E. This is similar to the No Free Lunch Theorem [54], but specifically for neural networks existing in the same search space.

Theorem 2 (Equi-Probable Input-Output Mapping [48]). *Let $\mathcal{N}$ be a minimal neural network with opposite networks $\Gamma(\mathcal{N})$. Without a-priori knowledge concerning $\mathbf{X}$ and for some $s^* \in S$,*

$$P(s^* = \min(S)) = \frac{1}{|S|}$$

where,

$$|S| = \prod_{l \in L} 2^{m_l},$$

where L corresponds to the number of layers which can utilize opposite transfer functions, each having m_l neurons and $S = \Gamma(\mathcal{N}) \cup \{\mathcal{N}\}$.

Therefore, before learning begins there is no inherent bias towards choosing a random network, or one of its opposites (with respect to the error function). While this holds for a random location in weight space, the rate at which this probability decays for each network during learning has not be determined analytically. In [48] we provide experimental evidence to support an exponential increase in probability using some common benchmark data sets.

Numerical Condition

Numerical condition is an important and fundamental concept affecting the speed and accuracy of neural network learning algorithms [4]. It refers to the sensitivity of the network output to changes in its weights and biases. Typically, conditioning is most important when initializing network weights since learning algorithms can rapidly determine a local minimum of a specific location on the error surface. Thus, an ill-conditioned network may require long training times or could converge to a poor solution. Therefore, techniques to alleviate or lessen the degree of ill-conditioning are important to neural learning, and numerical optimization in general.

Aside from the initialization phase of learning, the conditioning will change during training when nonlinear hidden or output units are employed [40]. So, it is also possible to improve conditioning during, especially early stages, of training. As such, some learning algorithms have been developed which are capable of adapting to an ill-conditioned situation, the most popular are resilient propagation (rProp) [38], quick propagation (quickProp) [18], conjugate gradient-based [23] and quasi-Newton-based [23] algorithms. Despite their ability to adapt to ill-conditioned situations, these techniques will still tend to yield more desirable results in well conditioned scenarios.

Common approaches to aid in alleviating the detriments of ill-conditioning include data preprocessing, weight initialization and regularization [4, 23, 35, 41]. Investigating the impact of transfer functions on the error surface has also been researched recently [13, 14, 15]. More importantly for this work is the consideration of adaptive transfer functions as a possible means to help alleviate ill-conditioning and improve accuracy and training times [29, 44].

One common measure related to conditioning is the rank of the Jacobian, **J**. This matrix is composed of first derivates of the residual error for each pattern $x \in \mathbf{X}$ with respect to the weights, **w**. For a network with one output neuron **J** is computed as

$$\mathbf{J} = \left[\frac{\partial e(\mathbf{X})}{\partial w_i} \right]_{i=0...,|\mathbf{w}|}, \tag{12.9}$$

where $e(x) = d(x) - \phi(x)$ is the residual error between the target output $d(x)$ and actual network output $\phi(x)$. The rank of an $m \times n$ matrix **J** represents the number of linearly independent rows or columns. Numerical condition occurs when rank$(\mathbf{J}) < \min(m, n)$.

Rank deficiency of the Jacobian is related to the concept of ill-conditioning [40, 46]. For backpropagation-like algorithms, a rank deficient Jacobian implies only partial information of possible search directions is known. This can lead to longer training times. Furthermore, many optimization algorithms such as steepest decent, conjugate gradient, Newton, Gauss-Newton, Quasi-Newton and Levenberg-Marquardt directly utilize the Jacobian to determine search direction [23, 40]. So, these algorithms will also likely exhibit slower convergence rates and possibly less desirable error than otherwise would be possible.

Since each opposite network represents a unique input/output mapping, it is likely that their respective Jacobians are also different. This will be true when the error surface itself is not symmetric about the origin (this symmetry is very rare in practice). Considering an error surface that is not symmetrical about the origin, the difference between the respective Jacobians of a random network $\mathcal{N}$ and one of its opposites $\check{\mathcal{N}}$ is computed according to

$$\mathbf{\Delta}^J = \mathbf{J}(\mathcal{N}) - \mathbf{J}(\check{\mathcal{N}}). \tag{12.10}$$

It must follow that $|\{\delta_{i,j}^{\mathbf{J}} \neq 0\}| \geq 1$ for $\delta_{i,j}^{\mathbf{J}} \in \mathbf{\Delta}^J$. Due to this property the rank$(\mathbf{\Delta}^J) > 0$, which may also lead to a difference in rank$(\mathbf{J}(\mathcal{N})) - $ rank$(\mathbf{J}(\check{\mathcal{N}}))$. If this value is negative, it may be beneficial to consider training $\check{\mathcal{N}}$ instead of $\mathcal{N}$.

The other popular conditioning measure is related to the Hessian matrix **H**, which represents the second derivatives of $Er(\mathbf{X})$ with respect **w**,

$$\mathbf{H} = \left[\frac{\partial^2 Er(\mathbf{X})}{\partial w_i w_j} \right]_{i,j=0...,|\mathbf{w}|}. \tag{12.11}$$

The Hessian is very important to nonlinear optimization as it reveals the nature of the curvature of the error surface. Most importantly, the eigenvalues of **H** have a large impact on the learning dynamics of backpropagation-like second-order algorithms [4, 23]. Also, the inverse $\mathbf{H}^{-1}$ has also been used in network pruning strategies [30].

For neural networks using a sum-of-squares based performance function it is common to computer

$$\mathbf{H} = \mathbf{J}^T \mathbf{J}. \tag{12.12}$$

We have already argued the expected change $\Delta^{\mathbf{J}}$. Since (12.12) directly uses $\mathbf{J}$, a change in the Jacobian will also lead to a change in the Hessian

$$\Delta^H = \mathbf{H}(\mathcal{N}) - \mathbf{H}(\check{\mathcal{N}}) \tag{12.13}$$

where there exists some $\delta_{i,j}^{\mathbf{H}} \in \Delta^{\mathbf{H}}$ such that $\delta_{i,j}^{\mathbf{H}} \neq 0$. Depending on the quantity and magnitude of each $\delta_{i,j}^{\mathbf{H}} \neq 0$, the difference between the two positions in weight space (i.e. $\mathcal{N}$ and $\check{\mathcal{N}}$) could be significant enough to warrant moving the search to that location.

As with rank($\mathbf{J}$), the conditioning of $\mathbf{H}$ has a profound impact on the learning time and accuracy of the learning algorithm. The most common measure of the condition of $\mathbf{H}$ is the condition number,

$$\kappa = \frac{\lambda_{max}}{\lambda_{min}} \tag{12.14}$$

where λ_{max} and λ_{min} are the largest and smallest nonzero eigenvalues of $\mathbf{H}$, respectively. The larger this ratio, the more ill-conditioned the network is. So, if some $\check{\mathcal{N}}$ is has a larger κ value than $\mathcal{N}$, it could be beneficial to consider training the opposite network instead.

Experiments concerning the numerical conditioning of opposite networks prior and during early stages of training have been conducted in [48]. It was shown that considering opposite networks can have a substantial influence on the Hessian, and a to a lesser degree the Jacobian. Nevertheless, results show that considering opposite networks seems promising and warrants further examination.

12.4   Proposed Framework

We present two frameworks for utilizing opposite networks during neural network training. The methods are presented in a general form and are flexible enough to work with most learning algorithms. As an example of the first famework, we discuss a method to improve backpropagation through time. The second example focuses on training feed-forward networks with backpropagation.

12.4.1   Global Decision-Based Framework

By a global decision-based algorithm we mean an approach which bases decisions on the entire neural network structure, its input/output mapping and related properties (such as the Jacobian and Hessian). The fundamental concept is that training is performed on a single network at each epoch, but to consider that another (opposite) network which may actually be more desirable with respect to some criteria, $\mathcal{F}$.

Algorithm 2 provides the general framework for the global decision-based approach. In line 3 the current network is trained using any learning algorithm. After a single epoch or iteration of the learning algorithm each opposite network is evaluated (line 4) on the training and/or validation data used to train $\mathcal{N}$. Finally, in line 5 we determine on which network, either $\mathcal{N}$ or one of $\gamma \in \Gamma(\mathcal{N})$ (since $S = \{\mathcal{N}\} \cup \Gamma(\mathcal{N})$), learning should continue. The function *argdes* returns the argument with the most desirable result with respect to $\mathcal{F}$, whether it be the minimum or maximum. It should be noted that $\mathcal{F}$ represents a (heuristic) decision as to which network training should continue on, and therefore could involve more variables than only the network error performance measure.

Algorithm 2. Global Decision-Based Framework

1: generate initial network $\mathcal{N}$
2: **while** termination criteria not satisfied **do**
3:    perform n iterations of learning on $\mathcal{N}$
4:    evaluate $\Gamma(\mathcal{N})$
5:    set $\mathcal{N} = \underset{s}{\text{argdes}} \, \mathcal{F}(s \in S)$
6: **end while**

There are three main concerns which must be addressed in order for this framework to be practical:

1. Minimizing the computational cost associated with examining $\Gamma(\mathcal{N})$.
2. Maximizing the usefulness of the decision function $\mathcal{F}$.
3. Minimizing the computational time for evaluating $\mathcal{F}$.

Minimizing the computational cost associated with the examination of the elements of $\Gamma(\mathcal{N})$ is very important. By Definition 2, $\Gamma(\mathcal{N})$ includes all possible opposite networks. If a large network with many hidden nodes and multiple hidden layers is being trained this set will be very large. So, evaluating each opposite network by processing all training data will be impractical, although for small networks may be possible. Therefore, to limit the computational cost, methods for determining the output of an opposite network given the original network output would be very beneficial. Since the relationship between $\mathcal{N}$ and each $\breve{\mathcal{N}}$ is known and strictly defined it seems reasonable to believe that a method should exist to directly determine the output for a given opposite network.

The design of an efficient and robust decision function $\mathcal{F}$ is vital because it determines the quality of all $s \in S$ and decides which one is the most desirable network to continue learning with. This function may be simple, for example a purely random decision or solely based on the evaluation of each network performed in line 4. Alternatively, more advanced information such as the Hessian, Jacobian or heuristic information based on the network architecture could be included. Nevertheless, the goal is to determine which network/location in weight space is most desirable to continue learning from.

As $\mathcal{F}$ becomes more complex, it will include more detailed information of each network. The gathering and evaluation of this information may also be time consuming and thus a tradeoff between information and computational efficiency may need to be made. Additionally, if the output of each opposite network can be determined without direct evaluation then it should also be possible to directly determine the Jacobian and Hessian matrices, further saving computation time.

12.4.2   Example: Backpropagation Through Time

The opposition-based backpropagation through time (OBPTT) algorithm [49] which extends the global decision-based framework. The general idea is based on utilizing probabilities to represent the state of each neuron transfer function, which in turn are employed when deciding which opposite networks should be considered. As learning progresses these probabilities are adjusted such that only the more likely networks are examined, until eventually only a single network remains (i.e. the probability of using $\phi(\cdot)$ converges to either 1 or 0).

The OBPTT pseudocode is presented in Algorithm 3. In line 2 the $\mathcal{Q} \subset \Gamma(\mathcal{N})$ networks to be considered are generated. The connection weights are all exactly the same as those of $\mathcal{N}$, however, the transfer function employed by each neuron is probabilistically decided when each $q \in \mathcal{Q}$ is generated. So, each hidden neuron $\eta_i \in \mathcal{N}$ has an associated probability $p(\eta_i)$, initially equal to 0.5 due to lack of information regarding the better of the two transfer functions, where each network transfer function $\phi_i(\cdot)$ is decided according to

$$\phi(\cdot)_i = \begin{cases} \phi(\cdot), & \text{if } r \geq p(\eta_i) \\ \breve{\phi}(\cdot), & \text{otherwise} \end{cases} \qquad (12.15)$$

where $r \in \mathcal{U}(0, 1)$ is a uniformly generated random number. For simplicity we also use the notation $\eta_i(\phi) = \textbf{true}$ or $\eta_i(\phi) = \textbf{false}$ if the transfer function for the ith neuron is in the original or opposite state, respectively.

Upon generating the subset of candidate opposite neurons, the backpropagation algorithm is then executed on each network in lines 3-6. In this manner, we can determine which network exists on a steeper area of the error surface. That is, given the current location in weight space represented by each network, which network will yield the lowest error after a single epoch of backpropagation. In lines 7 and 3, we greedily select this lowest error network.

After the best network $\mathcal{N}^*$ is selected, the transfer function probabilities need to be updated such that when generating the next $\mathcal{Q}$, the tendancy is towards networks which have already yielded low errors. This is accomplished in lines 9-14 using the user-defined control parameters τ_{ampl} and τ_{decay} which define the rate at which each $p(\eta_i)$ converges towards or away from ϕ or $\breve{\phi}$, respectively.

It is important to note that $|\mathcal{Q}| \to 0$ as the probabilities converge towards 1 or 0, respectively. Therefore, in line 2 when attempting to generate the set $\mathcal{Q}$, eventually with a high probability each $q \in \mathcal{Q}$ will equal $\mathcal{N}$, which is not permissible (i.e. a network cannot be equal to itself and its opposite simultaneously). In this situation, OBPTT degrades to traditional backpropagation, and in general towards the parent algorithm being employed.

Algorithm 3. Opposition-Based Backpropagation Through Time

1: **while** termination criteria not satisfied **do**
2:   generate $\mathcal{Q} \subset \Gamma(\breve{N})$

3:   backpropagate_one_epoch($\mathcal{N}$)
4:   **for all** $q \in \mathcal{Q}$ **do**
5:     backpropagate_one_epoch(q)
6:   **end for**

    {Find the network with minimum error}
7:   $S = \{\mathcal{N}\} \cup \mathcal{Q}$
8:   $\mathcal{N}^* = \underset{s \in S}{\operatorname{argmin}}\, E(s)$

    {Update probabilities}
9:   **for all** $\eta_i \in \mathcal{N}^*$ **do**
10:     **if** $E(\mathcal{N}^*) - E(\mathcal{N}) \leq 0$ **then**
11:       $p(\eta_i) = \begin{cases} \tau_{ampl} \cdot p(\eta_i) & \text{if } \eta_i(\phi) = \textbf{true} \\ 1 - \tau_{ampl} \cdot p(\eta_i) & \text{if } \eta_i(\phi) = \textbf{false} \end{cases}$
12:     **else**
13:       $p(\eta_i) = \begin{cases} \tau_{decay} \cdot p(\eta_i) & \text{if } \eta_i(\phi) = \textbf{false} \\ 1 - \tau_{decay} \cdot p(\eta_i) & \text{if } \eta_i(\phi) = \textbf{true} \end{cases}$
14:     **end if**
15:   **end for**

16:   $\mathcal{N} = \mathcal{N}^*$
17: **end while**

12.4.3   Local Decision-Based Framework

Local decision-based methods are based on decisions made at the local (i.e. neuron) level of the network. Here, each neuron decides whether to output the original transfer function value, $\phi(\cdot)$ or the opposite value $\breve{\phi}(\cdot)$.

The general idea behind the local framework is presented in Algorithm 4, where it is assumed to be utilized during the forward propagation step of backpropagation. The

---

Algorithm 4. Local Decision-Based Framework

1: {during forward propagation}
2: **for** each layer, l=1 to L **do**
3:   **for** each neuron, $j = 1 \ldots |l|$ **do**
4:     $v_j^l = \sum\limits_{i=0}^{|l-1|} w_{ij}^l y_i^{l-1}$
5:     $y_j^l = \begin{cases} \phi(v_j^l) & \text{if } \mathcal{G}(l, j) = \textbf{true} \\ \breve{\phi}(v_j^l) & \text{otherwise} \end{cases}$
6:   **end for**
7: **end for**

---

input to the neuron is computed, as usual, in line 4 and the output of each neuron is computed in line 5. The decision between the output of $\phi(\cdot)$ or $\breve{\phi}(\cdot)$ is made using the local decision function $\mathcal{G}(l, j)$. This function takes as input the current neuron, which can send a "message" to any neuron sending or receiving signals to/from it. Therefore, the neuron is restricted to local information between itself and the direct previous and post neurons.

The only difference between traditional forward propagation and this modified version is the addition of $\mathcal{G}(l)$. However, this function is very loosely defined and can have a profound impact on the learning process. Given that a neuron only has access to its internal parameters (total input signal, previous input/output signals, local gradient, etc) and to connected neurons, designing an efficient $\mathcal{G}(l)$ may be complicated. Furthermore, knowledge of the local influence on the input/output mapping of the network must also be accounted for such that a higher (global) decision function is not required to determine the quality of the local change.

12.4.4 Example: Backpropagation

As an example of a local decision-based learning approach we extend backpropagation for feedforward networks, we call this opposition-based backpropagation. Here, we are really only concerned with the decision $\mathcal{G}(l, j)$ employed by each neuron in line 5 of Algorithm 4, where j is the jth neuron in layer l. The rule presented here is based on the expected change in network output if a transfer function is changed to its opposite. It should be noted that if a transfer function is currently $\breve{\phi}$ then its opposite will of course be the original ϕ.

Algorithm 5 presents the pseudocode for the local decision rule $\mathcal{G}(l, j)$. To allow the learning process to customize to a transfer function configuration we define a minimum number of epochs between transfer function changes, λ. In line 1, the decision is considered if λ epochs have passed between the current epoch i and last transfer function change ω.

The first step of the decision is to determine the change in neuron output if the transfer function is changed (line 2). The idea behind this, is that if the output y is increasing then the associated input weights must have increased (since the data is constant). So, if a change in transfer function expedites the weight update process, then it should accelerate learning. To accomplish this, we compute the average input to a neuron according to

$$\bar{\mu} = \frac{1}{P} \sum_{p \in X} I \tag{12.16}$$

where I is the total input to (l, j) for each pattern.

To determine the current direction of weight updates, in line 3 we compute the sum of the difference in all outgoing neuronal outputs over the last two epochs. The change in output is Δy_o, where $o \in O$ and the set O corresponds to output neurons from the current neuron. In line 4 we then approximate the total change in output neuron response. That is, the estimated difference between the current output neuron signals and their output if neuron j changed its transfer function.

The decision in lines 5-7 whether to change the transfer function is based on using the results from lines 3 and 4. If both values have the same numerical sign, then a transfer

---

Algorithm 5. Local neuron decision rule $\mathcal{G}(l, j)$

Require: $\omega \equiv$ last epoch with transfer function change for neuron (l, j)
Require: $i \equiv$ current epoch
    {Ensure minimum number of epochs between transfer function change}
1: **if** $i - \omega > \lambda$ **then**
       {Determine difference in output between ϕ and $\check{\phi}$}

2:    $\Delta\phi(\bar{\mu}) = \begin{cases} \check{\phi}(\bar{\mu}) - \phi(\bar{\mu}) & \text{if } \eta_i(\phi) = \textbf{true} \\ \phi(\bar{\mu}) - \check{\phi}(\bar{\mu}) & \text{if } \eta_i(\phi) = \textbf{false} \end{cases}$

       {Compute the direction output neuron signals are moving in}

3:    $\psi = \sum_{o \in O} \Delta\phi_o$

       {Compute expected change in total output neuron response}

4:    $\Delta\psi = \sum_{o \in O} [\phi_o(\bar{\mu}_o) - \phi_o(\Delta\phi(\bar{\mu})w_{j,o})]$

       {Decide whether to change transfer function}
5:    **if** $\text{sign}(\psi) = \text{sign}(\Delta\psi)$ **then**
6:       change transfer function
7:    **end if**
8: **end if**

---

function change is made. Otherwise, the function is kept as it was. To limit transfer function changes which do not yield an improvement in overall network performance it is possible to require that the network error measure also show a beneficial affect. Furthermore, we also require that selecting a neuron to change its transfer function is based on a probabilistic decay.

12.5 Experimental Results

In this section we show how BP and BPTT can be improved by considering the two opposition-based frameworks. First the OBPTT approach is discussed and then we provide results for OBP. In both cases we show statistically signifant improvements over the original algorithms.

12.5.1 Opposition-based Backpropagation Through Time

We employ the OBPTT algorithm discussed in Section 12.4.2 and test it using the embedded Reber grammar benchmark problem [37]. The results of our approach are compared to traditional backpropagation through time over different hidden layer sizes and learning rate and momentum values. These findings are a summary of those found in [49].

    We restrict the experimental results to the Elman recurrent topology [17], as depicted in Figure 12.3. In this topology, the hidden layer outputs feed back into itself via a context layer, representing the previous time step. The context layer is thus a copy of the hidden layer from the previous time step. So, adding more context layers allows the

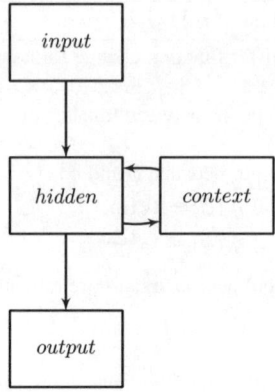

Fig. 12.3. The Elman topology in which the hidden layer has recurrent connections which feed back to itself

network to consider states further back in time. Due to this structure Elman networks can represent time-varying patterns, making them very useful for modeling dynamic problems.

The presented results assume a single hidden layer network, where only the hidden layer neurons are permitted to consider the logistic or opposite logistic transfer functions. The quality of the output is determined by the mean-squared error measure and the parameters $\tau_{decay} = 0.1$ and $\tau_{ampl} = 0.75$ were found to yield the best results for the Reber grammar problem.

To ensure a proper comparison between the approaches, both BPTT and OBPTT algorithms begin from the exact same starting weights, as determined by the Nguyen-Widrow rule [35]. In this manner, we can better show that any improvements must be due to the use of OTFs, and not a result of randomness in initial conditions. We train each network for 30 epochs, and average the results over 30 trials. During each epoch of OBPTT the maximum number of opposite networks under consideration was set to $|\mathcal{Q}| = 3$.

Embedded Reber Grammar

A Reber grammar [37] is a deterministic finite automaton representing some language, $\mathcal{L}$, where an element of the language is generated according to the rules which define the Reber grammar. We can also use the Reber grammar/automaton to determine whether a given string is invalid, with respect to the rules of the language.

It is possible to use an Elman network to represent the automaton of a grammar [11, 36]. The training data is composed of valid and optionally invalid strings. The final trained network should then be able to distinguish whether an unknown string is an element of $\mathcal{L}$ or not. Figure 12.4 shows a finite automaton representation of a Reber grammar having alphabet, $\Sigma = \{B, P, T, S, X, V, E\}$.

An embedded Reber grammar is a more difficult extension of the Reber grammar. These grammars are of the form $n\mathcal{L}m|a\mathcal{L}b$ where $n, m, a, b \in \mathcal{L}$ are unique strings.

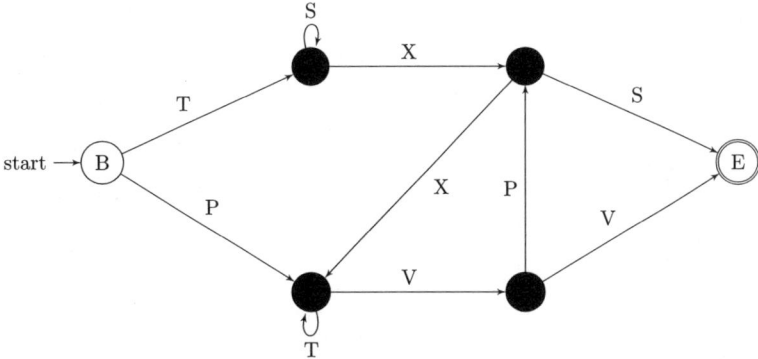

Fig. 12.4. An automaton representation of a Reber grammar which begins with the symbol B and ends with the symbol E

This tasks is much more difficult because the neural network must model the initial sequence over a greater number of time steps than the simple grammar shown in Figure 12.4.

The experiments shown in this chapter use the above language of 6 symbols. We convert each symbol to a 6-bit binary string where each bit represents one of the six symbols. So, each training string contains a single '1' bit, corresponding to one of the respective original symbols. In total our training data consists of 8000, 6-bit embedded Reber grammar strings. The goal of this learning task is to be able to predict the next symbol in the training data, as opposed to test whether a given string is indeed valid with respect to the rules of the grammar. Thus, the output layer also contains 6 neurons, representing each of the 6 symbols of Σ.

Considering Opposite Networks

At each iteration of OBPTT we train a maximum $|\mathcal{Q}| = 3$, as well as the original $\mathcal{N}$ network. So, there is an associated computational overhead for OBPTT that manifests itself into lost epochs which could have been performed by BPTT. However, since $|\mathcal{Q}| \to 0$ as discussed above, the practical number of additional epochs may be small. The additional computations are intended to search for better locations on the error surface, and thus may be acceptable.

The number of considered opposite networks is plotted against the number of epochs in Figure 12.5. We also present the average number of networks which were found to improve the overall network error.

From Figure 12.5, we see that on average about 20% of the networks in $\mathcal{Q}$ have a lower error than $\mathcal{N}$. Keeping in mind that $\mathcal{Q}$ is randomly selected from $\Gamma(\mathcal{N})$ this result is rather promising. A more intelligent strategy is likely to increase the usefulness of $\mathcal{Q}$, however, the current method still yields improvements which will be described below.

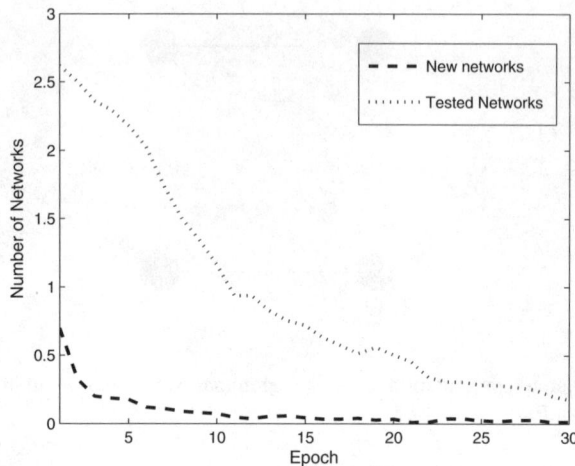

Fig. 12.5. The average usefulness of Q over 30 trials for a network with 7 hidden neurons

Learning Rate and Momentum

In this set of experiments we aim to examine the behavior of OBPTT for different settings of learning rate (α) and momentum (β). To better see the improvement we compare to traditional BPTT.

Table 12.1 shows the results of the both BPTT and OBPTT when the hidden layer size is fixed at 5 neurons. Each experiment varied the learning rate and momentum from values of $\{0.10, 0.25, 0.50, 0.75\}$ an $\{0.00, 0.10, 0.25, 0.50\}$, respectively. Also, we only consider the situation where $\alpha > \beta$. The results show the mean (μ) and standard deviation (σ) of the MSE averaged over the 30 trials.

Table 12.1. Varying learning rate and momentum, bolded values are statistically significant. All OBPTT values are found to be significant.

| | | OBPTT | | BPTT | |
α	β	μ	σ	μ	σ
0.10	0.00	**0.040**	0.002	0.045	0.004
0.25	0.00	**0.040**	0.002	0.045	0.004
0.25	0.10	**0.040**	0.002	0.043	0.003
0.50	0.00	**0.042**	0.003	0.046	0.003
0.50	0.10	**0.042**	0.003	0.046	0.004
0.50	0.25	**0.042**	0.001	0.046	0.004
0.75	0.00	**0.043**	0.002	0.047	0.004
0.75	0.10	**0.044**	0.003	0.047	0.005
0.75	0.25	**0.045**	0.004	0.050	0.005
0.75	0.50	**0.046**	0.003	0.051	0.006

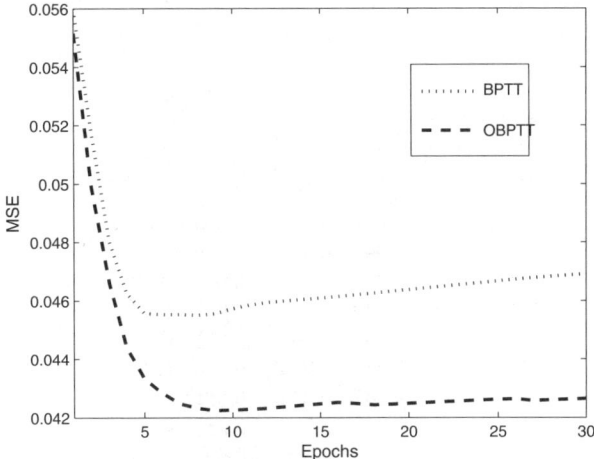

Fig. 12.6. Comparing BPTT and OBPTT over the 30 trials where $\alpha = 0.1$ and $\beta = 0.0$

For each experiment shown in Table 12.1, we see that the OBPTT approach is able to yield a lower MSE, as well as a lower standard deviation. Furthermore, all of the results significantly favor OBPTT according to a t-test at a 0.95 confidence level (bolded values). Another observation is that comparing the best results for each algorithm ($\alpha = 0.10, \beta = 0.0$ for OBPTT and $\alpha = 0.25, \beta = 0.10$ for BPTT), still reveals that OBPTT yields a statistically significant outcome at the 0.95 confidence level.

Figure 12.6 shows a characteristic plot of a network trained with learning parameters $\alpha = 0.10$ and $\beta = 0.0$. Not only does the OBPTT approach converge to a lower MSE value, but it also shows a more stable learning trajectory about the convergence point. This is important because it shows the network is not "unlearning" the data. It also indicates the network may have been better conditioned for the problem.

Hidden Layers

This experiment aims at examining the sensitivity of the approach to different hidden layer sizes. As with the previous experiment, we also consider different settings of learning rate for each size of hidden layer. The results for this experiment are presented in Table 12.2, which for comparison purposes, includes the findings from the previous section (for 5 hidden nodes).

For each set of learning parameters we find that the opposition-based algorithm outperforms the original one. Similarly to the aforementioned results, all except one of these is found to be statistically significant at the 0.95 confidence level. We also observe the standard deviation also tends to be lower, indicating more reliable results with the OBPTT algorithm. Additionally, for each layer size, comparing the best result obtained from each algorithm also shows a statistically significant result favoring OBPTT.

Table 12.2. Results varying the learning rate, momentum and hidden layer size. Values in bold are statistically significant.

		OBPTT		BPTT	
		Hidden Neurons=3			
α	β	μ	σ	μ	σ
0.10	0.00	**0.043**	0.003	0.047	0.006
0.25	0.00	**0.042**	0.003	0.045	0.005
0.25	0.10	**0.044**	0.004	0.049	0.006
0.50	0.00	**0.043**	0.003	0.049	0.008
0.50	0.10	**0.044**	0.005	0.049	0.006
0.50	0.25	**0.045**	0.003	0.051	0.005
0.75	0.00	**0.045**	0.004	0.050	0.005
0.75	0.10	**0.047**	0.005	0.051	0.006
0.75	0.25	**0.046**	0.004	0.052	0.005
0.75	0.50	0.051	0.006	0.052	0.004
		Hidden Neurons=5			
α	β	μ	σ	μ	σ
0.10	0.00	**0.040**	0.002	0.045	0.004
0.25	0.00	**0.040**	0.002	0.045	0.004
0.25	0.10	**0.040**	0.002	0.043	0.003
0.50	0.00	**0.042**	0.003	0.046	0.003
0.50	0.10	**0.042**	0.003	0.046	0.004
0.50	0.25	**0.042**	0.001	0.046	0.004
0.75	0.00	**0.043**	0.002	0.047	0.004
0.75	0.10	**0.044**	0.003	0.047	0.005
0.75	0.25	**0.045**	0.004	0.050	0.005
0.75	0.50	**0.046**	0.003	0.051	0.006
		Hidden Neurons=7			
α	β	μ	σ	μ	σ
0.10	0.00	**0.039**	0.001	0.042	0.001
0.25	0.00	**0.040**	0.001	0.042	0.002
0.25	0.10	**0.040**	0.001	0.042	0.002
0.50	0.00	**0.041**	0.001	0.044	0.004
0.50	0.10	**0.042**	0.002	0.046	0.004
0.50	0.25	**0.042**	0.001	0.046	0.003
0.75	0.00	**0.043**	0.002	0.045	0.004
0.75	0.10	**0.043**	0.002	0.046	0.004
0.75	0.25	**0.044**	0.002	0.047	0.006
0.75	0.50	**0.045**	0.002	0.049	0.005

12.5.2   Opposition-Based Backpropagation

To test the efficacy of the local decision-based OBP algorithm described in Section 12.4.4 we examine properties of accuracy and generalization ability as the network structure and learning parameters are varied. The experiments focus on feedforward networks for the task of function approximation, where the hidden layer neurons employ the hyperbolic tangent (tanh) and opposite tanh function $(\tanh(-x))$. We permit each neuron to change between these functions if $\lambda \geq 1$ iterations have passed since the last epoch the neuron transfer function changed.

The data for the experiments is obtained by training each network using 10 fold cross validation for 300 epochs. The results are averaged over 30 trials. As in the OBPTT experiments, we generate a random initial weight state and allow both BP and OBP to begin training from that identical location. This is done to eliminate bias in starting positions for each algorithm. Contrasting with the OBPTT experiments the initial weights are generated according to a uniform distribution over the interval $[-1, 1]$.

Function Approximation

Neural networks are extremely capable tools for performing the task of function approximation. According to the Universal Approximation theorem (Appendix A) a neural network can approximate any continuous function to arbitrary accuracy. Of course, this theorem merely states the existence of weights to satisfy this accuracy; it is up to learning algorithms to discover them.

Table 12.3 shows the 7 common benchmark functions we employ for this task [16]. These are simple problems having 2 dimensional input and a 1 dimensional output. However, they are sufficient to show the possible type of improvements OTFs can have over traditional backpropagation.

Table 12.3. The 7 benchmark functions

Function	Input Range
$f_1(x) = x_1 + x_2$	$x_1 \pm 100$
	$x_2 \pm 10$
$f_2(x) = x_1 \cdot x_2$	$x_1 \pm 10$
	$x_2 \pm 1$
$f_3(x) = \frac{x_1}{|x_2|+1}$	$x_1 \pm 100$
	$x_2 \pm 10$
$f_4(x) = x_1^2 - x_2^3$	$x_1 \pm 100$
	$x_2 \pm 10$
$f_5(x) = x_1^3 - x_2^2$	$x_1 \pm 100$
	$x_2 \pm 10$
$f_6(x) = 418.9829n + \sum_{i=1} n(-x_i \sin(\sqrt{|x_i|})$	$x_1 \pm 200.0$
	$x_2 \pm 100.0$
$f_7(x) = 0.26(x_1^2 + x_2^2) - 0.48x_1x_2$	$x_1 \pm 10.0$
	$x_2 \pm 1.0$

The training set for each function is generated by randomly generating 550 patterns. The same patterns are used by both networks. We do not perform any pre-processing of this data, which can influence the problem slightly, but since both networks are operating under this less than ideal scenario, they are on equal footing (in practice data is not always preprocessed, although it is usually recommended).

Consideration of Opposite Networks

To examine the usefulness and computational overhead involved with the local decision-based approach we will focus on determining the expected number of additional gradient calculations, as well as the number of neurons which considered changing their transfer functions in ratio to the number of changes which improved the network error.

Table 12.4 compares the number of times a transfer function change was considered by the parent algorithm, the number of neurons that considered making a change, and the number of these transfer function changes that were accepted because they improved the network MSE. For function f_1, for every iteration where a function change was considered, on average the impact of about 4 neurons was evaluated before one was accepted. This is skewed slightly, because the target function is very simple and so most attempts at improving the MSE were futile. However, this highlights the fact that using this type of decision rule for simple target concepts may not be useful. This is further enforced by noting that only 2 and 8 changes improved the network performance when using the logistic or tanh transfer functions.

The remaining six functions show an approximate ratio of Considered to Attempted is slightly less than 1:2, which is a large improvement over the 1:4 ratio for the f_1 function. However, this lower ratio is meaningless unless transfer function changes are actually accepted. As we can see, the ratio of Considered:Attempted average is about 1:1 which means that every time a network decides to allow neurons to change their functions, on average 2 neurons will attempt to do so and 1 of these will lead to an improved MSE. These improvements are a consequence of functions $f_2 - f_7$ having a more complicated error surface than that of function f_1. Therefore, the opposite networks were better conditioned to solving these problems.

Table 12.4. Average amount of additional function calls

Function		Frequency of procedure		
		Considered	Attempted	Accepted
f_1	logistic	12	54	2
	tanh	12	41	8
f_2	logistic	29	55	25
	tanh	29	54	25
f_3	logistic	38	43	38
	tanh	38	53	37
f_4	logistic	38	61	37
	tanh	38	49	36
f_5	logistic	38	75	33
	tanh	38	50	37
f_6	logistic	29	53	21
	tanh	29	40	27
f_7	logistic	38	88	28
	tanh	28	60	24

Another important observation is that the form of the transfer function (i.e. tanh or logistic) did not have a significant impact on the number of times a transfer function change was accepted. The likely reason for this is due to the similar shape of the respective functions and they have similar behavior on opposite sides of their inflection points.

Performance

In this section we will compare the final result with respect to the MSE measure between OBP and BP. We will additionally, examine the significant epoch which represents the first epoch where OBP is statistically significant to BP at a 0.95 confidence level, and remains significant. So, at every point after the significant epoch, the MSE at each successive epoch is also statistically superior. This measure will allow us to compare the average time (in epochs) required for opposition to yield improvements. As with the above experiments each network has a fixed number of 10 hidden neurons, and uses $\alpha = 0.1$ with no momentum value.

The results of this experiment are summarized in Table 12.5 and include both the logistic and tanh transfer functions. All the the results obtained by OBP are found to be statistically significant at a 0.95 confidence level. In all cases, this significance is reached by the 3rd epoch which shows that opposite networks have a very immediate impact on learning.

To further compare the improvements with respect to specific transfer function form we can compute the average improved error according to

$$\bar{r} = \frac{1}{7} \sum_{i=1}^{7} \frac{OBP(f_i)}{BP(f_i)}. \tag{12.17}$$

Table 12.5. Comparing performance of OBP versus BP

Function		BP		OBP		Sig. Epoch
		μ	σ	μ	σ	
f_1	logistic	506.344	103.158	**365.950**	124.474	3
	tanh	759.579	158.928	**487.974**	168.599	3
f_2	logistic	29928.644	6927.741	**24337.836**	7002.546	3
	tanh	45922.595	10733.058	**35334.398**	12803.404	3
f_3	logistic	90.473	6.967	**70.287**	6.940	3
	tanh	179.967	14.954	**78.165**	9.802	3
f_4	logistic	558.956	53.448	**376.568**	42.427	3
	tanh	789.224	82.806	**470.703**	63.647	3
f_5	logistic	524.381	69.982	**421.764**	85.143	3
	tanh	676.859	90.202	**450.181**	104.757	3
f_6	logistic	1284.312	41.198	**1080.652**	103.901	3
	tanh	2139.096	125.796	**1364.533**	232.263	3
f_7	logistic	43.213	4.862	**29.619**	4.584	3
	tanh	60.808	6.532	**40.288**	9.672	3

This measure shows the relative average value obtained by OBP compared to BP. For the logistic function we find that $\bar{r} \approx 0.66$ and for the hyperbolic tangent function we calculate $\bar{r} \approx 0.63$. Both these improvements are rather substantial and highlight the type of impact that can be achieved using OTFs, with even a simple implementation strategy. Furthermore, in these cases the improvement seems to be invariant with respect to the transfer function being employed.

Generalization Ability

The ability for a neural network to generalize lies in its structure (number of hidden nodes, connections, etc). However, for a specific data set and error surface, each neural network configuration will represent its own input/output mapping, assuming the properties described above. Due to this, each of these networks will yield a different generalization ability. In this section we will compare this characteristic of generalization between BP and OBP which will be calculated from the testing subset of data used by the 10-fold cross-validation technique used to train the networks.

Table 12.6 presents the results for both logistic and tanh transfer functions. From these results we find that 6/7 tanh and 3/7 logistic function experiments showed a statistically significant difference in generalization ability. In total 9/14 experiments with OBP trained networks had a lower generalization error and thus seems less prone to problems of over/underfitting.

Table 12.6. Comparing the generalization ability between BP and OBP trained networks using testing subset of cross-validation data

Function		BP		OBP	
		μ	σ	μ	σ
f1	logistic	386.134	77.314	370.589	89.760
	tanh	1029.007	451.438	**519.156**	219.262
f2	logistic	30317.572	12171.253	**24158.174**	9029.284
	tanh	45238.370	28918.824	35734.524	20214.202
f3	logistic	80.383	38.327	71.551	34.783
	tanh	122.942	55.640	**79.020**	43.652
f4	logistic	407.556	149.557	386.865	142.802
	tanh	570.027	166.172	**502.224**	131.787
f5	logistic	706.207	231.591	**484.803**	179.726
	tanh	1116.769	437.079	**476.906**	240.339
f6	logistic	1098.062	186.130	1095.329	175.426
	tanh	1589.940	377.450	**1393.687**	298.162
f7	logistic	46.578	14.887	**31.753**	11.661
	tanh	88.483	29.722	**44.398**	14.625

Using the metric of Equation 12.17, we calculate improvements of $\bar{r} \approx 0.85$ and $\bar{r} \approx 0.66$ for the logistic and tanh transfer functions. Both of these results are very impressive, especially for the tanh function. This experiment shows that considering opposite networks can not only improve the MSE measure, but also the input/output mapping seems to be more representative of the true distribution of the data.

12.6   Conclusions and Future Work

In this chapter we discussed previous results regarding the influence of adaptive transfer functions on feedfoward and recurrent neural networks. While these results are very promising, we focus on a specific type of adaptive transfer function, namely the opposite transfer function. The opposite function allows for a rapid transformation of a given network to one of its opposite networks, each having some degree of correlation in their output.

In order to be useful in neural network learning we propose two frameworks which can be used to inject the usefulness of opposite networks into existing learning algorithms. The first framework took a global decision-based perspective on this task. Before each epoch a subset of opposite networks is randomly selected and a single epoch of learning performed on each one. Learning then proceeds with the network which exhibited the lowest error at that epoch. The second framework took a local decision-based approach where each neuron decides, based only on local information, whether a change in transfer function is likely to yield a lower error than the network currently being trained. If so, the change is accepted and learning proceeds with the new network, if indeed an improvement was observed. Each hidden neuron can be selected as a candidate to change its function.

Using these frameworks we showed extensions of the backpropagation and backpropagation through time learning algorithms. While the frameworks should work for any gradient-based learning algorithm, we chose these because of their simplicity and familiarity as a proof of concept. Our experimental results showed an improvement in accuracy and generalization abilities, at the expense of a small amount of computational overhead. Since the error was generally lower for opposition-based approaches, it can be said that they also converged to the lower error faster, but convergence rate is not necessarily higher.

Future work involves examining different learning algorithms such as rProp, quick-Prop, Levenberg-Marquardt, etc as parent algorithms. Hopefully, similar improvements can also be observed for those techniques as well. An examination of different transfer functions is also an important investigation, as is considering more flexible adaptive transfer functions where its opposite is dynamically decided at each epoch. More practical experiments for high dimensional, more complex real world problems is also an important experiment that should be conducted.

On the theoretical side some possible future works involve convergence conditions and proofs for each framework. Additionally, the development of more intelligent criteria for deciding amongst which opposite network is most likely to yield a higher quality input/output mapping is a very important direction. This will likely involve the use of higher order information, such as the Hessian matrix. Finally, developing a heuristic method to decide candidate networks which are most likely to show improvements would improve not only the input/output mapping quality, but also lower the computational requirements of the frameworks.

Appendix

A   Universal Approximation Theorem

Theorem 3 (Universal Approximation [23]). *Let $\phi(\cdot)$ be a nonconstant, bounded, and monotone-increasing continuous function. Let I_{m_0} denote the m_0-dimensional unit hypercube $[0,1]^{m_0}$. The space of continuous functions on I_{m_0} is denoted by $C(I_{m_0})$. Then, given any function $f \ni C(I_{m_0})$ and $\epsilon > 0$, there exist an integer m_1 and sets of real constants α_i, b_i, and w_{ij}, where $i = 1, \ldots, m_l$ and $j = 1, \ldots, m_0$ such that we may define*

$$F(x_1, \ldots, x_{m_0}) = \sum_{i=1}^{m_1} \alpha_i \phi \left(\sum_{j=1}^{m_0} w_{ij} x_j + b_i \right)$$

as an approximate realization of the function $f(\cdot)$; that is,

$$|F(x_1, \ldots, x_{m_0}) - f(x_1, \ldots, x_{m_0})| < \epsilon$$

for all $x_1, \ldots, x_{m_0}$ that lie in the input space.

References

1. Albertini, F., Sontag, E.: For Neural Networks, Function Determines Form. Neural Networks 6, 975–990 (1993)
2. Barakova, E.I., Spaanenburg, L.: Symmetry: Between Indecision and Equality of Choice. Biological and Artificial Computation: From Neuroscience to Technology (1997)
3. Bishop, C.: Neural Networks for Pattern Recognition. Oxford University Press, Oxford (1995)
4. Bishop, C.: Pattern Recognition and Machine Learning. Springer, Heidelberg (2007)
5. Chandra, P., Singh, Y.: An Activation Function Adapting Training Algorithm for Sigmoidal Feedforward Networks. Neurocomputing 61 (2004)
6. Chandra, P., Singh, Y.: A Case for the Self-Adaptation of Activation Functions in FFANNs. Neurocomputing 56, 447–545 (2004)
7. Charalambous, C.: Conjugate Gradient Algorithm for Efficient Training of Artificial Neural Network 139(3), 301–310 (1992)
8. Chen, A., Hecht-Nielsen, R.: On the Geometry of Feedforward Neural Network Weight Spaces. In: Second International Conference on Artificial Neural Networks, pp. 1–4 (1991)
9. Chen, A.M., Lu, H., HechtNielsen, R.: On the Geometry of Feedforward Neural Networks Error Surfaces. Neural Computation 5(6), 910–927 (1993)
10. Chen, C.T., Chang, W.D.: A Feedforward Neural Network with Function Shape Autotuning. Neural Networks 9 (1996)
11. Cleeremans, A., Servan-Schreiber, D., McClelland, J.L.: Finite State Automata and Simple Recurrent Networks. Neural Computation 1(3), 372–381 (1989)
12. Drago, G.P., Morando, M., Ridella, S.: An Adaptive Momentum Back Propagation (AMBP). Neural Computation and Applications 3(4), 213–221 (1995)
13. Duch, W., Jankowski, N.: Optimal Transfer Function Neural Networks. In: 9th European Symposium on Artificial Neural Networks, pp. 101–106 (2001)
14. Duch, W., Jankowski, N.: Transfer Functions: Hidden Possibilities for Better Neural Networks. In: 9th European Symposium on Artificial Neural Networks, pp. 81–94 (2001)

15. Duch, W., Jankowski, N.: New Neural Transfer Functions. Applied Mathematics and Computer Science 7, 639–658 (1997)

16. El-Fallahi, A., Mart, R., Lasdon, L.: Path Relinking and GRG for Artificial Neural Networks. European Journal of Operational Research 169, 508–519 (2006)

17. Elman, J.L.: Distributed Representations, Simple Recurrent Networks and Grammatical Structure. Machine Learning 7(2), 195–226 (1991)

18. Fahlman, S.E.: Faster-Learning Variations on Backpropagation: An Empirical Study. Morgan Kaufmann, San Francisco (1988)

19. Funahashi, K.: On the Approximate Realization of Continuous Mappings by Neural Networks. Neural Networks 2(3), 183–193 (1989)

20. Goh, S.L., Mandic, D.: Recurrent Neural Networks with Trainable Amplitude of Activation Functions. Neural Networks 16 (2003)

21. Hagan, M.T., Menhaj, M.B.: Training Feedforward with the Marquardt Algorithm. IEEE Transactions on Neural Networks 5(6), 989–993 (1994)

22. Hanna, A., Mandic, D.: A Complex-Valued Nonlinear Neural Adaptive Filter with a Gradient Adaptive Amplitude of the Activation Function. Neural Networks 16(3), 155–159 (2003)

23. Haykin, S.: Neural Networks: A Comprehensive Foundation, 2nd edn. Prentice-Hall, Englewood Cliffs (1998)

24. Hoffmann, G.A.: Adaptive Transfer Functions in Radial Basis Function (RBF) Networks. In: Bubak, M., van Albada, G.D., Sloot, P.M.A., Dongarra, J. (eds.) ICCS 2004. LNCS, vol. 3037, pp. 682–686. Springer, Heidelberg (2004)

25. Hornik, K., Stinchcombe, M., White, H.: Multilayer Feedforward Networks are Universal Approximators. Neural Networks 2(3), 359–366 (1989)

26. Hu, Z., Shao, H.: The Study of Neural Network Adaptive Control Systems. Control and Decision 7 (1992)

27. Jacobs, R.A.: Increased Rates of Convergence Through Learning Rate Adaptation. Neural Networks 1(4), 295–308 (1988)

28. Jordan, F., Clement, G.: Using the Symmetries of Multilayered Network to Reduce the Weight Space. In: IEEE Second International Joint Conference on Neural Networks, pp. 391–396 (1991)

29. Lagaros, N., Papadrakakis, M.: Learning Improvement of Neural Networks used in Structural Optimization. Advances in Engineering Software 35(1), 9–25 (2004)

30. Le Cun, Y., Denker, J.S., Solla, S.A.: Optimal Brain Damage. In: Advances in Neural Information Processing Systems, pp. 598–605 (1990)

31. Liang, Y.: Adaptive Neural Activation Functions in Multiresolution Learning. In: IEEE International Conference on Systems, Man, and Cybernetics, vol. 4, pp. 2601–2606 (2000)

32. Liu, Y., Yao, X.: Evolutionary Design of Artificial Neural Networks with Different Nodes. In: International Conference on Evolutionary Computation, pp. 670–675 (1996)

33. Magoulas, G.D., Vrahatis, M.N., Androulakis, G.S.: Improving the Convergence of the Backpropagation Algorithm Using Learning Rate Adaptation Methods. Neural Computation 11, 1769–1796 (1999)

34. Moller, M.F.: A Scaled Conjugate Gradient Algorithm for Fast Supervised Learning 6, 525–533 (1993)

35. Nguyen, D., Widrow, B.: Improving the Learning Speed of 2-Layer Neural Networks by Choosing Initial Values of the Adaptive Weights. In: IEEE Proceedings of the International Joint Conference on Neural Netowrks, vol. 3, pp. 21–26 (1990)

36. Omlin, C.W., Giles, C.L.: Constructing Deterministic Finite-State Automata in Recurrent Neural Networks. Journal of the ACM 43(6), 937–972 (1996)

37. Reber, A.S.: Implicit Learning of Synthetic Languages: The Role of Instructional Set. Journal of Experimental Psychology: Human Learning and Memory 2(1), 88–94 (1976)

38. Riedmiller, M., Braun, H.: A Direct Adaptive Method for Faster Backpropagation Learning: The RPROP Algorithm. In: IEEE Conference on Neural Networks, pp. 586–591 (1993)
39. Rumelhart, D., Hinton, G., Williams, R.: Learning Internal Representations by Error Propagation. In: Parallel Distributed Processing, ch. 8. MIT Press, Cambridge (1986)
40. Saarinen, S., Bramley, R., Cybenko, G.: Ill-Conditioning in Neural Network Training Problems. SIAM Journal on Scientific Computing 14(3), 693–714 (1993)
41. Schraudolph, N.: Centering Neural Network Gradient Factors. In: Orr, G., Muller, K.R. (eds.) Neural Networks: Tricks of the Trade, pp. 207–226. Springer, Heidelberg (1998)
42. Sussmann, H.J.: Uniqueness of the Weights for Minimal Feedforward Nets with a Given Input-Output Map. Neural Networks 5(4), 589–593 (1992)
43. Tezel, G., Özbay, Y.: A New Neural Network with Adaptive Activation Function for Classification of ECG Arrhythmias. In: Apolloni, B., Howlett, R.J., Jain, L. (eds.) KES 2007, Part I. LNCS (LNAI), vol. 4692, pp. 1–8. Springer, Heidelberg (2007)
44. Thome, A.G., Tenorio, M.F.: Dynamic Adaptation of the Error Surface for the Acceleration of the Training of Neural Networks 1, 447–452 (1994)
45. Trentin, E.: Networks with Trainable Amplitude of Activation Functions. Neural Networks 14(4), 471–493 (2001)
46. van der Smagt, P., Hirzinger, G.: Solving the Ill-Conditioning in Neural Network Learning. In: Orr, G., Muller, K.R. (eds.) Neural Networks: Tricks of the Trade, pp. 193–206. Springer, Heidelberg (1998)
47. Vecci, L., Piazza, F., Uncini, A.: Learning and Approximation Capabilities of Adaptive Spline Activation Function Neural Networks. Neural Networks 11, 259–270 (1998)
48. Ventresca, M., Tizhoosh, H.: Numerical Condition of Feedforward Networks with Opposite Transfer Functions. In: International Joint Conference on Neural Networks (to appear, 2008)
49. Ventresca, M., Tizhoosh, H.R.: Opposite Transfer Functions and Backpropagation Through Time. In: IEEE Symposium on Foundations of Computational Intelligence, pp. 570–577 (2007)
50. Ventrescda, M., Tizhoosh, H.: Improving the Convergence of Backpropagation by Opposite Transfer Functions. In: International Joint Conference on Neural Networks, pp. 9527–9534 (2006)
51. Duch, W., Jankowski, N.: Survey of Neural Transfer Functions. Neural Computing Surveys 2, 163–213 (1999)
52. Werbos, P.J.: Backpropagation Through Time: What it Does and how to do it. Proceedings of the IEEE 78, 1550–1560
53. Werbos, P.J.: Back-propagation: Past and future. In: Proceedings of the IEEE International Conference on Neural Networks, pp. 343–353 (1988)
54. Wolpert, D.H., Macready, W.G.: No Free Lunch Theorems for Optimization. IEEE Transactions on Evolutionary Computation 1(1), 67–82 (1997)
55. Xu, S., Zhang, M.: Adaptive Higher-Order Feedforward Neural Networks. In: International Joint Conference on Neural Networks, vol. 1, pp. 328–332 (1999)
56. Xu, S., Zhang, M.: Justification of a Neuron-Adaptive Activation Function. In: International Joint Conference on Neural Networks, vol. 3, pp. 465–470 (2000)

Part V

Real World Applications

13

Opposite Actions in Reinforced Image Segmentation

Farhang Sahba[1] and H.R. Tizhoosh[2]

[1] Department of Electrical and Computer Engineering, University of Toronto, Canada
fsahba@uwaterloo.ca
[2] Department of Systems Design Engineering, University of Waterloo, Canada
tizhoosh@uwaterloo.ca

Summary. In many vision-based applications we need to segment an object of interest in digital images. For methods which rely on a learning process, the lack of sufficient number of training samples is usually an obstacle, especially when the samples need to be manually prepared by an expert. In addition, none of the existing methods uses online feedback from the user in order to evaluate the generated results and continuously improve them. Considering these factors, a new algorithm based on reinforcement learning is discussed in this chapter. The approach starts with a limited number of training samples and improves its performance in the course of time.

A potential obstacle when we apply reinforcement learning into image-based applications is the large number of state-action pairs. In such cases, it is usually difficult to evaluate the state-action information especially when the agent is in the exploration mode. The opposition-based leaning is one of the methods that can be applied to converge to a solution faster in spite of a large state space. Using opposite actions we can update the agent's knowledge more rapidly. The experiments show the results for a medical application.

13.1 Introduction

Image segmentation plays a pivotal rule in many computer vision applications. The demand for higher robustness, reliability and automation of segmentation algorithms has been steadily increasing in recent years. Segmentation is partitioning of an image into meaningful regions based on characteristics such as intensity, texture, color and gradient. Techniques used for segmentation are highly dependent on the particular situation and the specifications of the problem at hand. Many, if not all, segmentation methods usually require some user interaction to adjust critical parameters for optimal object extraction. There is not a general segmentation algorithm that could generate acceptable results for all cases. But there are numerous methods specialized for particular applications which can usually give better results by taking into account expert or a-priori knowledge.

In many applications we need to segment only one object in the image. This simplifies the segmentation process. However, we still face difficulties due to some factors such as poor image contrast, noise and missing or diffuse boundaries. In methods which rely on learning techniques the lack of a sufficient number of training samples is another obstacle, especially when the samples are being manually prepared by an expert.

The segmentation of prostate in trans-rectal ultrasound (TRUS) images is an example for extremely challenging medical image segmentation [2, 6]. The detection of the

H.R. Tizhoosh, M. Ventresca (Eds.): Oppos. Concepts in Comp. Intel., SCI 155, pp. 287–297, 2008.
springerlink.com © Springer-Verlag Berlin Heidelberg 2008

prostate boundary in such images is crucial for computer-aided cancer diagnosis and classification. However, due to a very low signal-to-noise ratio, it is difficult to extract the accurate boundaries such that improvements of segmentation process are still desirable [8]. Many methods have been proposed in literature to facilitate more accurate automatic or semi-automatic segmentation of the prostate boundaries in ultrasound images [1, 3, 9, 10, 12, 22].

Recently, some works have been published that show the applications of reinforcement learning for image-based problems [13, 14, 15, 16, 17, 18, 19]. In this chapter a segmentation system based on opposition-based reinforcement learning (ORL) is introduced. To validate the findings, the system is applied to segment the prostate in TRUS images. Generally, a reinforcement learning agent can learn from interactions with an offline simulation system and its online model [20, 23]. Due to the exploratory policy used in reinforcement learning (RL), it would be well-suited to demonstrate the usefulness of considering opposite alternatives compared to pure randomness. The goal of this ORL system is to identify an object of interest in an image and separate it from the background. The system can work in offline or online mode. The offline mode is performed using the manually segmented samples to acquire the fundamental information about the specifications of the problem at hand. In the online mode, the segmentation is mainly based on the fact that the object of interest has a meaningful shape with respect to the application at hand.

13.2   Problem Statement and Proposed Approach

A framework for adjusting the parameters of a multi-stage segmentation system by using an RL agent is discussed in this section. The framework is depicted in Figure 13.1.

As shown, an intelligent agent is employed to find the appropriate parameters for image processing tasks. The system contains a series of image processing tasks with parameters that must be adjusted to manipulate the input images in some desired way. The goal is to choose a final set of parameters for various tasks such that an object of interest can accurately be extracted. An RL agent operates based on *states* which are the features that describe the image content in various stages (after each processing task). To change the result of segmentation, the RL agent takes *actions* which have the capability to change the critical parameters. The learning is guided through *rewards* which could be based on subjective or objective measures.

As depicted in Figure 13.1, there are a series of N image processing tasks T_1, $T_2, \ldots, T_N$ with corresponding parameters that must be adjusted. The system adaptively computes the set of parameters that increase its performance for a given segmentation problem. Limiting ourselves to the model previously described, we must first define the states s, actions a and reward r. To generate the states, features of the input and output images for each processing task are used. These features must describe the status (content) of these images. It is desirable to use such features with a proper level of discretization not leading to extremely large state spaces.

Actions can be easily defined as modification of the parameters of the processing tasks. After the agent adjusts these parameters properly, it receives a reward. The reward is an external reinforcement signal and must correctly reflect the goal of the agent.

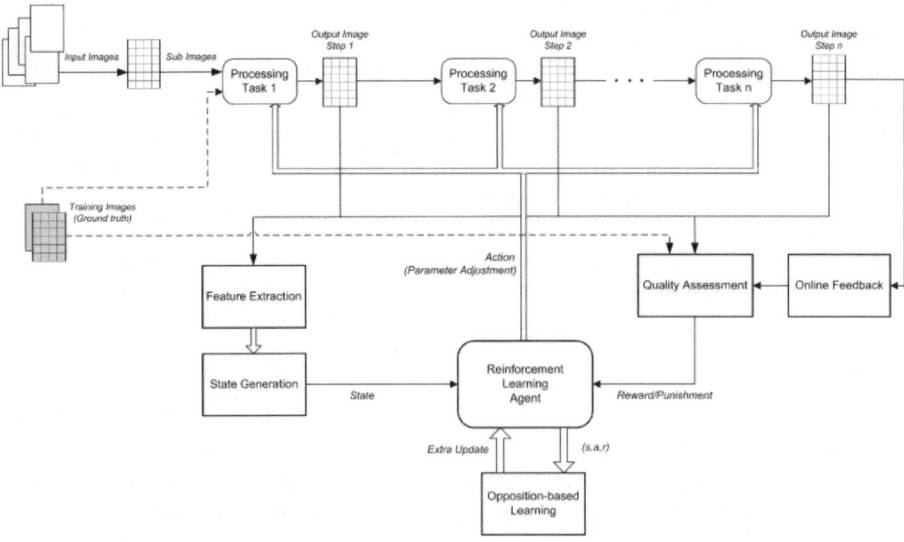

Fig. 13.1. The general model of the proposed ORL system

The ORL method for object segmentation has two modes: offline and online. The offline training is conducted by using the manually segmented samples represented as ground-truth images. In this mode, the agent is adopted in a simulation environment and interacts with training samples to acquire the fundamental information necessary to segment the object of interest. Once the agent's behavior is acceptable in terms of established criteria (e.g. accuracy), the agent switches to the online mode. In this mode, the RL agent operates in real-time to segment new (unseen) images. With continuous learning in this mode, the agent can adapt to changes in the input images.

In the proposed system, an $R \times C$ input image is divided into $R_S \times C_S$ sub-images. The RL agent works on each of these sub-images separately. Local processing on sub-images is carried out to find the best segmentation parameters for each of them. To construct the processing task chain, the sub-images are thresholded using local values. Due to disturbing factors such as poor contrast, noise, or non-uniform illumination, artifacts exist after thresholding. Therefore, morphological operators are applied to post-process each thresholded sub-image. The RL agent determines two factors: 1) the local threshold value, and 2) the size of the structuring elements for each individual sub-image.

During the *offline mode* where the desired output image is available, the agent works on each sub-image and explores the solution space until a sufficient number of actions have been taken. In this mode, the agent tries different actions in different states via an exploratory procedure. The Q-learning algorithm is used to implement the reinforcement scheme (see Chapter 11). After the RL agent changes the parameters for each sub-image, it receives a reward/puishment for that state-action pair and updates the corresponding value in the Q-matrix, a matrix storing accumulated rewards (see Chapter 11). After the offline mode, the agent has already explored many actions and is capable of exploiting the most rewarding ones for segmenting new images. Needless to say that the performance of the system after the offline mode depends on the number of samples that have been learned.

In the *online mode*, the system operates episodically. During each iteration in each episode, the agent works on each sub-image and completes the iteration by using the knowledge previously gained and stored in the Q-matrix. The rewards are not calculated immediately, and the agent waits until the whole image is scanned and processed. The reward is then provided for each sub-image objectively and/or subjectively based on the quality of the segmented object before and after the action is taken.

Components required to construct the RL agent, namely states, actions and rewards, are defined in the following sub-sections.

13.2.1   States

We considered the idea of the state consisting of some features representing the quality of the image. These features can be chosen based on various shape and/or boundary properties reflecting the quality of each sub-image after thresholding and post-processing. Generally, the feature selection is highly dependent on the specifications of the problem at hand. For the TRUS images, the following features extracted from the main object in each sub-image are used to define the states:

Feature 1 – *Area*
   In each sub-image, the normalized area with respect to the total area of the sub-image is calculated and used as the first feature:

$$A_{norm} = \frac{A_{subimage} - A_{SO}}{A_{subimage}}. \tag{13.1}$$

where $A_{subimage}$ and A_{SO} are the area of the sub-image and the area of its object, respectively.

Feature 2 – *Compactness*
   The compactness Ψ of the object after thresholding is defined as:

$$\Psi = \frac{4\pi A_{SO}}{P_{SO}^2}, \tag{13.2}$$

where A_{OS} and P_{OS} are the area and perimeter of the object in the sub-image, respectively [4].

Feature 3 – *Relative Position*
   By using the geometric center (x_c, y_c) of the prostate given by the user via a single mouse click (or by calculating the center of gravity of the extracted segment), the relative distance ρ and angle ϕ of the sub-image with respect to the geometric center is adopted as a state parameter:

$$\rho = \sqrt{(x_s - x_c)^2 + (y_s - y_c)^2}, \tag{13.3}$$

$$\phi = \tan^{-1}\left(\frac{y_s - y_c}{x_s - x_c}\right),$$ (13.4)

where x_s and y_s are the coordinates of the center of the current sub-image.

Feature 4 – *Gray Level Information*
A histogram prototype on the edge of the prostate can be calculated in the ground-truth images, which shows the gray level distribution along true edgy areas. The histogram distance between each sub-image and the prototype is then calculated and used to define a feature. One of the most popular histogram distances, the χ^2 distance, is selected for this purpose [24]:

$$D_{\chi^2}(h_1, h_2) = \sum_{b=1}^{M_h} \frac{(h_1(b) - h_2(b))^2}{(h_1(b) + h_2(b))},$$ (13.5)

where M_h is the number of gray levels and h_1 and h_2 are the normalized histograms (gray level probability density functions) of each sub-image and the prototype, re-spectively. To incorporate the spatial gray level information, one may extract tex-ture information which has frequently been used as a feature in ultrasound image analysis. However, this may lead to an increase in computational expense.

Feature 5 – *Number of the Objects*
The last feature used for state definition is the number of revealed objects, N_O, after the morphological opening.

13.2.2   Actions

To extract the object of interest, the actions are defined as "changing of the parameters" for processing of each sub-image. The assigned values are increased or decreased, or chosen from the predefined values to control the effect of each processing task. In order to extract the object of interest, all sub-images are binarized using local thresholds. Due to disturbances such as speckle noise or poor contrast, irrelevant structures exist after thresholding. Hence, morphological opening and closing are employed to post-process each thresholded sub-image. The RL actions are defined as a change in the threshold value, and the size of the structuring elements for each sub-image.

The assigned local threshold values are increased or decreased by a specific amount Δ. Using this definition, the values τ_{g_1} and τ_{g_2} from a predefined set $(\tau_1, \tau_2, ..., \tau_n)$, equally spaced between the local maximum gray level g_{lmax} and local minimum gray level g_{lmin}, are selected. For the opening and closing morphological operators, the size of the structuring elements are chosen from predefined values to control the effects of these operators. Thus, the set of all actions A can be presented as follows:

$$A = \{\tau_{g_1} \pm \Delta, \tau_{g_2} \pm \Delta\} \cup \{v_j\} \cup \{v_u\},$$ (13.6)

where v_j and v_u are the sizes of the structuring elements for the morphological operators.

13.2.3   Reward/Punishment

The rewards and punishments are defined according to the quality criteria that quantify how well the object has been segmented in each sub-image. In the offline mode, a straightforward method is to compare the results with the ground-truth image after each action. To measure this value for each sub-image, we note how much the quality has changed after each action was taken. In each sub-image, for improving the quality of the segmented object, the agent receives a reward otherwise a punishment. A general form for the reward function is represented by

$$r = \begin{cases} \kappa_1 & D_\Delta \geq 0 \\ \kappa_2 & D_\Delta \leq 0, \end{cases} \tag{13.7}$$

where D_Δ is a measure indicating the difference between the quality before and after taking the action and is calculated by the normalized number of misclassified pixels in the segmented sub-images. In Eq. 13.7, κ_1 and κ_2 are constant values.

13.2.4   Opposition-Based Computing

A potential obstacle when we apply RL agents to image-based applications, is the large number of state-action pairs involved. Also we are operating on two-dimensinal data and each action will take time to produce a new result. Therefore, we need to speed up the learning process to increase the practicality of the system. Opposition-based computing is one of the methods that can be applied for this purpose (see Chapter 2).

Employing opposition within learning for image segmentation means we have to introduce opposite states and/or opposite actions and simultaneously explore their effect when we use the original states and actions. This, however, is highly dependent on how the state-action pairs are defined in context of the application at hand. Because the RL agent must visit all sub-images, it takes too long to try various actions especially when the agent is rather in exploration mode. Using opposition-based computing we can update the knowledge of the RL agent more frequently resulting in shortening the exploration time. For image segmentation, this can be easily done for actions. For example if the action a is to increase the threshold, $\tau_{g_1} + \Delta$, the opposite action $\breve{a}$ can be defined as decreasing it, $\tau_{g_1} - \Delta$. Analogously, if the action is choosing a specific value among some predefined values, the opposite action can be defined based on the relative distance of others with respect to the current value. Generally speaking, we can define the degree of oppositeness based on the distance to the current state.

Whereas the offline reward r_{OFF} is calculated from the difference between the available ground-truth image and the binary image produced by the agent, in the online mode the object contour is considered in a global manner to identify the image regions that correspond to the object of interest. This serves as a guidance from the higher level of perception to update the agent's information. The RL agent then uses this feedback as online reward r_{ON} for the action taken on each sub-image. If a sub-image is segmented correctly in terms of the global feedback, the parameters remain unchanged and learning process is terminated.

Objective Evaluation

Objective evaluation is applied as a primary assessment. There are several methods for implementing this evaluation. One way is to compute the distance between two shapes. Typically, these techniques compute the best global match between two shapes by pairing each element of shape Γ_1 with each element of shape Γ_2. The technique introduced in [11] is adopted in a 2-D plane to measure the object similarity. In this technique there is an unknown curve, C^1 and a prototype curve C^2 corresponding to shapes Γ_1 and Γ_2, respectively. If the characteristics of the points located on the two curves are represented as string $\tilde{x}$ and $\tilde{y}$, the editing distance between the two strings is defined as the shortest way (minimum cost) to transform $\tilde{x}$ into $\tilde{y}$. The idea is to find a minimum cost path as a sequence of editing operations that transform the string corresponding to the curve C^1 into curve C^2 [11]. The algorithm not only gives the distance $\Delta(\tilde{x}, \tilde{y})$ between two strings $\tilde{x}$ and $\tilde{y}$, but also the way in which the symbols of $\tilde{x}$ correspond to those of $\tilde{y}$.

The signature of a shape is used to generate its string. The shape signature is defined as a 1D function representing a 2D area or boundary of the shape. In the proposed system, the central distance is calculated as the distance from the points on the boundary to the central point of the object (given by the expert). It is represented as a 2π periodic function [5]. The signature of the extracted object is then compared with the standard signature of the object of interest prior or during the offline mode using the ground-truth images. Finally, the significant deviations are estimated and the objective reward (/punishment) r_{ON}^O for the corresponding sub-image is calculated.

Subjective Evaluation

Another alternative to provide the agent with a reinforcement signal is subjective evaluation from the online feedback. The user considers the results of the segmentation system for each image. If he/she is not satisfied, he/she can change the results manually (manual editing of the extracted contour). These changes are evaluated as subjective punishment for the agent, where the manually corrected result is used as a new ground-truth image to improve the agent's knowledge. The agent then proceeds to the next image with its updated information. By adopting this method, the agent can be further trained online by a subjective evaluation as r_{ON}^S. The Q-matrix is updated and the segmentation system can follow the changes in the new input image. During all previous stages, the object of interest (in our case, the prostate) is extracted by using the position of the central point (the position of the click).

13.3 Results and Discussions

To evaluate the performance of the segmentation system, validation experiments were performed. The results of the segmentation system on an image data set containing 80 TRUS image slices from five patients are represented in this section. To train the agent, 12 manually segmented images from the data set were chosen.

The ϵ-greedy policy is implemented to explore/exploit the solution space when the RL agent is in the offline mode [21]. The size of the TRUS images are 460×350 pixels.

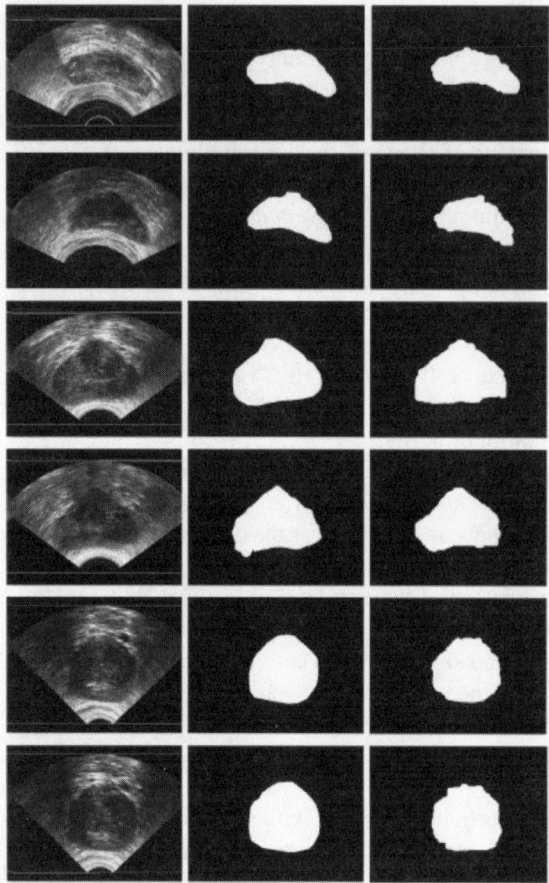

Fig. 13.2. Prostate segmentation in TRUS images. From left to right: original, manually segmented and result of the proposed system.

The size of the sub-images $R_S \times C_S$ is empirically set to 11×14. The threshold action is defined by changing the two values between the local maximum and minimum gray levels in each sub-image by $\Delta = 1/10$. For the post-processing tasks, the size of a disk shape structuring element for the morphological opening is changed by 5 in the set $\{0, ..., 15\}$ and for the morphological closing by 1 in the set $\{0, ..., 5\}$. The criterion to terminate the process for each sub-image is to reach a pixel difference less than 5% by comparing it with the ground-truth image. The reward (Eq. 13.7) is calculated with $\kappa_1 = 10, \kappa_2 = 0$.

The average time for training was $61s$ per patient set and $7.9s$ for test images. After the offline learning, the Q-matrix is filled with the appropriate values and the agent has gained enough knowledge to recognize the optimal values for the sub-images in new (unseen) images. The method was consequently applied on the remaining sample images. Figure 13.2 displays the results of the proposed approach for six sample images.

Table 13.1. The average area overlap AO, μ, between the segmented and the ground-truth images for the training samples (for training samples manually segmented versions are available)

	$AO(\%)$ Patient 1	$AO(\%)$ Patient 2	$AO(\%)$ Patient 3	$AO(\%)$ Patient 4	$AO(\%)$ Patient 5
μ	**93.56**	**95.30**	**94.07**	**94.58**	**93.83**

The choice of criteria to measure the accuracy depends on the application and can be derived from the region or boundary information. For all images, the accuracy of the final segmented object is defined as the *area overlap* of the segmented images and the ground-truth images, a commonly used metric [2]. Table 13.1 shows the results for training images.

Then, the algorithm was applied again using the opposition-based reinforcement learning. When employing opposition-based computing, we observed that the training time is reduced during the exploration mode. This is because the Q-matrix is filled more rapidly using extra opposition-based updates in each iteration. Table 13.2 shows the average Learning Time Reduction (LTR) with respect to the standard Q-Learning. As we can see, a reduction in learning time has been achieved.

Table 13.2. Learning Time Reduction (LTR%) using opposition-based computing comparing to the strandard Q-learning for patients

	Patient 1	Patient 2	Patient 3	Patient 4	Patient 5
LTR%	17.2%	18.2%	19%	16.7%	21.5%
$\mu = 18.52$			$\sigma = 1.88$		

The results in terms of visual appearance and accuracy can be employed as suitable coarse level estimations to serve a fine-tuning segmentation algorithm [2]. For instance, these results can be used as the initial points for the well-known snake method introduced in [7]. In some cases, the results can even be regarded as the final segmentation. It is important to note that our proposed approach is not designed to compete with the existing segmentation approaches. As a proof of concept, it is introducing a new class of knowledge-based methods that require a limited number of samples for training and can improve their performance in an online learning procedure. To compare with other techniques and understand the advantages of the proposed method, we note that static methods such as active contours have static performance; they do not have any ability to be trained. On the other hand, in the methods that use learning algorithms, we usually need a large number of training samples. They do not have online training unless we "retrain" them.

13.4 Conclusion

This chapter intended to present the concept of opposition-based reinforcement learning in the field of image segmentation by providing some experimental results. This system finds the appropriate local values for image processing tasks and segments an object. Opposition-based computing was employed to reduce the training time. During an offline stage, the agent uses some images and their manually segmented versions to gather knowledge. The agent is provided with scalar reinforcement signals to explore/exploit the solution space. It can then use the acquired knowledge to segment similar images. Also, it can improve itself using online feedback. We have to notice that our proposed method (as the first version) is a proof of concept of a new learning-based method and is not designed to compete with the existing segmentation approaches in terms of accuracy.

NOTE: The subject matter of this work is covered by a US provisional patent application.

References

1. Betrounia, N., Vermandela, M., Pasquierc, D., Maoucheb, S., Rousseaua, J.: Segmentation of abdominal ultrasound images of the prostate using a priori information and an adapted noise filter. Computerized Medical Imaging and Graphics 29, 43–51 (2005)
2. Nanayakkara, N.D., Samarabandu, J., Fenster, A.: Prostate segmentation by feature enhancement using domain knowledge and adaptive region based operations. Physics in Medicine and Biology 51, 1831–1848 (2006)
3. Chiu, B., Freeman, G.H., Salama, M.M.A., Fenster, A.: Prostate segmentation algorithm using dyadic wavelet transform and discrete dynamic contour. Physics in Medicine and Biology 49, 4943–4960 (2004)
4. Gonzalez, R.C., Woods, R.E.: Digital Image Processing, 2nd edn. Prentice-Hall, Englewood Cliffs (2002)
5. Holt, R.J., Netravali, A.N.: Using Line Correspondences In Invariant Signatures For Curve Recognition. Image and Vision Computing 11(7), 440–446 (1993)
6. Insana, M.F., Brown, D.G.: Acoustic scattering theory applied to soft biological tissues, Ultrasonic Scattering in biological tissues. CRC Press, Boca Raton (1993)
7. Ladak, H.M., Mao, F., Wang, Y., Downey, D.B., Steinman, D.A., Fenster, A.: Prostate boundary segmentation from 2D ultrasound images. Medical Physics 27, 1777–1788 (2000)
8. Noble, J.A., Boukerroui, D.: Ultrasound Image Segmentation: A Survey. IEEE Transactions on Medical Imaging 25, 987–1010 (2006)
9. Pathak, S.D.V., Chalana, D.R., Haynor, Kim, Y.: Edge-guided boundary delineation in prostate ultrasound images. IEEE Transactions on Medical Imaging 19, 1211–1219 (2000)
10. Prater, J.S., Richard, W.D.: Segmenting ultrasound images of the prostate using neural networks. Ultrasound Imaging 14, 159–185 (1992)
11. Rodriguez, W., Lastb, M., Kandel, A., Bunked, H.: 3-Dimensional curve similarity using string matching. Robotics and Autonomous Systems 49, 165–172 (2004)
12. Shen, D., Zhan, Y., Davatzikos, C.: Segmentation of Prostate Boundaries From Ultrasound Images Using Statistical Shape Model. IEEE Transactions on Medical Imaging 22, 539–551 (2003)
13. Sahba, F., Tizhoosh, H.R.: Filter Fusion for Image Enhancement Using Reinforcement Learning. In: Canadian Conference on Electrical and Computer Engineering, pp. 847–850 (2003)

14. Sahba, F., Tizhoosh, H.R., Salama, M.M.A.: Using Reinforcement Learning for Filter Fusion in Image Enhancement. In: The Fourth IASTED International Conference on Computational Intelligence, Calgary, Canada, pp. 262–266 (2005)
15. Sahba, F., Tizhoosh, H.R., Salama, M.M.A.: A Reinforcement Learning Framework for Medical Image Segmentation. In: The IEEE World Congress on Computational Intelligence (WCCI), Vancouver, Canada pp. 1238–1244 (2006)
16. Sahba, F., Tizhoosh, H.R., Salama, M.M.A.: Increasing object recognition rate using reinforced segmentation. In: The IEEE International Conference on Image Processing (ICIP), Atlanta, pp. 781–784 (2006)
17. Shokri, M., Tizhoosh, H.R.: Q(λ)-based Image Thresholding. In: Canadian Conference on Computer and Robot Vision, pp. 504–508 (2004)
18. Taylor, G.A.: A Reinforcement Learning Framework for Parameter Control in Computer Vision Applications. In: Proceedings of the First Canadian Conference on Computer and Robot Vision, pp. 496–503 (2004)
19. Tizhoosh, H.R., Taylor, G.W.: Reinforced Contrast Adaptation. International Journal of Image and Graphic 6, 377–392 (2006)
20. Singh, S., Norving, P., Cohn, D.: Introduction to Reinforcement Learning. Harlequin Inc. (1996)
21. Sutton, R.S., Barto, A.G.: Reinforcement Learning. MIT Press, Cambridge (1998)
22. Wang, Y., Cardinal, H., Downey, D., Fenster, A.: Semiautomatic three-dimensional segmentation of the prostate using two dimensional ultrasound images. Medical physics 30, 887–897 (2003)
23. Watkins, C.J.C.H., Dayan, P.: Q-Learning. Machine Learning 8, 279–292 (1992)
24. Brunelli, R., Mich, O.: Histogram analysis for image retrieval. Pattern Recognition 34, 1625–1637 (2001)

14

Opposition Mining in Reservoir Management

Masoud Mahootchi, H.R. Tizhoosh, and Kumaraswamy Ponnambalam

Department of Systems Design Engineering, University of Waterloo, Canada
mmahootc@engmail.uwaterloo.ca, tizhoosh@uwaterloo.ca,
ponnu@vlsi.uwaterloo.ca

Summary. Water resource management is one of the important issues for most governmental and private agencies around the globe. many mathematical and heuristic optimization or simulation techniques have been developed and applied to capture the complexities of the problem; however, most of them suffered from the curse of dimensionality. Q-learning as a popular and simulation-based method in Reinforcement Learning (RL) might be an efficient way to cope well practical water resources problems because of being model-free and adaptive in a dynamic system. However, it might have trouble for large scale applications. In this chapter, we are going to introduce a new type-II opposition of Q-Learning technique in a single and multiple-reservoir problem. The experimental results at the end of chapter will confirm the contribution of the opposition scheme in speeding up the learning process especially at the early stage of learning and making it more robust at the end. These results are promising for large-scale water resources applications in the future.

14.1 Introduction

In the modern world where the population is on the rise, the efficient utilization of limited natural resources such as water plays a crucial role in human life. There are vital questions how to efficiently control, use, and manage different water resources around the world such that demands are met and minimum cost and waste are achieved. Sometimes, these questions for some specific resources are not local concerns of one community or government, but strategic issues for several interrelated communities.

The surface water resources or *reservoir management* is one of these issues involving various equipments, facilities, and substantial budgets. For the planning or designing step of a single reservoir or a set of reservoirs, different objectives such as flood control, irrigation needs, drinking and industrial supplies, power generation, recreation, navigations, and environmental issues should be considered. In real-world, the optimization of these objectives might be non-linear and non-convex which makes the problem more complex. Moreover, embedding the stochasticity of the parameters such as inflow, evaporation, and demand into the simulation model of the reservoir creates further complexities. Existing computational limitations cause various physical constraints.

In this chapter, we would like to apply Q-Learning along with type-II opposition schemes to a single- and multi-reservoir problem. In the next section, we will describe the nature of reservoirs, potential objectives and important constraints. In section 14.3, we will demonstrate how to design Q-Learning algorithm to be applied for single

H.R. Tizhoosh, M. Ventresca (Eds.): Oppos. Concepts in Comp. Intel., SCI 155, pp. 299–321, 2008.
springerlink.com   © Springer-Verlag Berlin Heidelberg 2008

and multi-reservoir applications. Section 14.4 introduces a specific type-II opposition scheme. In section 14.5, we will demonstrate some experimental results for a single reservoir case study. In the last chapter, we will summarize the final results and point to future research areas.

14.2   Problem Description

In the following subsections, we will demonstrate schematic figures of the single- and multi-reservoir case studies and establish some important terminologies. A typical optimization model including objective function and some common constraints will be introduced at the end of this section.

14.2.1   Components of a Single Reservoir

Figure 14.1 represents a general sketch of a single reservoir. The total storage volume behind a reservoir is usually categorized in four main levels: 1) flood control storage is a space used to keep the excessive water behind the reservoir, prevent it from spilling, and reduce the harmful and devastative effects in downstream, 2) water supply storage or the active storage which is used for different demands such as irrigation, drinking and industrial water supplies etc., 3) buffer storage which is a safety level of storage volume set up for every period in a cycle, 4) dead storage or inactive storage which is usually required for power generation because the pumping level to drain the water for generating power has been placed at this point. This is also considered for sediment collection and recreation. There is also a spillway which releases the overflow downstream.

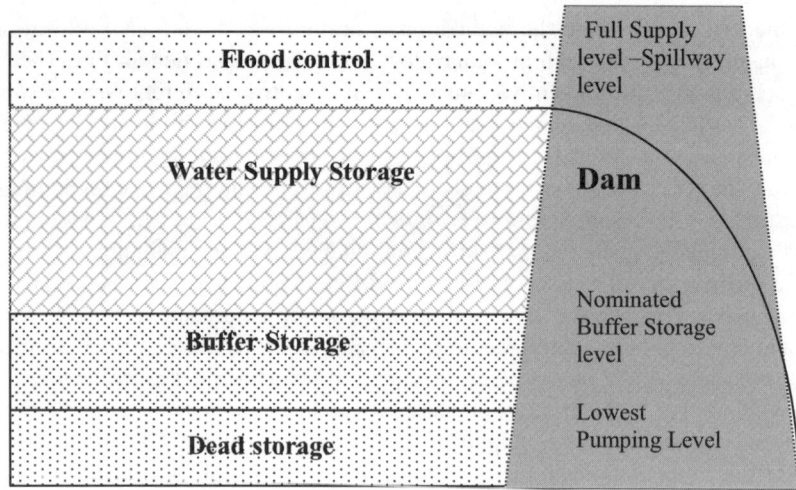

Fig. 14.1. Schematic representation of a single reservoir

14.2.2 Multi-reservoir Configuration

Depending on the purpose of the environment, different configurations of reservoirs can be constructed. The interrelation between these reservoirs could be serial, parallel, or a combination of both (Figure 14.2). Generally, the whole structure of reservoirs is to satisfy a specific demand in the downstream reservoirs. This means that the releases of these reservoirs are only utilized to meet the total demand including irrigation, municipal water supply, etc. However, in some cases each individual upstream reservoir has to meet a portion of the whole demand as well. For example, there is a powerplant corresponding to each reservoir having responsibility to generate power. It is clear that the water released to generate hydroelectric power in each reservoir flows again to downstream reservoirs for other purposes [11]. It is worth mentioning that a portion of release in some reservoirs in the configuration can be assigned for other purposes because of existing water-sharing agreements between different states or governments [4].

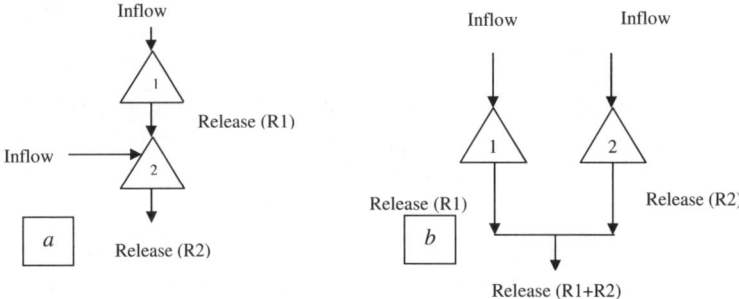

Fig. 14.2. Different configurations of two reservoirs: (a) serial, (b) parallel

14.2.3 Reservoir Modeling

The purpose in the management of surface water resources is to find an operating policy such that the maximum benefit or minimum cost is achieved in a short or long period of time. This operating policy should map the periodic water releases to system states (which are usually predefined discrete values) indicating the different storage volumes at the beginning of each period of a cycle (e.g., year). In optimization of some applications, when the respective model is in the situation of non-stationary Markov decision process, the preceding inflow in every time period can be considered as a new state variable [16]. This kind of state variable, along with other state variables for different reservoirs in the system, form a vector quantifying the system state. It should be noted that the decision (action) of the system can be a vector representing separate decisions for each demanding point, corresponding to each reservoir. In other words, if there is M reservoirs and each reservoir has N demanding points, the number of elements in the respective vector will become $M \times N$. **Objective function** - Assuming that the storage volumes of each reservoir form the vector of system state, the objective function and important constraints can be considered as follows

$$Z = \max \sum_{t=1}^{T} f^t(\mathbf{s}^t, \mathbf{a}^t, \mathbf{d}^t), \qquad (14.1)$$

where T is the number of periods, $\mathbf{s}^t$ is a vector presenting the values of state variables at the time t, $\mathbf{a}^t$ and $\mathbf{d}^t$ are the vectors of actions taken and the given demands, respectively. For example,

$$f^t(\mathbf{s}^t, \mathbf{R}^t) = \max \sum_{t=1}^{T} (\mathbf{R}^t \times \mathbf{c}^t + \mathbf{c}_s^t \times \mathbf{s}^t). \tag{14.2}$$

Here, $\mathbf{R}$ is the release matrix, $\mathbf{c}^t$ is the vector of unit prices of water released, and $\mathbf{c}_s^t$ is the vector of constant values for the respective storage volume of each reservoir. The second term in the objective function is related to power generation. In fact this is a linear form of the nonlinear function as described in [8]. This linear objective function is the zeroth order approximation of Taylor's series for energy function, which is presented by Loucks et al. [8]. **Balance Equation -** This equality constraint indicates the conservation of mass with respect to net inflow and outflow of the reservoir:

$$\mathbf{s}_i^{t+1} = \mathbf{s}_i^t + \mathbf{I}_i^t - \mathbf{R}_i^t + \sum_{\forall j} u_{i,j} \quad \forall t = 1 \cdots T, \ i = 1 \cdots N. \tag{14.3}$$

Here, $\mathbf{I}_i^t$ is the amount of inflows to reservoir i. The variable $u_{i,j}$ is determined based on the routing matrix established by the configuration of reservoirs [7] and demonstrates the volume of water released from reservoir j to reservoir i. For short term optimization, the vector of initial storage volume $\mathbf{s}_i^0$ is assumed to be given.

Minimum and maximum storage - Every element of vector $\mathbf{s}^t$ should satisfy the corresponding minimum and maximum storage level at period t in the respective vectors $\mathbf{s}_{MIN}^t$ and $\mathbf{s}_{MAX}^t$.

Minimum and maximum releases - All possible values for every element of decision vector $\mathbf{R}^t$ should be between the corresponding minimum and maximum release levels at period t in the respective vectors $\mathbf{R}_{MIN}^t$ and $\mathbf{R}_{MAX}^t$. The purpose of this set of constraints is to provide suitable water quality for the existence of wildlife and fish, as well as preventing floods in downstream.

14.3   Reinforced Techniques for Reservoir Management

Stochastic Dynamic Programming (SDP) is a powerful and efficient algorithm for reservoir management in which the original problem is broken down into different subproblems. The optimal solutions for all sub-problems constitute the optimal policy for the main problem. The different techniques in reinforcement learning (RL) such as Q-learning, Sarsa-Learning, Q(λ), and Sarsa(λ) mostly use the idea or some mathematical formulations of SDP. For instance, in Q-Learning the Bellman optimality equation [1] and Robbins-Monro algorithm [13] both contribute to create the main formulation. This technique is model-free which means it does not need the probability distribution of the transition from one state to another state. This advantage makes this technique very attractive for large-scale applications of water resources. In the following subsections, we will explain how to define/configure the state variables, admissible actions, reward/punishment, the optimal policy, and update action-value functions for Q-Learning.

14.3.1 System States

Assuming the storage level is the only state variable in our system, we should descretize the storage volume before the learning is started. There exist some methods for discretizing the storage level (e.g., [5], [6], [2], and [12]); however, the following formulation is a practical way for a single reservoir which is extendable to multi-reservoir applications:

$$A_p = \left\{ s_i^t = s_{min}^t + \frac{(s_{max}^t - s_{min}^t)}{K-1}(i-1) \right\} \text{ for } i = 1 \cdots K, \qquad (14.4)$$

where K is an integer value showing the number of discretizations and A_p is the set of all possible actions. One can find a mathematical way to create a label for each possible state such that the discrete storage levels pertinent to every reservoir in the state vector are extracted. This could be quite beneficial when the state space is very large. Therefore, this information is not needed as a reference during the learning process.

14.3.2 Admissible Actions

At each time period, an RL agent should choose an action among alternative actions. In other words, possible actions may be defined as a set of all discrete values of the control variable (release) within the range between minimum and maximum release volumes (R_{min}^t and R_{max}^t). Given the maximum number of discrete release volumes at each period as H, one may determine the set of discrete possible actions as follows:

$$A_p = \left\{ a_i^t | a_i^t = R_{min}^t + \frac{(R_{max}^t - R_{min}^t)}{H-1}(i-1) \right\} \text{ for } i = 1 \cdots H. \qquad (14.5)$$

The agent may pick an action from this set at each time period based on the policy. It should be noted that in problems with a small state-action space, more simulation periods could be performed to experience all possible actions at different states to find the optimal policy. However, the convergence may be quite slow and perhaps intractable in multi-reservoir systems with numerous state variables. On the other hand, some of those actions are physically infeasible.

Since inflows into reservoir are random variables, the set of admissible actions corresponding to each specific storage volume as state variable might be different for every new inflow. Therefore, in order to have a unique set of admissible actions for every state, a fixed value for inflow should be chosen in the balance equation (Equation 14.3). Depending on whether the lower or upper bound of the inflow bound is taken, two different ways in defining the set of admissible actions associated with any storage volume may be used: 1) optimistic, and 2) pessimistic.

In the optimistic case, the inflow is supposed to be high enough and close to its upper bound based on historical data available or based on a probability distribution function. However, in many time periods of the simulation and as the real inflow would be less than the optimistic inflow chosen, many of those actions might result in infeasible end-of-period storage volumes in simulation and therefore the control system has to punish them. A similar procedure is followed in the pessimistic case, except that a low inflow value is used to specify the set of admissible actions for any state. That low inflow

---

Algorithm 1. Finding admissible actions - single reservoir

1: Find all possible actions with respect to maximum and minimum release

$$R_{min}^t = a_1^t, a_2^t, \ldots a_H^t = R_{max}^t.$$

2: Discretize storage levels of reservoir to K values with respect to maximum and minimum storage levels in each period

$$s_{min}^t = s_i^t, s_2^t, \ldots s_K^t = s_{max}^t.$$

3: Compute minimum and maximum inflow, I_{min}^t and I_{max}^t, from the historical data or from the respective probability distribution

$$P(I^t \leq I_{min}^t) = \epsilon,$$
$$P(I^t \leq I_{max}^t) = 1 - \epsilon,$$

where ϵ is a small positive value close to zero.

4: Compute minimum and maximum evaporation, e_{min}^t and e_{max}^t, from the historical data or from the respective probability distribution

$$P(e^t \leq e_{min}^t) = \epsilon_1,$$
$$P(e^t \leq e_{max}^t) = 1 - \epsilon_1,$$

where ϵ_1 is a small positive value close to zero.

5: Find the end-period-storage
   * Pessimistic:

$$s_i^{t+1} = \bar{s}_j^t + I_{min}^t - e_{max}^t - a_h^t,$$
$$\text{for } j = 1 \cdots K, \quad h = 1 \cdots H.$$

   * Optimistic:

$$s_i^{t+1} = \bar{s}_j^t + I_{max}^t - e_{min}^t - a_h^t,$$
$$\text{for } j = 1 \cdots K, \quad h = 1 \cdots H.$$

6: Find the set of admissible actions (A_i^t) with respect to storage values computed in step 5 (those actions are feasible if the related end-of-period storage s^{t+1} volume in balance equation does not violate the minimum allowable storage volume, s_{min}^{t+1})

$$(A_i^t) = \{a_h^t | s_i^{t+1} \geq s_{min}^{t+1}\} \; for \; h = 1 \cdots H.$$

---

value can be easily determined given the probability distribution function of the random inflow, can be seen in Algorithm 1.

For a stationary system (with respect to time), if the agent is able to maximize the total benefit obtained by system operations over a long time, it may receive less expected benefit with lower variance compared to the optimistic case with more expected benefit and higher variance of the simulated objective function [9].

Algorithm 2. Finding admissible actions - multi-reservoir

1: Compute minimum and maximum inflow, $I_{m,min}^t$ and $I_{m,max}^t$, and minimum and maximum evaporation, $e_{m,max}^t$ and $e_{m,min}^t$ for all reservoirs ($m = 1 \cdots M$)

$$\text{pessimistic} \begin{cases} P(I_m^t \leq I_{m,min}^t) & = & \epsilon_1, \\ P(e_m^t \leq e_{m,max}^t) & = & 1 - \epsilon_2, \end{cases}$$

$$\text{optimistic} \begin{cases} P(I_m^t \leq I_{m,max}^t) & = & 1 - \epsilon_1, \\ P(e_m^t \leq e_{m,min}^t) & = & \epsilon_2. \end{cases}$$

2: Find all possible actions

$$a_{m,h_m}^t = R_{m,min}^t + \frac{(R_{m,max}^t - R_{m,min}^t)}{H_m - 1}(h_m - 1), \text{ for } h_m = 1 \cdots H_m,$$

where a_{m,h_m}^t is the h_m possible action in reservoir m, $R_{m,min}^t$ and $R_{m,max}^t$ are the minimum and maximum release for reservoir m, respectively, and H_m is the number of discretizations for release in reservoir m.

3: In an optimistic scenario, find the maximum release flowing from one reservoir to another, $u_{m',m}^{*t}$, with respect to routing matrix [7]

$$\begin{cases} s_m^{t+1} & = & \bar{s}_{m,k_m}^t + I_{m,max}^t - e_{m,min}^t - a_{m,h_m}^t + \sum_{m'=1,m'\neq m}^{M} u_{m',m}^{*t}, \\ u_{m',m}^{*t} & = & max\{a_{m',h_{m'}'}^t | s_{m'}^{t+1} \geq s_{m',min}^{t+1}, \bar{s}_{m',k_{m'}}^t\}, \\ & & \text{for } m = 1 \cdots M, \quad k_m = 1 \cdots K_m, \quad h_m = 1 \cdots H_m \end{cases}$$

where $\bar{s}_{m,k_m}$ is k_m^{th} discrete value of storage in reservoir m, K_m is the number of discretizations for storage level in reservoir m, and s_m^{t+1} is the end-of-period storage volume.

4: In a optimistic scenario, find admissible actions

$$A(m, k_m) = \{a_{m,h_m}^t | s_m^{t+1} \geq s_{m,min}^{t+1}, \bar{s}_{m,k_m}^t\},$$
$$\text{for } m = 1 \cdots M, \quad k_m = 1 \cdots, K_m, \quad h_m = 1 \cdots H_m.$$

5: In a pessimistic scenario, perform steps 3, 4 after changing $I_{m,max}^t$ and $e_{m,min}^t$ to $I_{m,min}^t$ and $e_{m,max}^t$; respectively, and change maximization to minimization for finding $u_{m',m}^{*t}$.

In multi-reservoir applications, in addition to precipitation and run off water flowing to each reservoir, water might be released from one reservoir to another depending on the structures of reservoirs given by the routing matrix [7]. Algorithm 2 demonstrates the process for finding the admissible action in pessimistic and optimistic schemes.

14.3.3 Reward/Punishment

To find the immediate or delayed reward/punishment, we should first define a new terminology, action taken and actual decision. As the environment is stochastic and because of existing constraints of maximum and minimum storage volume, it is clear that the action taken by the agent might not be the same as the release performed during the respective period. This release is called the actual decision and is a basis to calculate the immediate reward and punishment. In other words, because of stochasticity in reservoir management, there might be different rewards/punishments for one specific action in every interaction within the environment. For instance, if the action and the actual decision are a^t and d^t and the objective function is total income earned from selling water for r^t per unit, the reward and punishment will be $a^t \times r^t$ and $(a^t - d^t) \times r^t$, respectively. Therefore, the final reward in this simple example will be $d^t \times r^t$.

14.3.4   Optimal Policy and Updating Action-Value Function, Q-Learning

Using a greedy, ϵ-greedy, or Softmax policy, an agent takes the next action based on updated action-value functions. In other words, after receiving the reward from environment, the agent updates the action-value function related to the previous state and takes a new action. For the starting point, arbitrary values should be considered for all action-value functions $Q(i, a)$ (e.g., $Q(i, a) = 0$, for all states i, and actions a). In each iteration at least one value should be updated. Assuming that the agent places storage s^t $= i$ and takes action a^t, based on the balance equation (Equation 14.3) and the value of the stochastic variables, the next storage becomes $s^{t+1} = j$. Suppose this storage meets the minimum level of reservoir, then the action-value functions are updated as follows:

$$
Q_t^{(n)}(i, a) = \begin{cases} Q_t^{(n-1)}(i, a) + \alpha \times [r_e^t + \gamma \max_{b \in A(j,t')} Q_{t'}^{n-1}(j, b) - Q_t^{(n-1)}(i, a)], \\ \qquad \text{If } a = a^t \text{ and } i = i^t \\ Q_t^{(n-1)}(i, a), \quad \text{Otherwise} \end{cases}
\tag{14.6}
$$

where n is the number of iterations performed from the the starting point of the learning and r_e^t is the immediate reward in time period t.

Although the transition from one storage to another is continuous and is controlled by the balance equation (Equation 14.3), there is no guarantee that the current or next storage is exactly one of the discrete values. There are two different storages in each iteration that should be approximated to discrete values:

1. Storage, which the agent currently redobserves (current state), and
2. Storage, which the agent will observe on in the next iteration (next state).

An easy way to tackle this problem is to find the closest discrete value to the current or next storage. However, this seems to be error-prone because the respective storage only partially belongs to this state. To increase the accuracy, we might use a linear or non-linear interpolation. To interpolate linearly, firstly, we have to find two successive values s^{-t}, s^{+t}, which are the closest discrete values to the current storage. Finally, the degrees of closeness of the storage to these boundaries should be computed based on the following formulation:

$$
w_1 = 1 - \frac{s^t - \bar{s}^{-t}}{\bar{s}^{+t} - \bar{s}^{-t}}, \qquad w_2 = 1 - \frac{\bar{s}^{+t} - s^t}{\bar{s}^{+t} - \bar{s}^{-t}}.
\tag{14.7}
$$

Algorithm 3. Updating action-value functions using interpolation (single-reservoir)

1: Take the closest discrete value of storage in period t, $\bar{s}^{-t} = i$ or $\bar{s}^{+t} = i$, to the actual current storage, s_i^t

2: Compute the next storage volume using the balance equation (Equation 14.3) and other boundary constraints

3: Find two successive discrete values which are the closest values to the next storage ($\bar{s}^{-(t+1)} = j$, $\bar{s}^{+(t+1)} = j + 1$)

4: Find w_1 and w_2 using Equation 14.7

5: Perform interpolation for the next storage

$$Q_z^{(n)} = Q_t^{(n-1)}(i, a) + \alpha[r_e^t + \gamma \max_{b \in A(j', t')} Q_{t'}^{n-1}(j', b) - Q_t^{(n-1)}(i, a)],$$
$$j' = (j - 1) + z \quad \& \quad z = 1, 2$$

where r_e^t is the immediate reward/punishment.

6: Update the action-value function

$$Q_t^{(n)}(i, a) = \sum_{z=1}^{2} w_z \times Q_z^{(n)}.$$

7: Perform interpolation for the current storage as needed. Consider ($\bar{s}^{-t} = i$, $\bar{s}^{+t} = i + 1$) as two consecutive boundaries for current storage and w_1' and w_2' as respective weights calculated using Equation 14.7

$$Q_{zz'}^{(n)} = Q_t^{(n-1)}(i', a) + \alpha w_{z'}'[r_e^t + \gamma \max_{b \in A(j', t')} Q_{t'}^{n-1}(j', b) - Q_t^{(n-1)}(i', a)],$$
$$i' = (i - 1) + z', \quad j' = (j - 1) + z, \quad z = 1, 2 \quad \& \quad z' = 1, 2.$$

8: Update action-value function

$$Q_t^{(n)}(i', a) = \sum_{z=1}^{2} w_z \times Q_{zz'}^{(n)} \quad for \quad z' = 1, 2.$$

As seen in Algorithm 3, since the current storage partially belongs to both upper and lower state boundaries, the learning rate α in each step should be multiplied by a new weight obtained from Equation 14.8 for the closest state to the current state ($\alpha \times w'$). The new learning rate is smaller than the previous one, therefore, the process of learning of action-value functions is slower and presumably more accurate than the situation in which only one action-value function pertinent to lower or upper discrete storage is updated.

In multi-reservoir case, the analogous process should be performed to update action-value functions but the weighting and some kind of normalization are needed. In a reservoir application composed of M reservoirs, after taking an action in the current state, the storages in all reservoirs become $s_1^{t+1}, \cdots, s_M^{t+1}$. Each of these storages could be exactly the same discrete value considered for the respective state variable or can be between two successive representatives of state variable:

$$
\begin{array}{l}
\bar{s}_{1,k_1}^{(t+1)} \le s_1^{t+1} \le \bar{s}_{1,k_1+1}^{(t+1)} \\
\bar{s}_{2,k_2}^{(t+1)} \le s_2^{t+1} \le \bar{s}_{2,k_2+1}^{(t+1)} \\
\cdots\cdots\cdots\cdots\cdots \\
\bar{s}_{M,k_M}^{(t+1)} \le s_M^{t+1} \le \bar{s}_{M,k_M+1}^{(t+1)}
\end{array}
\longrightarrow
\begin{bmatrix}
\bar{s}_{1,k_1}^{(t+1)} & \bar{s}_{1,k_1+1}^{(t+1)} \\
\bar{s}_{2,k_2}^{(t+1)} & \bar{s}_{2,k_2+1}^{(t+1)} \\
\vdots & \vdots \\
\bar{s}_{M,k_M}^{(t+1)} & \bar{s}_{M,k_M+1}^{(t+1)}
\end{bmatrix}.
\tag{14.8}
$$

So there are at most 2^M combinations to be considered if none of them is equal to their boundaries:

$$
\bar{S} =
\begin{matrix}
1 \\ 2 \\ 3 \\ \vdots \\ 2^M
\end{matrix}
\begin{bmatrix}
\bar{s}_{1,k_1}^{(t+1)} & \bar{s}_{2,k_2}^{(t+1)} & \cdots & \bar{s}_{M,k_M}^{(t+1)} \\
\bar{s}_{1,k_1+1}^{(t+1)} & \bar{s}_{2,k_2}^{(t+1)} & \cdots & \bar{s}_{M,k_M}^{(t+1)} \\
\bar{s}_{1,k_1}^{(t+1)} & \bar{s}_{2,k_2+1}^{(t+1)} & \cdots & \bar{s}_{M,k_M}^{(t+1)} \\
\vdots & \vdots & \vdots & \vdots \\
\bar{s}_{1,k_1+1}^{(t+1)} & \bar{s}_{2,k_2+1}^{(t+1)} & \cdots & \bar{s}_{M,k_M+1}^{(t+1)}
\end{bmatrix},
\tag{14.9}
$$

where $\bar{S}$ is a matrix showing all possible combinations of storage volumes in the boundaries of the respective state in each reservoir. If the current or next storage of one reservoir becomes one of the values of its boundaries, the possible combinations are decreased to 2^{M-1}. Therefore, if all storages in the state vector become equal to one of the boundaries, there is only one combination to be considered as a state of the system, then no interpolation is required.

Analogously, the closeness of each element of state could be computed as follows:

$$
w_{j,k_j} = 1 - \frac{s_j^{t+1} - \bar{s}_{j,k_j}^{t+1}}{\bar{s}_{j,k_j+1}^{t+1} - \bar{s}_{j,k_j}^{t+1}} \quad , \quad w_{j,k_j+1} = 1 - \frac{\bar{s}_{j,k_j+1}^{t+1} - s_j^{t+1}}{\bar{s}_{j,k_j+1}^{t+1} - \bar{s}_{j,k_j}^{t+1}}.
\tag{14.10}
$$

To obtain the weight for each combination, we have to normalize these values as follows:

$$
\bar{w}_{j,k_j} = \frac{w_{j,k_j}}{2^{M-1} \sum\limits_{j}^{M} w_{j,k_j} + w_{j,k_j+1}},
\tag{14.11}
$$

$$
\bar{w}_{j,k_j+1} = \frac{w_{j,k_j+1}}{2^{M-1} \sum\limits_{j}^{M} w_{j,k_j} + w_{j,k_j+1}}.
\tag{14.12}
$$

The summation of normalized weights for each combination is considered as a final weight of the respective state (w_l, $l = 1 \cdots 2^M$). This method can be implemented for the current state after taking an action. The weight for all possible combinations of current state is denoted by $w'_{l'}$ ($l' = 1 \cdots 2^M$).

14.4   Opposition-Based Schemes in Reservoir Management

As mentioned in the previous section, a reinforcement agent attempts to acquire knowledge through interactions with the environment. However, when the system of reservoirs is large or the number of possible state-action pairs is huge, these interactions might be time-consuming. Moreover, a physical limitation in terms of computer memory may exist. Therefore, the question may arise how to achieve an optimal or near-optimal solution with fewer interactions. Opposition-Based Learning (OBL) scheme, introduced by Tizhoosh [14], and applied to extend different artificial intelligent (AI) techniques such as reinforcement learning (RL) [15] could be a suitable answer to the mentioned question because it uses the current knowledge to update other action-value functions without taking a new action and visiting a new situation.

14.4.1   Opposite Action and Opposite State

As described in Chapter 2 (pp. 17), we distinguish between type I and II opposition. For type II opposition, the action-value functions play a key role in finding the opposite action and opposite state. However, the majority of action-value functions have zero value at the very beginning of the learning process or may have not been sufficiently observed. Therefore, they cannot be reliable values for finding the opposites. To increase the accuracy of finding the actual opposite action and opposite state, some kind of function approximation such as a feed-forward Multi-Layer Perceptron networks (MLP) or a Fuzzy Rule-Based modeling (FRB) can be used. In this type of opposition mining (see Chapter 2), regular learning is performed in order to obtain initial information about the behavior of the system and specifically about action-value functions. The information related to action-state pairs which are visited during learning can be empolyed as training data for the function approximations. Using other action-state pairs as test data, new knowledge or action-value functions are extracted. In other words, we can introduce an opposite agent which is not actually taking any action. However, the knowledge provided by the function approximation can be used to create a set of new action-value functions which then are used for finding the opposites.

Furthermore, because training an MLP or creating FRB is time-consuming and might not be practical during the learning process, the training step could be performed only once every several episodes. The distance between two consecutive trainings is considered as a model parameter and should be specified at the beginning of the learning or dynimically decided based on a heuristic.

As a result, there are two sets of action-value functions: one is obtained from the direct interactions of the agent with the environment, and the second is extracted from the function approximation. This knowledge is kept and updated using the action taken by the agent and the opposite action introduced by the opposite agent until another function approximation is run and new knowledge is generated. This might increase the accuracy of finding the actual opposites between two runs of function approximation. It is quite clear that as the learning process continues, more accurate opposite actions and states can be achieved. The training and test data can be obtained with respect to

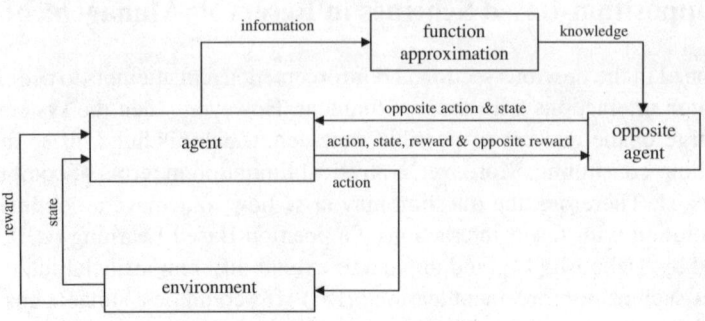

Fig. 14.3. Type-II oppositional learning using function approximation

the number of observations for action-state pairs or the rate of change in action-value functions in every iteration.

Figures 14.3 illustrates the opposition-based learning based on this scheme.

14.4.2   Reward and Punishment

As previously mentioned, the opposite agent is not really taking the opposite action. Given the value of stochastic parameters such as inflow and evaporation after taking an action, the real opposite decision $\breve{R}^t$ pertinent to opposite action $\breve{a}$ or action a in the opposite state $\breve{s}$ can be extracted using the balance equation (Equation 14.3) and the boundaries of storage volume. This is illustrated in the equation 14.14 when the opposite action is introduced:

$$\breve{s}^{t+1} = s^t + I^t - e_t - \breve{a}^t, \tag{14.13}$$

$$\breve{R}^t = \begin{cases} \breve{a}^t & \text{if } s_{min}^{t+1} \leq \breve{s}^{t+1} \leq s_{max}^{t+1} \\ s^t - s_{min}^{t+1} + I^t - e_t & \text{if } \breve{s}^{t+1} < s_{min}^{t+1} \\ \breve{a}^t & \text{otherwise} \end{cases} \tag{14.14}$$

where $\breve{R}^t$ is the actual opposite release at time period t. It should be noted that the next state pertinent to the opposite action is set to the boundary values s_{min} or s_{max} if they are out of limit.

14.4.3   Proposed Opposition Mining Algorithm

Using the above definitions, the pseudo-code of the opposition-mining/type-II oppositional Q-learing, is presented through the Algorithm 4. The analogous algorithm can be designed for opposite state. Moreover, this algorithm is easily extendable to other RL techniques such as $Q(\lambda)$ and $sarsa(\lambda)$.

Algorithm 4. Q-Learning with Type-II Opposition

1: Initialize action-value functions, e.g., $Q_{i,a}^t = 0$
2: Find admissible actions, $A^t(i)$
3: Determine the total number of episodes *"noepisode"* and the number of years *"noyears"* in each episode
4: Set the starting episode sp, a counter showing the number of training $n = 1$, and the interval between two consecutive trainings, c
5: Determine the starting and ending o_s and o_e episode for using OBL
6: Start episode with $mm = 1$
7:    Choose initial state: $i = s$ and time t (e.g., randomly)
8:    Take an admissible action using policy π (e.g., greedy,...) and receive next state $j=s'$ and reward r
9:    Find the closeness weights w_1 and w_2
10:   **If** $mm \leq o_e$
11:      ** **If** $o_s \leq mm < sp$
12:         * Use type-I scheme to find the opposite action as follows
$$\breve{a}^t = R_{max}^t + R_{min}^t - a^t$$
13:      ** **If** $mm = sp + c \times (n - 1)$
14:         * $n = n + 1$
15:         * Initialize approximated action-value functions
16:            $\hat{Q}^t(i, a) = Q^t(i, a)$ for all i and a
17:         * Choose action-state pairs which are visited enough as training data
18:         * Train the function approximation, MLP, FRB,...
19:         * Update $\hat{Q}^t(i, a)$ for the test data using approximated functions, $\hat{Q}^t(i, a)$
20:      ** **If** $mm \geq sp$ & $mm \geq o_s$
21:         * Compute the type-II opposite action using the following equation

$$\breve{a} \in \{\hat{a} \mid \hat{Q}_{(i,\hat{a})}^t \approx \max_{(b \in A_i)} \hat{Q}_{(s_i,b)}^t + \min_{(b \in A_i)} \hat{Q}_{(s_i,b)}^t - \hat{Q}_{(s_i,a)}^t\}.$$

22:      ** Find the closeness weights $\breve{w}_1, \breve{w}_2$ for next state when the opposite action is taken,
23:   Use Algorithm 3 to update original action-value functions and their approximates corresponding to the action and the opposite action, $Q^t(i, a), Q^t(i, \breve{a}), \hat{Q}^t(i, a)$, and $\hat{Q}^t(i, \breve{a})$
24:   **If** $t \neq T$, **then** $t = t + 1$, $i = s'$, and go to step 8; **otherwise,** go to the next step
25:   **If** $year \leq noyears$, **then** $year = year + 1$, $t = 1$, $i = s'$, and go to step 8; **otherwise,** go to the next step
26: **If** $mm \leq noepisode$, **then** $mm = mm + 1$, $year = 1$, and go to step 7; **otherwise,** go to the next step
27: Find the best decisions for all states

14.4.4 Performance Criteria

One way to demonstrate the performance of an operating policy is to run a simulation using sufficient historical or synthetic data to calculate the various criteria such as average cyclic gain

$$\mu = \sum_{i=1}^{Y} \mu_i / Y, \tag{14.15}$$

cyclic variance

$$\sigma = \sum_{i=1}^{Y} (\mu_i - \mu) / Y - 1, \tag{14.16}$$

and coefficients of variation $cov = \sigma/\mu$ where Y is the number of years in the simulation and μ_i is the benefit/cost in every cycle. It is quite possible to have different operating policies at the end of each complete run of the learning program for the same number of episodes.

Another criterion to compare two approaches is to calculate the deviation of achieved policies if there is an optimal solution. In a multiple-solution application, we can measure the deviation of value functions with respect to every policy. As a result, a lower deviation in operating policies or value functions at the end of all experiments represents a more robust learning method.

To find a performance criterion measuring the robustness of learning and its opposition versions, we can define the distance between all possible policy pairs or action-value functions as follows:

$$L_{(d,d')} = \sum_{t=1}^{T} \sum_{i=1}^{K} \left(|\pi_{(i,d)}^t - \pi_{(i,d')}^t| \right),$$
$$\text{for } d = 1 \cdots D - 1, \ d' = d + 1 \cdots D, \ J = \frac{D!}{2! \times (D-2)!} \tag{14.17}$$

where D is the number of experiments for each set of parameters, K is the number of state variables, $\pi_{(i,d)}^t$ and $\pi_{(i,d')}^t$ are two different action policies which map the ith state and the tth period to the optimal action a, and J is the number of pairwise combinations of these action policies.

Given these distances, the following criteria called the mean and variance of distance, $\bar{L}$ and σ_L^2, are introduced and utilized to measure the robustness of learning methods [10]:

$$\bar{L} = \frac{\sum_{d=1}^{D-1} \sum_{d'=d+1}^{D} L_{(d,d')}}{J},$$
$$\sigma_L^2 = \frac{\sum_{d=1}^{D-1} \sum_{d'=d+1}^{D} (L_{(d,d')} - \bar{L})^2}{J-1}. \tag{14.18}$$

14.5   Experimental Results

We have chosen an application of single reservoir case study according to Fletcher [3]. The detailed information of respective data including minimum and maximum storage and release in different periods of a cycle are given in Table 14.1. One cycle in this problem is one complete year with 12 periods (months).

Table 14.1. Maximum and minimum storage and release, average inflow, and demand

Value	Month											
m^3	1	2	3	4	5	6	7	8	9	10	11	12
Max. Storage	8.0	8.0	8.0	8.0	8.0	8.0	8.0	8.5	8.0	8.0	8.0	8.0
Min. Storage	1.0	1.0	1.0	1.0	1.0	1.0	1.0	1	1.0	1.0	1.0	1.0
Max. release	4.0	4.0	6.0	6.0	7.5	12.0	8.5	8.5	6.0	5.0	4.0	4.0
Min. release	0.0	0.0	0.0	0.0	0.0	0.0	0.0	0.0	0.0	0.0	0.0	0.0
Average inflow	3.4	3.7	5.0	5.0	7.0	6.5	6.0	5.5	4.3	4.2	4.0	3.7

Table 14.2. The benefit of release per unit for each month of a year

benefit	Month											
($)	1	2	3	4	5	6	7	8	9	10	11	12
Release	1.4	1.1	1.0	1.0	1.2	1.8	2.5	2.2	2.0	1.8	2.2	1.8

Table 14.3. Performance criteria of the the learning methods with variable learning rate ($\alpha = A/visit(i, a)$, $A = 1$)

Methods	Criterion	number of episodes								
		20	40	60	80	100	200	300	400	500
QL	μ	84.72	91.27	94.10	96.17	98.36	99.14	99.32	99.57	99.89
	σ	38.78	52.37	55.09	60.11	60.03	80.20	82.08	82.80	82.28
	Cov	0.073	0.079	0.079	0.080	0.079	0.090	0.091	0.091	0.091
OBL QL type I	μ	92.43	96.60	98.44	99.26	99.23	99.30	99.57	99.88	100.18
	σ	54.50	60.83	65.72	70.43	75.01	81.59	82.65	80.52	79.68
	Cov	0.080	0.081	0.082	0.085	0.087	0.091	0.091	0.090	0.089
OBL QL type II	μ^*	96.46	98.17	98.51	98.99	99.09	99.35	99.73	99.84	100.01
	σ	55.75	65.07	73.91	77.34	78.32	82.48	82.20	83.02	82.29
	Cov	0.077	0.082	0.087	0.089	0.089	0.091	0.091	0.091	0.091

To simplify the problem, inflow to the reservoir is normally distributed and the monthly averages are presented in Table 14.1. The coefficient of variation ($\frac{\sigma_{It}}{\mu_{It}}$) is used for finding the variance in all periods and is given as an input parameter. Moreover, the evaporation from the reservoir is neglected.

Power generation is considered as a main objective function. However, in order to make the objective simple and linear, the benefit of release per/unit is approximated in a single value for each month and given in Table 14.2.

Table 14.4. The result of t-test to compare the average gain of regular Q-learning and its OBL versions ($\alpha = \frac{A}{visit(i,a)}$, $A = 1$)

comparison t_v & p_v		number of episodes								
		20	40	60	80	100	200	300	400	500
QL vs.	t_v	13.35	13.11	14.30	16.18	4.37	1.14	1.79	2.46	2.10
OBL I	p_v^*	0	0	0	0	0	0.13	0.040	0.01	0.02
OBL I vs.	t_v	8.71	5.18	0.30	-1.44	-0.70	0.33	1.12	-0.29	-1.22
OBL II	p_v^*	0	0	0.38	0.92	0.76	0.37	0.13	0.62	0.89

$* \; p_v = Pr(t \geq t_v)$.

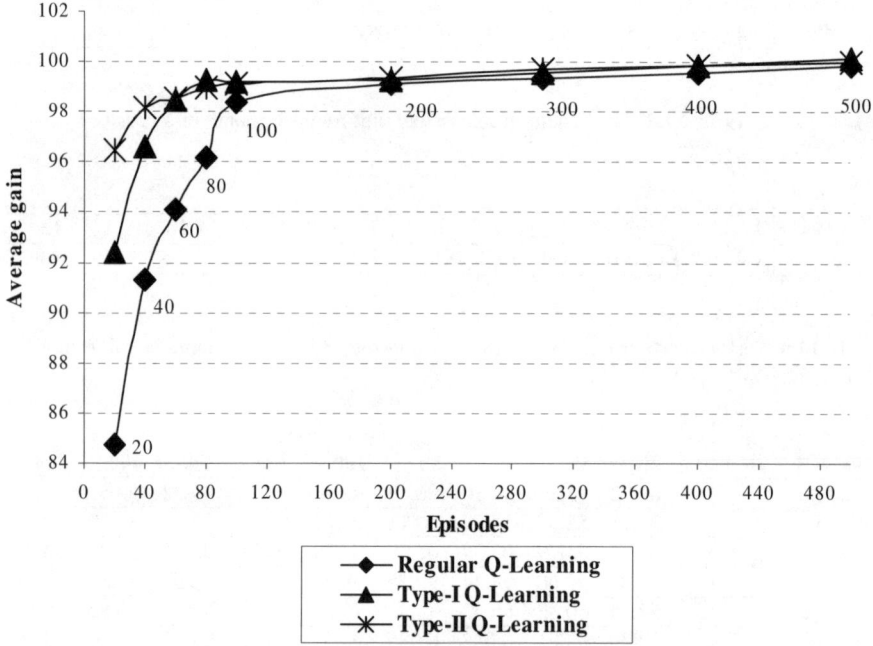

Fig. 14.4. The average annual benefit for different episodes with variable learning rate

The following configuration was used to generate the results:

- Storage volume and water release were discretized into 55 and 7 equal intervals, respectively.
- The number of years in each episode is equivalent to 30.
- We have only used opposite actions (no opposite states).
- A multi-layer perceptron (MLP) with one hidden layer and 20 nodes is used to approximate the respective action-value functions. The number of trainings is set as a constant value to 2. Moreover, the episode number, c, in which the training

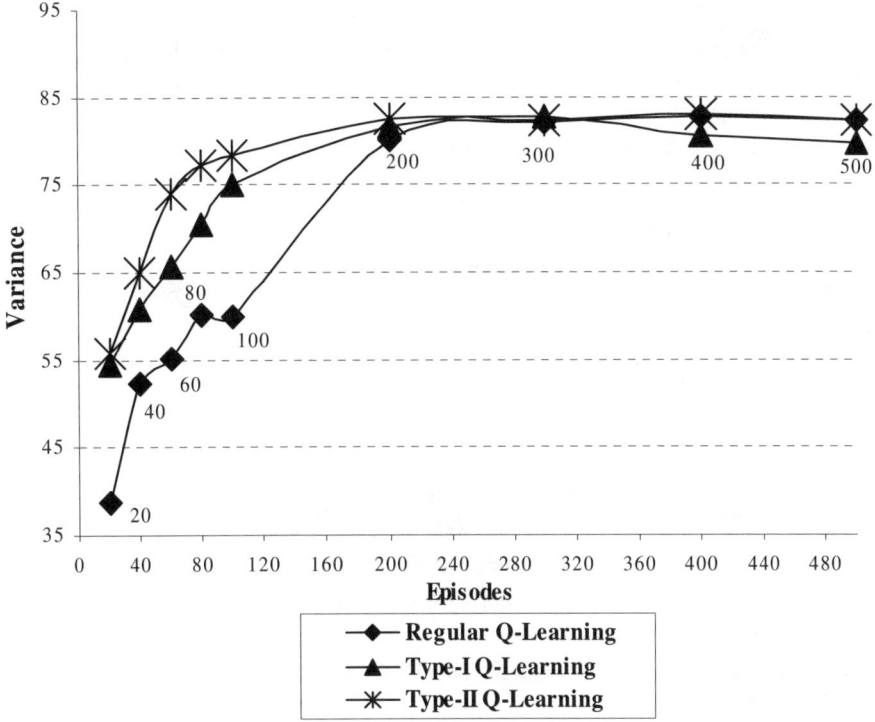

Fig. 14.5. The average of variance of annual benefit for different episodes with variable learning rate

is started should be determined at the beginning of each experiment. This number varies for every set of parameters.

- The results of Stochastic Dynamic Programming (SDP) are used as gold standard to show how accurate the learning methods are.
- The extracted operating policy from the SDP or the learning parts were simulated for 2000 years and the mean, variance, and coefficient of variations of benefit in one year were obtained. However, each experiment was performed 30 times and the average and the variance of all criteria were used for comparison.
- In order to investigate the efficiency of our approach in a highly stochastic situation, we have chosen a large coefficient of variations equivalent to 0.3 which creates reasonable variabilities in inflow during the learning.
- The optimistic scheme for finding admissible action was used.
- Both constant and variable learning rates were considered (α=0.5 or $\alpha = A/visit$ (i,a) where $A = 1$ and $visit(i, a)$ is the number of times that state i and action a have been observed or considered).
- We have chosen the ϵ-greedy policy with $\epsilon = 0.1$ for the agent to take action [9].

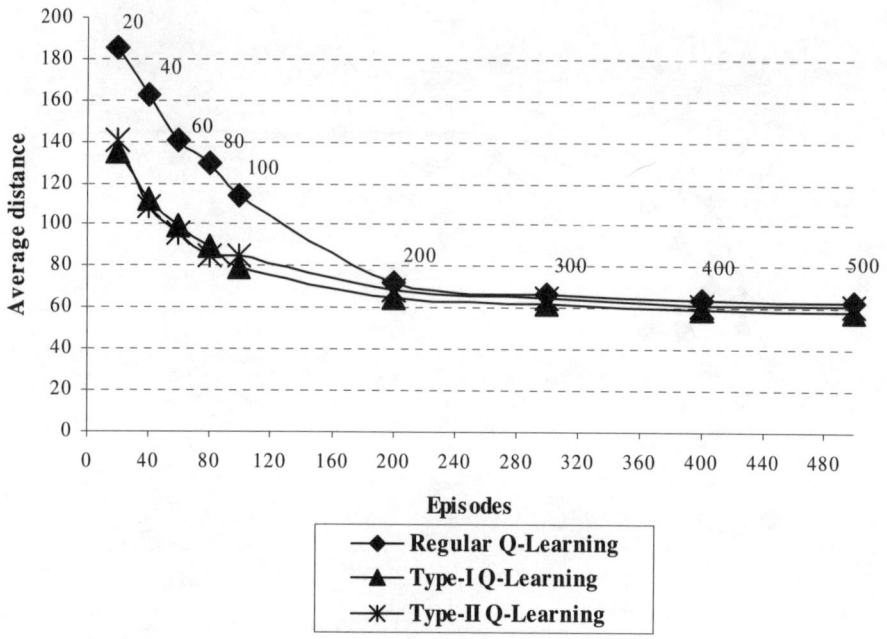

Fig. 14.6. The mean of distance between two policies for different episodes in the learning methods with variable learning rate

- In contrary to those experiments performed in [10], the staring point of the learning in our experiments was exactly after the training of first function approximation, and there is no oppositional learning before this point.
- The target performance criteria in terms of average benefit, variance, and coefficient of variation in this application are 102.27, 80.42, and 0.088, respectively. These evaluations are obtained through the simulation of the optimal policy which is extracted from SDP approach.

Table 14.3 compares the performance of regular Q-learning and its opposition-based versions where the learning rate α is a variable parameter: $\alpha = \frac{1}{visit(s^t, a^t)}$. Both OBL versions make significant improvements for episodes from 20 to 100 as the total interactions of the agent within the environment before the operating policy is extracted. However, as the number of episodes increases, the performance criteria of all algorithms get closer together such that there is almost no difference between these values when the number of episode is sufficiently large. Moreover, the type II scheme of the OBL learning shows slightly better results compared to type I version for some experiments with few episodes in terms of average annual gain (Figures 14.4). For instance, in the case with 20 episodes, the average benefit and the average variance for the simulated policy pertinent to OBL type II scheme are 96.46 and 55.75, respectively. However, these values are 92.43 and 54.50 for the OBL type I. In order to verify these results statistically, assuming the student distribution for average gains, we can use the t-test

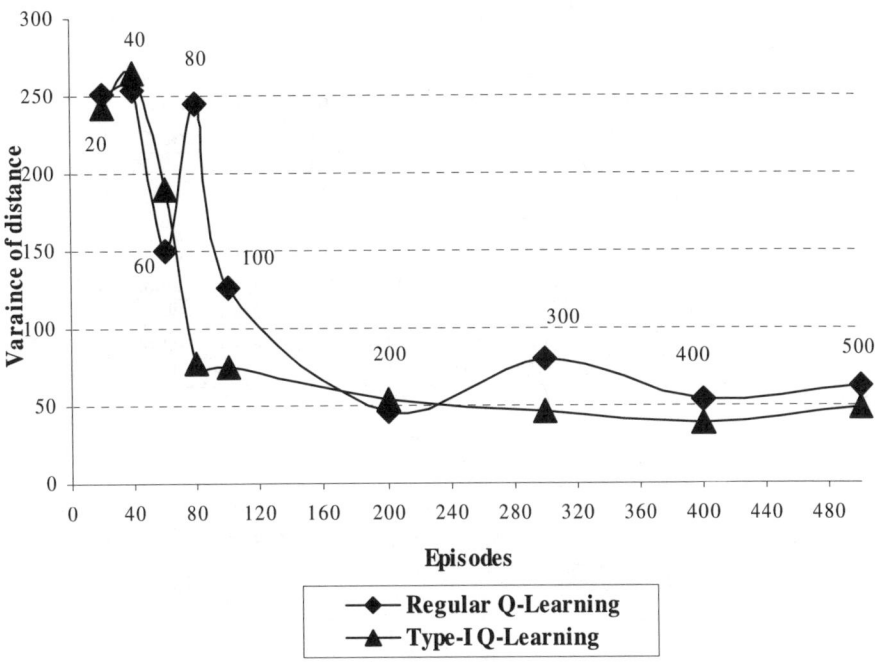

Fig. 14.7. The variance of distance between two policies for different episodes in the learning methods with variable learning rate

with 58 degrees of freedom to measure the superiority of the learning method in terms of different criteria. We set null hypothesis as average benefit in regular learning equivalent to its OBL versions versus the alternative hypothesis in which the OBL versions are more appropriate for different episodes. We can also repeat the respective t-test to compare both OBL versions with each other.

As it can be seen in Table 14.4, there is strong evidence that the averages in OBL versions are higher than regular learning for all early episodes (e.g., $p_v = 0$ for all episodes lower than 100 in the comparison of type I and regular learning) while in the rest the evidence is not strong enough (e.g., $p_v = 0.13$ where the number of episodes is equivalent to 200 in the comparison of type I and regular learning). Furthermore, the test confirms the previous results for both OBL versions.

In terms of the robustness criteria explained in Equation 14.18, the OBL learning for both type I and type II have better results compared to regular learning specifically for small number of episodes (Figures 14.6-14.7). It is worth mentioning that we use the policy $\pi_{i,a}^t$ instead of action-value functions $Q_{i,a}^t$ to derive these criteria because the operating policies extracted at the end of different experiments, subject to sufficient iterations, are almost the same. In other words, it seems that this problem has a unique solution and it therefore can be used to measure the robustness of learning method.

Table 14.5. Performance criteria of the learning methods with constant learning rate ($\alpha = 0.5$)

Methods	Criterion	number of episodes								
		20	**40**	**60**	**80**	**100**	**200**	**300**	**400**	**500**
QL	μ	71.13	77.28	80.49	83.42	85.92	89.57	93.45	95.57	96.59
	σ	35.43	38.59	39.29	43.73	47.84	49.17	48.71	49.35	54.38
	Cov	0.081	0.080	0.077	0.079	0.080	0.078	0.075	0.073	0.076
OBL QL type I	μ	90.62	94.02	95.40	96.36	96.70	98.42	99.32	99.42	99.48
	σ	40.90	42.48	39.06	45.33	46.32	55.11	62.62	65.50	64.38
	Cov	0.070	0.069	0.065	0.070	0.070	0.075	0.080	0.081	0.081
OBL QL type II	μ^*	83.88	87.45	89.76	92.38	94.14	99.15	99.70	99.76	99.64
	σ	45.3	48.28	50.79	53.84	56.42	62.74	64.18	63.69	66.28
	Cov	0.080	0.079	0.080	0.079	0.080	0.080	0.080	0.080	0.082

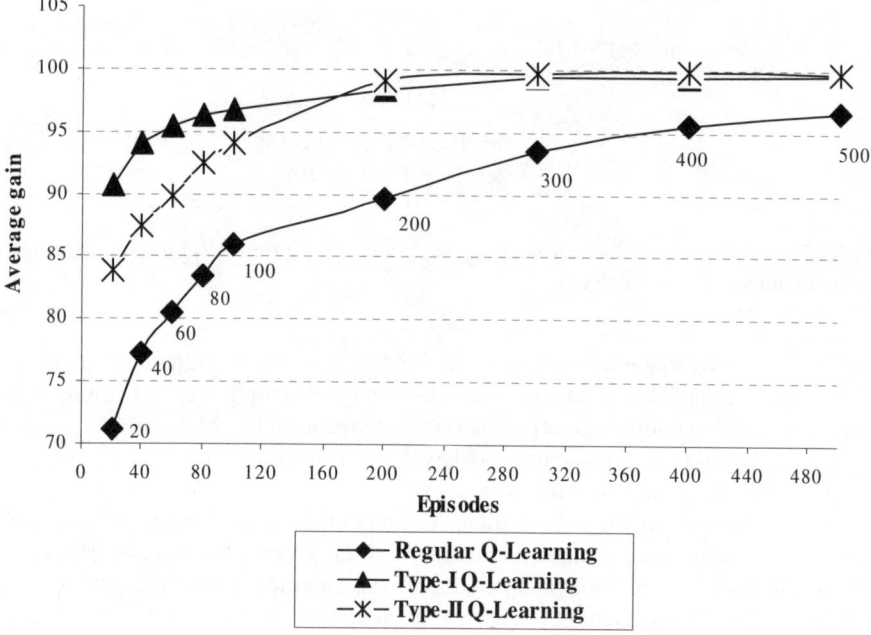

Fig. 14.8. The average annual benefit for different episodes in the learning methods with constant learning rate ($\alpha = 0.5$)

Furthermore, it seems that the way of determining the learning rate α might affect the efficiency of the OBL. In order to investigate this issue, we have used the same set of criteria achieved for this application as reported in [10] with a constant learning rate $\alpha = 0.5$ (Table 14.5). As observed in this table, the variation between the average benefits in all learning methods vanishes as the number of episodes increases. However, the respective variances of OBL remain almost unchanged in all types of learning for

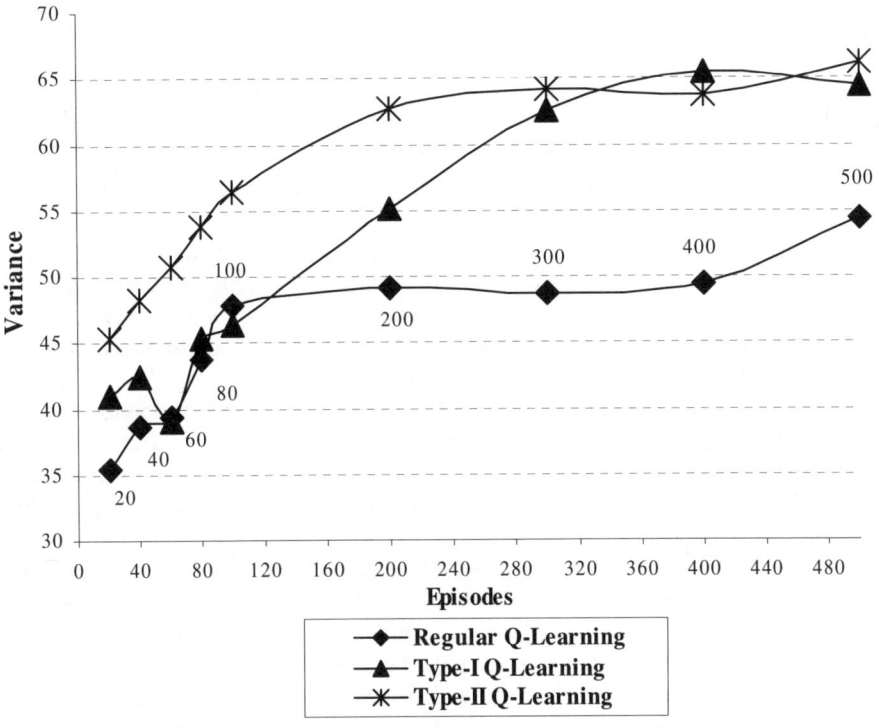

Fig. 14.9. The average of variance of annual benefit for different episodes in the learning methods with constant learning rate ($\alpha = 0.5$)

episodes higher than 300. This implies that there is no justification to continue oppositional learning after early episodes and we can switch to regular Q-Learning thereafter. However, as previously illustrated, if the learning rate is variable, the results could be slightly different. Moreover, the contribution of OBL learning for both versions with a constant learning rate are better compared to the situation in which the learning rate is variable specifically when the learning process is performed for a set of small number of episodes such as 20 or 40.

14.6 Summary and Conclusions

In this chapter, we have demonstrated how to utilize Q-Learning as a popular technique in Reinforcement Learning (RL) to obtain the operating policies in reservoir management and showed some experimental results for a single reservoir case study. However, in order to overcome the potential computational burden for very large-scale problems and speed up the learning process, we introduced and tuned a new type-II opposition-based scheme.

In type II oppositional RL, we have defined an opposite agent to assist the main agent to find the opposite action and/or state. This opposite agent receives the knowledge from a trained function approximation and updates itself using the information coming from the main agent until receiving new knowledge from the function approximator. The respective function approximation which is a multi layer perceptron in our case study is trained using the information captured by the main agent. An equal interval between two training times is a parameter and has to be determined before the learning is started.

In order to show that the OBL versions are beneficial in finding near-optimal operating policies, we have chosen a single reservoir problem in which the target value of the optimal policy can be easily obtained by a Stochastic Dynamic Programming (SDP) algorithm. Furthermore, to evaluate the extracted policies, some performance criteria including average annual benefit, variance, and coefficient of variation have been calculated using the respective policies through the simulation.

Both types of OBL-extended algorithms have better results where the total number of iterations are not sufficiently for regular learning. This means that we could compute the near-optimal operating policy in a very small number of iterations. This fact creates a promising perspective to efficiently cope with large-scale applications in the field of reservoir management.

It seems that the policy extracted in type II learning method is approaching an optimal policy faster than OBL type I. We found that the closeness of operating policy to optimal one is really dependent on the efficiency of the function approximation as an input to the opposite agent. The training data fed to the function approximation should be determined with caution such that it embodies different situations. In other words, the boundary limit used to separate the training and test data can be specified with respect to the number of observations for all state-action pairs. it is also possible to set the feeding data based on the rate of change in action-value functions.

The action-value functions conveyed to the opposite agent could be composed of actual values which are obtained by the regular learning process and considered as labels for training data, and approximation values which are calculated by the trained networks. Since in the early stage of the learning, there are not sufficient visited action-state pairs and the most action-value functions have zero values, the respective inputs to the function approximation are not accurate enough. It might be also misleading for the opposite agent to find the actual opposites using these values. One simple solution to tackle this problem and improve the accuracy of opposites is that the opposite agent can use the approximated action-value functions for all training and test data. We found this solution effective specially when the function approximation is trained for the first time at the early stage of learning. Of course, more research work is needed in the future to investigate this issue.

In our case study, the statistical information of stochastic parameters (inflow) is not changing with the time. Therefore, considering a constant learning rate for all interactions might create a slight oscillation which in turn influences the operating policy at the end.

Type-II algorithm like type I can significantly contribute in speeding up the learning process especially during the early stage of the learning. However, the respective operating policy might create more variations during the simulation.

References

1. Bellman, R.: Dynamic Programming, 1st edn. Princeton University Press, Princeton (1957)
2. Doran, D.G.: An efficient transition definition for discrete state reservoir analysis: the divided interval technique. Water Resou. Res. 11(6), 867–873 (1975)
3. Fletcher, S.G.: A new formulation for the stochastic control of systems with bounded state variables: an application to reservoir systems. PhD thesis, Dept. of Systems Design Engineering, University of Waterloo, Canada (1995)
4. Ponnambalam, K., Adams, B.J.: Stochastic optimization of multireservoir systems using a heuristic algorithm: case study from India. Water Resou. Res. 32(3), 733–742 (1996)
5. Klemes, V.: Discrete representation of storage for stochastic reservoir optimization. Water Resou. Res. 13(1), 149–158 (1977a)
6. Klemes, V.: Comment on An efficient transition definition for discrete state reservoir analysis: the divided interval technique by D.G. Doran. Water Resou. Res. 13(2), 481–482 (1977b)
7. Labadie, J.W.: Optimal operation of multireservoir systems: State-of-the-art review. J. Water Resou. Plan. and Manage. 130(2), 93–111 (2004)
8. Loucks, D.P., Stedinger, J.R., Haith, A.D.: Water resource systems planning and analysis. Prentice-Hall, New Jersey (1981)
9. Mahootchi, M., Tizhoosh, H.R., Ponnambalam, K.: Reservoir operation optimization by reinforcement learning. In: Proceedings of international conference on storm water and urban water systems, pp. 165–184 (2006); James, Irvine, McBean, Pitt, Wright (eds.) Monograph 15, Ch. 8 in Contemporary Modeling of Urban Water Systems
10. Mahootchi, M., Tizhoosh, H.R., Ponnambalam, K.: Opposition-based reinforcement learning in the management of water resources. In: Proceedings of IEEE Symposium Series on Computational Intelligence, Hawaii, U.S.A., pp. 217–224 (2007)
11. Mays, L.W., Tung, Y.K.: Hydrosystems engineering and management. McGraw-Hill, U.S.A. (1992)
12. Ponnambalam, K.: Optimization of the Integrated Operation of Multi-Reservoir Irrigation Systems. PhD thesis, Dept. of Civil Engineering, University of Toronto (1987)
13. Robbins, H., Monro, S.: A stochastic approximation method. Annals of Mathematical Statistics 22(3), 400–407 (1951)
14. Tizhoosh, H.R.: Opposition-based learning: a new scheme for machine intelligence. In: Proceedings of the International Conference on Computational Intelligence for Modeling Control and Automation, CIMCA, Vienna, Austria, vol. I, pp. 695–701 (2005)
15. Tizhoosh, H.R.: Opposition-based reinforcement learning. Journal of Advanced Computational Intelligence and Intelligent Informatics (JACIII) 10(4), 578–585 (2006)
16. Wang, D., Adams, B.J.: Optimization of real-time reservoir operations with markov decision processes. Water Resou. Res. 22(3), 345–352 (1986)

Index

Author Index

Printing: Krips bv, Meppel, The Netherlands
Binding: Stürtz, Würzburg, Germany